Social Relationships and Cognitive Development

Edited by

Robert A. Hinde, Anne-Nelly Perret-Clermont,
and Joan Stevenson-Hinde

A Fyssen Foundation Symposium

CLARENDON PRESS · OXFORD
1985

Oxford University Press, Walton Street, Oxford OX2 6DP
Oxford New York Toronto
Delhi Bombay Calcutta Madras Karachi
Kuala Lumpur Singapore Hong Kong Tokyo
Nairobi Dar es Salaam Cape Town
Melbourne Auckland
and associated companies in
Beirut Berlin Ibadan Nicosia

Oxford is a trade mark of Oxford University Press

Published in the United States
by Oxford University Press, New York

British Library Cataloguing in Publication Data

Social relationships and cognitive development.—
(A Fyssen Foundation symposium)
1. Cognition   2. Social interaction
I. Hinde, Robert A.   II. Perret-Clermont,
Anne-Nelly   III. Stevenson-Hinde, Joan   IV. Series
153.4   BB311
ISBN 0–19–852155–3
ISBN 0–19–852167–7 Pbk

Library of Congress Cataloging in Publication Data
Main entry under title:
Social relationships and cognitive development.
(A Fyssen Foundation symposium)
Bibliography: p.
Includes indexes.
1. Interpersonal relations—Congresses. 2. Cognition
—Congresses. 3. Cognition and culture—Congresses.
4. Socialization—Congresses.   I. Hinde, Robert A.
II. Perret-Clermont, Anne Nelly.   III. Hinde, J. S.
(Joan Stevenson)   IV. Series.
HM132.S572 1985   302   85–13587
ISBN 0–19–852155–3
ISBN 0–19–852167–7 (pbk.)

Set by Promenade Graphics Ltd., Cheltenham
Printed in Great Britain at the University Press, Oxford
by David Stanford
Printer to the University

This volume is dedicated to the memory of

H. Fyssen

Preface

This volume consists of a report on the first symposium sponsored by the Fyssen Foundation. In opening the symposium, Madame Fyssen said:

About 5 years ago, my husband and myself decided to create a Foundation, whose aim is to *encourage all forms of scientific enquiry into cognitive mechanisms, including thought and reasoning, underlying animal and human behaviour, and their ontogenetic and phylogenetic development*. Until now, the Foundation's activities have been directed primarily towards the awarding of fellowships to young researchers. Since the creation of the Foundation, 90 fellowships have been granted, both to French and to foreign researchers.

In addition, each year the Foundation awards an international prize. This has been presented, successively, to Professors André Leroi-Gourhan (1980), William H. Thorpe (1981), Vernon B. Mountcastle (1982), Harold C. Conklin (1983), and Roger W. Brown (1984).

Finally, the bye-laws of the Foundation provides for the organization of specialized symposia on topics related to the scientific objectives of the Foundation about once per year.

The first symposium was held at the Trianon Palace Hotel, Versailles, from the 16 to the 20 November 1984. Draft manuscripts had been pre-circulated, in order to allow ample time for discussion. Final papers were submitted after the meeting so that authors could, if they wished, take up points made in the discussions. Many of these points were submitted also in writing, either at the meeting or soon afterwards. Many of the issues raised in discussion and not included in the authors' revisions appear, with the contributors' approval, in an edited form after each chapter in this volume.

In the initial planning of the conference we were helped by the scientific committee of the Fyssen Foundation, and especially by Dr Perriault. The immediate organization of the meeting was undertaken by Madame Colette Leconte, whose foresight and indefatigable devotion was largely responsible for its success. Madame Colette Kouchner also helped in the pre-conference planning and in the arrangements for publication. The successive drafts of the editorials and discussions, the sorting out of the references, and the correspondence with authors necessary for the production of the final manuscript were in the capable and imperturbable hands of Mrs Ann Glover. To all of these we are duly grateful.

Finally we would especially like to express our gratitude to Madame Fyssen who played an active part throughout the planning, chose the location, attended all the meetings, and gave us much encouragement.

Cambridge
March 1985

R. A. H.
A.-N. P.-C.
J. S.-H.

Contents

List of participants

Jean-Louis Adrien
Service de Pédopsychiatrie, Professeur Lelord, C.H.R. de Tours, 49, Boulevard Béranger, 37000 Tours, France.

Grazia Attili (contributor—absent)
Istituto di Psicologia del CNR, via dei Monti Tiburtini 509, 00157 Roma, Italy.

Monique Ballion
Institut National de Recherche Pédagogique, 29, rue d'Ulm, 75230 Paris Cédex 05, France.

Bernadette Bresard
Groupe de Recherche sur les Anthropoïdes, Musée National d'Histoire Naturelle, 57, rue Cuvier, 75231 Paris Cédex 05, France.

Alain Brossard (contributor)
Séminaire de Psychologie, Université de Neuchâtel, Maladière 10, 2000 Neuchâtel, Switzerland.

Peter E. Bryant (contributor)
Department of Experimental Psychology, University of Oxford, South Parks Road, Oxford, OX1 3UD, United Kingdom.

Michael J. Chandler (contributor)
Department of Psychology, University of British Columbia, 2075 Wesbrook Mall, Vancouver, B.C. V6T 1 W5, Canada.

Jean-Pierre Changeux
Unité de Neurobiologie Moléculaire Institut Pasteur, 28, rue du Docteur Roux, 75724 Paris Cédex 15, France.

Jean Chavaillon
Maître de Recherches, CNRS–5ème Circonscription, 1, Place Aristide Briand, 92190 Meudon-Bellevue, France.

Dorothy Cheney (contributor)
Department of Anthropology, University of California, 405 Hilgard Avenue, Los Angeles, California 90024, USA.

Verena Dasser (contributor)
Zoologisches Institut, Universität Zurich-Irchel, Gebäude 25, Winterthurerstrasse 190, Ch-8057 Zurich, Switzerland.

Willem Doise (contributor)
Faculté du Psychologie et des Sciences de l'Education, Université de Genève, 24, rue Général-Dufour, 1211 Genève 4, Switzerland.

Mme A. H. Fyssen
194, rue de Rivoli, 75001 Paris, France.

Willard W. Hartup (contributor)
Institute of Child Development, University of Minnesota, 51 East River Road, Minneapolis, Minnesota 55455, USA.

Robert A. Hinde (chairperson and co-editor)
MRC Unit on the Development and Integration of Behaviour, Cambridge University, Madingley, Cambridge, CB3 8AA, United Kingdom.

Antonio Iannacona
Istituto di Psicologia dell' Università di Salerno, Facoltà di Lettere, via Irno 136, Salerno, Italy.

Nada Ignjatovic-Savic (contributor)
Filozofski Fakulted, Cika Ljubina 18–20, 1100 Beograd, Yugoslavia.

Lothar Krappmann (contributor)
Max-Planck-Institut für Bildungsforschung, Lentzeallee 94, 1 Berlin 33–Dahlem, RFA.

Helgard Kremin
Laboratoire de Pathologie du Langage, INSERM U 111, 2 ter, rue d'Alésia, 75014 Paris, France.

Hans Kummer
Ethologie und Wildforschung, Universität Zurich-Irchel, Winterthurer-strasse 190, 8057 Zurich, Switzerland.

Marie-Christine Lacour
Etudiante de 3ème cycle en Psycholinguistique à Paris VII Option Ethologie, 18, rue Oudry, 75013 Paris, France.

Bruno Latour
Maître de Recherches, Ecole National Supérieure des Mines, Centre de Sociologie de l'Innovation, 62, Boulevard Saint-Michel, 75006 Paris, France.

Alain Legendre
Laboratoire de Psychobiologie de l'Enfant, 41, rue Gay-Lussac, 75005 Paris, France.

Marie-Thérèse Lenormand
INSERM U 3, Professeur Scherrer, Hôpital de la Salpétrière, 47, Boulevard de l'Hôpital, 75634 Paris Cédex, France.

Anne-Nelly Perret-Clermont (contributor, co-chairperson, and co-editor)
Séminaire de Psychologie, Université de Neuchâtel, Maladière 10, CH. 2000 Neuchâtel, Switzerland.

Jacques Perriault
Directeur de Département, Institut National de Recherche Pédagogique, 91, rue Gabriel Péri, 92120 Montrouge, France.

Marian Radke-Yarrow (contributor)

Chief, Laboratory of Developmental Psychology, Department of Health and Human Services, National Institutes of Health, Bethesda, Maryland 20205, USA.

Michael Rutter (contributor)

Department of Child and Adolescent Psychiatry, Institute of Psychiatry, De Crespigny Park, Denmark Hill, London, SE5 8AF, United Kingdom.

Robert L. Selman (contributor)

Department of Psychiatry, Judge Baker Guidance Center, 295 Longwood Avenue, Boston, Massachusetts 02115, USA.

Robert Seyfarth (contributor—absent)

Department of Anthropology, University of California, 405 Hilgard Avenue, Los Angeles, California 90024, USA.

Tracy Sherman (contributor)

Laboratory of Developmental Psychology, Department of Health and Human Services, National Institutes of Health, Bethesda, Maryland 20205, USA.

Myrna B. Shure (contributor)

Hahnemann University, Preventive Intervention Research Center, Broad & Vine, Philadelphia, Pennsylvania 19102–1192, USA.

Mira Stambak (contributor)

Institut National de Recherche Pédagogique, 29, rue d'Ulm, 75230 Paris Cédex 05, France.

Joan Stevenson-Hinde (co-editor)

MRC Unit on the Development and Integration of Behaviour, Cambridge University, Madingley, Cambridge, CB3 8AA, United Kingdom.

Serge Stoleru

Institut Universitaire des Sciences Psychosociales et Neurobiologiques, 74, rue Marcel Cachin, 93012 Bobigny Cédex, France.

Barbara Tizard (contributor)

Thomas Coram Research Unit, 41 Brunswick Square, London WC1N 1AZ, United Kingdom.

Jacques Vauclair

Département de Psychologie Animale, CNRS–INP 9, 31, Chemin Joseph-Aiguier, 13402 Marseille Cédex 9, France.

Lawrence Weiskrantz

Department of Experimental Psychology, University of Oxford, South Parks Road, Oxford, OX1 3UD, United Kingdom.

James V. Wertsch (contributor)

Department of Linguistics, Northwestern University, College of Arts and Sciences, 2016 Sheridan Road, Evanston, Illinois 60201, USA.

Introduction: The dialectics between levels of social complexity

Most children grow up in a nuclear or extended family, interacting first with one or two parents, and then with siblings, with relations, and with friends. Later, through schools and peer groups, the number and variety of their interactants increase (e.g. Foot *et al.* 1980; Rubin and Ross 1982). Whilst the details differ, such a pattern of an expanding network of relationships is virtually ubiquitous in human societies (e.g. Whiting and Whiting 1975) and also in most non-human primates (Berman 1983), and was almost certainly present in our environment of evolutionary adaptedness (Alexander 1974; Hinde 1984).

This network of relationships constitutes the most important part of the child's environment. In any social group, individuals must adjust their behaviour according to whom they are with. Abilities to assess the capacities and predict the behaviour of others are invaluable, and it has been argued that the need for managing interpersonal relationships provided the selective forces that moulded the evolution of the cognitive capacities of higher mammals and of man (Chance and Mead 1953; Jolly 1966; Humphrey 1976).

Yet the study of the properties of interpersonal relationships has until recently been neglected by psychologists. Until the seventies, most students of personality focused primarily on the individual, with little attention to variations in behaviour between different social contexts. Developmental psychologists have been concerned with the effect of others (usually the mother, but more recently also the father and siblings) on the development of personality, but only in the last few years have they directed much attention to children's friendships (e.g. Foot *et al.* 1980; Asher and Gottman 1981; Duck 1983). Most social psychology texts treat individuals' perceptions of social phenomena, and the bases of their attraction to others, but then leap to group phenomena, neglecting the dyad. Although the way was paved by earlier work (e.g. Blau 1964; Homans 1961; Thibaut and Kelley 1959), only in the last few years has the necessity for a science (in the sense of an ordered body of knowledge) about long-term dyadic and triadic relationships been widely recognized (e.g. Duck 1973; Hinde 1979; Kelley *et al.* 1983).

The nature of relationships poses special problems to the psychologist. The nature of the constituent interactions depends upon the natures of the participant individuals, whilst the characteristics that individuals display depend in part on the nature of the interaction and relationship in which they are involved, and in the longer term the characteristics that they *can* display are influenced by the interactions and relationships they have experienced. The nature of relationships depends on those of the constituent interactions, but the nature of those interactions depends on the participants' perceptions of the nature of the relationship. The nature of any relationship is affected by that of the social group in which it is embedded, whilst the nature of that social group depends in part on its constituent dyadic (and higher order) relationships.

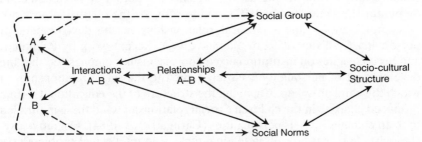

Fig. Int. 1 The dialectics between successive levels of social complexity. (Discontinuous lines represent dialectics perhaps of less importance than the continuous ones.) (Modified from Hinde 1984.)

Interactions and relationships are affected also by the norms and values of the participants—norms and values which are in part created, transmitted, and transmuted through the agency of dyadic relationships. And the sociocultural structure, used here for the system of institutions and beliefs, and the relations between them, shared by the members of the group, in turn both influences, and is influenced by, the relationships between individuals (Fig. Int. 1). And beyond that, each group is juxtaposed with other groups, contact with which affects diverse aspects of the social behaviour of its individuals. And finally each group is set in a physical environment, which affects and is affected by the group members. The social scientist must therefore come to terms with a series of dialectics between successive levels of social phenomena—relationships, social structure, sociocultural structure, and intergroup relationships, each of which has emergent properties not relevant to the level below. And at the same time he must remember that each level represents not an entity but a process in continuous creation through the agency of the dialectics (Mead 1934; Doise 1980, 1982; Hinde 1979, 1984).

Whilst some of the social sciences seem to be concerned with one or other of these levels, in practice the dialectics always obtrude. Thus students of personality, finding that the cross-situational consistency of supposed 'traits' tended to be low, were forced to recognize that behaviour may be affected (to differing extents according to the nature of the individual and the behaviour) by the context (Bem and Allen 1974; Bem and Funder 1978; Kenrick and Stringfield 1980). And the most important aspect of the context is the interactional and relationship one, including the meaning that the individual attributes to them (the sense he makes out of the situation) according to his sociocultural scheme of reference (systems of understanding and beliefs), and his personal past experience. Developmental psychologists, concerned first with the growing child, have had to come to terms with the interacting influences of parent on child and child on parent (Bell and Harper 1977), and to consider the relative importance of complementary and reciprocal relationships in the development of personality (Sullivan 1953a,b; Youniss 1980). Recent work by Palmonari and his colleagues on institutionalized adolescents has emphasized that the development of personality is closely linked to the set of relationships in which the child grows up. Changing the structure of the educational setting produced changes in the children's social relationships and these in turn led to deep changes in their personalities (Carugati *et al.* 1984). Cognitive psychologists find that how an individual tackles an intellectual problem may change radically with the social context (Carraher *et al.* 1985). Educationalists have found that students' school performance is affected by their teachers' expectations and these are affected by the teachers' own career and identity (Marc 1984) and by the organization of the institution (Gilly 1980). Anthropologists, concerned with the sociocultural structure, seek to understand the ways in which beliefs, myths, and legends are created and passed on by individuals and affect the lives of individuals and their mode of establishing relationships and social groups (e.g. Herdt 1981; Verdier 1979; Hainard and Kaehr 1983).

This volume is concerned with one aspect of these dialectics—namely the interplay between an individual's social interactions and relationships on the one hand, and his or her cognitive development on the other. As we proceed we shall see both that relationships cannot be considered independently of the individual or social group, and that individuals cannot be considered independently of the relationships, social group, norms, values, and socio-historical context in which they are embedded. We shall see also that cognitive development is closely related to other aspects of the individual, including the emotions. And we shall be forced to bear in mind that the concepts we use—relationships, cognition, emotion, stage, and so on—are at the same time essential tools for understanding and blinkers that constrain our vision.

Section A

Section A

1

Methodology and the concept of cognition
EDITORIAL

Our verbal language depends upon, and is critical for, many of our own cognitive abilities, and we use those abilities to form, and in the presence of, the norms, values, beliefs, and institutions of our society. We shall see later that there is a close interdependence between our cognitive abilities and our social situation. It is thus helpful to seek a different perspective by considering first the abilities of monkeys and apes, which lack verbal language and anything comparable to human culture.

Two issues raised by the animal data permeate much of the rest of the volume, and may be introduced here. The first is methodological. Upon what sorts of data can we base conclusions about something as apparently nebulous as cognitive abilities, and even more, how could we obtain hard evidence that relationships affect cognitive abilities or vice versa? The chapters in this volume represent diverse approaches to this problem, none of which is fully adequate on its own but each of which can be seen as a contribution towards a solution:

1 *Description of what individuals do in the real social world*. This can provide evidence about both cognitive capacities and performance, and about factors that may affect them. The absence of experimental control poses difficulties for the interpretation of process and the long-term consequences of a given experience may seem probable but are likely to remain unproven (e.g. in this volume chapters by Dasser; Cheney and Seyfarth; Attili; Tizard; Stambak *et al.*).

2 *Laboratory experiments to assess performance*. These can provide hard evidence about performance and the factors that can affect it, but not necessarily about maximum capacities, since the individual may not show optimum performance in the inevitably artificial test conditions (Dasser, this volume). They may also reveal abilities which may not be apparent in nature.

3 *Correlational studies showing associations between life experiences and cognitive performance*. These provide important evidence for the existence of (direct or indirect) causal links, especially with additional evidence concerning for instance the sequence of events. However, in general correlational data cannot prove the direction of causal links,

whether they are direct or indirect, or their potency (e.g. Rutter; Krapp-mann).

4 *Laboratory experiments to assess the effects of changes in conditions or of the nature of subjects.* These can provide strong evidence about factors that can affect performance and/or the nature of process, especially in the short-term (e.g. chapters by Radke-Yarrow and Sherman; Bryant; Chandler; Wertsch and Sammarco; Doise; Perret-Clermont and Brossard). The relevance to real life of factors demonstrated in the laboratory must of course be assessed.

5 *Real life intervention studies.* These can produce strong evidence for the existence of effects, but it is often difficult to pin down the factors responsible (e.g. chapters by Rutter; Shure; Selman; Bryant). Bryant argues for a combination of (3) and (5).

None of these approaches is likely to be successful on its own; even laboratory experiments usually depend on earlier observational or correlational studies.

A second issue concerns the relation between the cognitive abilities we use to manipulate the physical environment and those we use in social interaction. In the past developmental psychologists often assumed that, within limits, cognition was unitary and applicable to any domain, so that aspects of cognitive development proceeded in parallel across areas of differing behavioural content (e.g. Piaget 1929). This does not mean that an infant performs at precisely the same level in different tasks—for instance conservation of volume occurs later than conservation of weight or substance (the phenomenon of horizontal décalage). However performance progresses in the same way in the different areas, and at any one age discrepancies between an individual's performance in the different tasks will be limited.

However this view has been challenged from two directions. One involves a primarily biological argument, that learning is guided by constraints and predispositions, which differ according to species, age, and context (Seligman and Hager 1972; Hinde and Stevenson-Hinde 1973). Many animals show complex learning abilities in limited areas, whilst lacking them in others. For instance hunting wasps (*Ammophila campestris*) can remember the location of several nest holes, and rats associate accurately past tastes with digestive discomfort, although each may behave with apparent stupidity in other contexts. Indeed Rozin (1976) has gone so far as to suggest that animals initially evolved not so much general learning abilities as adaptive specializations which enabled them to solve particular problems. On this view the subsequent evolution of intelligence involved the harnessing of abilities originally specific to one problem or set of problems for other purposes. Rozin suggested that a similar extension of access is involved in development, for instance learning to read may depend on the difficult task of gaining access to phonetic representation of what is heard (Rozin and Gleitman 1976).

The second source of doubt about the unitary nature of cognition comes from comparisons between the cognitive abilities shown by human subjects in different contexts. Thus:

1 Piaget himself claimed that the abilities to deal with tasks in different cognitive domains emerge in a definite order within each stage. In general, cognitive development is more piecemeal than has sometimes been supposed, especially in infancy (Harris and Heelas 1979; Gelman and Baillargeon 1983). Some décalages persist into adulthood (Keating, 1979). Some skills are first acquired in specific contexts, and subsequently become decontextualized (Vygotsky 1978).

2 Even within social cognition, an individual's judgements do not form a unified system, and there are differences between domains (e.g. Turiel 1983). Age changes are not always uniform across different contextual domains of social cognitive reasoning (though see data reviewed by Hartup, this volume), and changes in social reasoning are not clearly reflected in social behaviour (review Bearison 1982).

3 Consistent high correlations between skills in the socio-cognitive and physical domains are seldom found (e.g. Shantz 1975).

4 There are many obvious differences between physical and social domains, which imply that physical and social cognitions must be based on different premises. For instance the way in which an object moves can be predicted from knowledge of the physical forces acting on it, but the movement of people cannot (Glick 1978; Gelman and Spelke 1981). And acting on or manipulating another person requires techniques (cries, smiles, gestures, language) different from those used for acting on objects. Furthermore strong emotions are more likely to be involved in dealings with people than with things.

5 Because the behaviour of people is more complex than that of objects, one might expect social cognitive development to lag behind that of physical cognition. Yet this seems not to be the case: for instance children seem able to accommodate to another's perspective when only two years old (Lempers et al. 1977) and to carry an internal representation of a person before they can that of an object (Bell 1970; Uzgiris and Hunt 1966. See also Dunn and Kendrick 1982; Gelman 1979; Stambak, this volume). Whether this would still be the case if the object were as complex and familiar as the person is questionable (Jackson et al. 1978), but the fact is that people are more complex than objects in the real world.

6 Social performance may depend on factors in addition to cognitive skills—for instance norms inherent in the culture and social knowledge acquired through experience (e.g. Higgins and Parsons 1983; Costanzo and Dix 1983). Adults often use intuitive strategies, seemingly far removed from cognitive judgement (Nisbett and Ross 1980). Furthermore social cognition may involve competencies not relevant in the

physical domain, such as the ability to assess people's internal states and intentions (Damon 1981, 1983a).

None of these arguments is conclusive. It could be that there are logical competences common to domains, but they are not critical for deciding performance in the conditions tested. The classic view, whilst supposing that cognitive structures might be shared between domains, also supposed that structures were actively created through interaction with the physical and social environments, so that cognitive structures could contribute in different ways to different domains. Furthermore social judgements may depend on social knowledge as well as, or more than, upon cognitive abilities. Kohlberg (1969) supposed that certain stages of cognitive development were necessary but not sufficient for stages of moral judgement.

The apparent relative precocity of socio-cognitive abilities can be explained in a number of ways. Quite apart from the possibility that children by nature soon develop an ability to enter into the worlds of others intersubjectively (Trevarthen 1979), social cognition may be facilitated by the emotional involvement the child has with people, and by the feedback people provide for children, thereby facilitating the correction of their inadequate or inaccurate interpretations of behaviour. Furthermore, because people are fundamentally similar to each other, basically egocentric adaptions may be adaptive (Hoffman 1983). And social relationships may affect socio-cognitive functioning in diverse ways—for instance through the acquisition of social knowledge (Higgins and Parsons 1983) or norms which themselves affect social perceptions (Costanzo and Dix 1983) which may affect future relationships. 'In a general sense, social development is a process of mutual adaptation between society and the child' (Damon 1983a, p. 386).

Little could be gained here by entering the argument as to whether all knowledge is basically social or basically non-social (e.g. Chandler 1976; Damon 1981); indeed to the foetus or very young infant the distinction scarcely matters. However, there are two points to be made. First, the reader of this book must be aware that some authors use 'cognitive' mostly in a narrow sense, to refer to the manipulation of mental structures (Rutter, Bryant, Chandler, Wertsch and Sammarco); others use 'cognitive' or 'social cognitive' to refer to a wider (to varying degrees) range of abilities and knowledge used in social interaction (Dasser, Cheney and Seyfarth, Attili, Hartup, Radke-Yarrow and Sherman, Shure, Stambak *et al.*, Krappmann, Selman); whilst yet others regard cognitive coordinations as elaborated by certain kinds of social interactions and their actualization as often affected by the social context (Doise, Perret-Clermont). Here the term social refers to the observations that these skills are produced socially, and are sometimes specific to particular interactions before being interiorized in the individual and thereby becoming available for use in other interactions (see also Perret-

Clermont *et al.* 1984a,b). Whether or not it is useful to make a clear distinction between, for example, social attributions based on a social inference process 'guided by logical structures and characterized by a systematic and logical consideration of relevant social information' and attributions based on social norms and more content-specific social beliefs (Costanzo and Dix 1983, p. 72–73) is still an open issue. Turiel (1983) has suggested that much confusion could be avoided if the terms *social* and *cognitive* always had referents—such as social environment, social experience, social cognitive, and non-social cognitive. However, it could be argued that a difficulty of definition would still remain: mathematics may be non-social, but mathematical ability depends on socio-cultural activity.

The second point is that this is an area in which we must differentiate the questions rather carefully. First, do social cognitive abilities depend in part on previously established skills acquired in interchange with the physical world? The answer here seems clear—there may be structures common to the two domains, but performance in each depends also on specific factors (see also Selman 1980). Indeed the distinction between physical and social domains is itself too crude, for each is complex and abilities may be specific to subdivisions of either. In any case our perception of the physical world is largely symbolic, with social meanings attached to it. Much of the physical world is a world of objects, with social rules on how to deal with them. Second, do physical and social cognitive abilities depend on identical basic operations, with differences between people and things accounting for differences between the ways in which skills are exhibited in the social and physical world? Here it seems that some of them may, but the social world certainly soon calls for skills not demanded by the physical world. Furthermore, as we shall see (Radke-Yarrow and Sherman, this volume), affect, knowledge, and cognitive skills all play a part in social-cognitive performance.

Third, does the possession or exercise of one type of skill affect the ability to display the other? This seems certain. Objects in the physical world often provide a focus for social interaction and, as we shall see in later chapters, social interaction may provide a context for acquiring skills for dealing with the physical world. This does not mean that the one is *always* necessary for the other. The earlier view (Mueller and Vandell 1979) that toys provided a necessary focus for the interactions of young children was certainly an overstatement—even very young children, and especially siblings, may take part in complex interchanges which are purely social in character (Dunn and Kendrick 1982). And conversely, individuals brought up in near-total social isolation nevertheless acquire some abilities in dealing with the physical world.

Fourth, in the course of evolution, have social and non-social cognitive abilities become adapted through different selection pressures? This could be in line with Rozin's views, summarized above. In the next two chapters Dasser, and Cheney and Seyfarth, consider the abilities shown by monkeys in laboratory situations and in natural social contexts. Both find apparent differences between the performances of monkeys in social and non-social situations. Dasser is cautious, concluding that the data she reviews provide no evidence that non-human primates have concepts for abstract entities like relationships, and acknowledges the possibility that aspects of the social behaviour of monkeys could be explained in terms of relatively simple psychological principles. Cheney and Seyfarth believe that their data go farther than this, and stress that vervet monkeys show complex cognitive abilities in situations where there might be a marked selective advantage in their doing so, but not in apparently comparable situations which are perhaps less important to them. They thus suggest that social and physical cognition evolved independently.

2

Cognitive complexity in primate social relationships
VERENA DASSER

Introduction

Classical ethologists, and many early field workers in primatology, were concerned with the study of species-typical behaviour and mostly with the observation of wild animals. The term 'cognition' was not often used, and only recently have primatologists began to ask what animals know about each other and their environment (Premack and Woodruff 1978; Cheney and Seyfarth 1980; Kummer 1982). In animal psychology, where many researchers have found the stimulus-response approach of behaviourism inadequate, there has been a longer but muted tradition of a more mentalistic conception of animal behaviour (e.g. Hulse *et al.* 1978; Mellgren 1983). However definitional problems pose difficulties in applying cognitive concepts unequivocally to the everyday life of monkeys and apes. Thus to qualify as cognitive, a process must bring about a change in the internal state of an individual that is neither a simple stimulus-response association nor solely motivational (e.g. Staddon 1983); and it must also involve distancing from the here and now, from immediate sensation and perception (Honig 1978). A key concept for both animal and human cognition is representation (Cooper 1982). We need not be concerned here with the precise nature of representation in animals, but it must be postulated if an animal does not merely respond to stimuli present at the time of the response, but rather its behaviour is influenced by stimuli no longer present.

In this paper findings from formal laboratory tests and relevant observations on the behaviour of caged and free-ranging primate groups are reviewed. The discussion is restricted for the most part to lower primates.

1 Cognitive processes: evidence from learning tests

It is the purpose of this section to review some findings in those cognitive areas that seem to be most relevant for the topic of this paper (for a more extensive review see Jarrard 1971).

The following paragraphs give examples of cognitive processes found in non-human primates. As such tests have not yet been carried out for the cognitive processes involved in interindividual relationships, the examples are drawn from non-social contexts.

Visual memory

The ability to store information and to retrieve it from memory is basic to all cognitive functions. Most formal testing has been done on the visual memory of non-human primates. In humans, the use of language in encoding non-linguistic material may substantially support the memorizing process. Monkeys, although lacking this aid, recognize complex visual stimuli remarkably well. Overman and Doty (1980) have shown that pigtail macaques (*Macaca nemestrina*) identify images seen 24 hours previously with an accuracy of 90 per cent. One subject correctly remembered 76 per cent of the pictures after a delay of 96 hours. Longer intervals were not tested. When tested on their memory capacity for lists of pictures, monkeys show retention curves strikingly similar to those of humans, suggesting similar mechanisms in the visual memory system of both species (Sands and Wright 1980).

Perceptual categories

The performance in the experiments mentioned above appears to be based on mere association learning—the subject simply remembers individual pictures. Sorting stimuli into categories might reflect higher cognitive functioning. Do monkeys show a disposition to put like items together? The method used by Sands *et al.* (1982) assesses some complex visual categories in rhesus monkeys (*Macaca mulatta*) that are only in part predetermined by the experimenter. Two monkeys were trained to identify two successively presented pictures as either identical or different. The pictures represented five categories: monkey faces, human faces, trees, fruits, and flowers. The monkeys' natural categories were identified through an analysis of the errors made by the subjects when responding to novel combinations of pictures, e.g. 'same' responses to different pictures. A multidimensional scaling procedure revealed clear clusters of faces and fruits. The monkeys treated these stimuli categorically, lumping together pictures that were physically dissimilar but represented a common conceptual category. Thus, monkeys encode certain sensory inputs along more dimensions than the directly perceived ones.

Concepts

The concepts of monkeys are not limited to non-relational absolute categories (Premack 1978); monkeys are also able to form more abstract, relational concepts. In complex learning tasks, monkeys can learn to use oddity or sameness as a cue for picking the correct of three novel stimuli (e.g. Noble and Thomas 1970; Thomas and Kerr 1976, *Saimiri sciureus*) or base their response on the relative novelty of a stimulus (Essock-Vitale 1978; Massel *et al.* 1981, rhesus monkeys). When rhesus monkeys are trained on numerous different discrimination problems, they become progressively more efficient in solving new problems (Harlow 1949; Miles 1965). While it seems probable that performance in such learning set experiments is based on a win-stay, lose-shift strategy shown by some non-primate species as well, only primates have so far been shown to generalize this strategy across large changes in specific stimuli (Mackintosh 1974).

No sharp boundary can be drawn between perceptual categories and abstract concepts. 'Natural' concepts, first extensively studied in pigeons by Herrnstein and his co-workers, assume a position in between the extremes. These are concepts of classes of things that occur in nature. Pigeons have been found to be very adept at identifying photographic slides showing any possible instance of such naturally occurring things as trees, bodies of water, a particular human (Herrnstein *et al.* 1976; Herrnstein 1979), humans in general (Herrnstein and Loveland 1964), fish (Herrnstein and deVilliers 1980), and other pigeons (Poole and Lander 1971). It is still not clear what exactly the stimulus features were that controlled the discrimination behaviour of the pigeons. Similar experiments have recently been carried out with monkeys (*Macaca arctoides*) by Schrier *et al.* (1984). The concepts studied were humans, monkeys in general, and the non-natural concept of the letter A. The monkeys appeared to have more difficulty in learning the concepts than the pigeons. Although the monkeys in these tests gave evidence that they formed the concepts, their performance was somewhat below that of the pigeons. Further experiments will have to show what the nature of such natural concepts in different species really is and how they might work in nature.

Cross-modal perception

Cross-modal perception refers to the capacity to compare information about stimuli perceived by different sense modalities. Monkeys have been shown to recognize novel objects they could feel but not see following visual exposure to the stimuli, or vice versa (Cowey and Weiskrantz 1975; Weiskrantz and Cowey 1975; Jarvis and Ettlinger 1977; Malone *et al.* 1980). Their performance reached levels comparable to that of chimpanzees. Our ability to recognize the equivalence of spoken and written words rests on this cognitive function and it was long postulated that non-humans lack this prerequisite for language (Reynolds 1981). If monkeys can match across modalities, they may also be able to match objects or events on the bases of their effects on themselves and other individuals rather than on mere physical similarity.

Self-recognition in mirrors

Monkeys and gorillas, in contrast to chimpanzees and orangutans, do not recognize themselves in mirrors (Gallup 1983). This has been interpreted by Gallup (1975, 1983) as an absence of self-awareness or consciousness in the former but not in the latter species. If this is true, it could also indicate a qualitative difference in how monkeys and gorillas on the one hand and chimpanzees and orangutans on the other perceive social partners and their relations with them. However, from their performance in visual-haptic matching tasks and their ability to use mirror cues to manipulate objects that they could not see otherwise (Brown *et al.* 1965) one would expect a monkey to be able to recognize its mirror image. In spite of numerous unsuccessful efforts to elicit self-directed responses in monkeys (e.g. Benhar *et al.* 1975; Gallup 1977), their failure may still reflect motivational rather than cognitive restrictions—the highly emotional other-directed reaction to the mirror image of a 'conspecific' may prevent learning of mirror-contingencies and cross-modal comparison, both processes that can be applied to non-social objects.

In a comparable experimental set-up, pigeons mimicked the behaviour of the chimpanzees (Epstein, Lanza, and Skinner 1981). This should caution us against interpreting positive findings in mirror recognition tests in terms of such rich concepts as self-awareness (Gallup 1970, 1975, 1983). While it seems highly plausible that the overtly similar behaviour of such widely different species as pigeons and chimpanzees is based on different underlying processes, chimpanzee behaviour may 'only' demonstrate their ability to form a concept of their own body rather than a 'person-concept' (Kummer, pers. comm.).

Inferential ability

Little is known about a monkey's capacity for reasoning. This may be largely because it is difficult to devise reasoning tests for a non-linguistic organism. Reasoning problems have nonetheless been successfully presented to non-language-trained chimpanzees (e.g. Gillan 1981; Premack 1983). McGonigle and Chalmers (1977) evaluated the performance of squirrel monkeys (*Saimiri sciureus*) in a transitive inference test. A series of elements, A–B–C, is said to be transitive if a relation r between A–B and B–C also holds for A and C. In the test, the monkeys were given information only about the relation of all adjacent elements of a series of 5 elements A to E, i.e. A heavier than B, and B heavier than C, and so on, and were rewarded for choosing the heavier element. As only two weight values were used, no specific weights could be associated with B, C, and D because these elements were equally often heavy and light, depending on the element they were paired with. In subsequent critical trials, the subjects were confronted with pairs of non-adjacent elements and had to infer the relation between them. The choices of the monkeys on these trials were consistent with the notion that they used transitive inference, and was equivalent to the data reported for 4-year-old children. A binary sampling model proposed by the authors and not requiring deductive reasoning does however also fit the data. Whatever the precise mechanism that helped the monkeys to solve the problem, it is clear that they used some kind of mental representation of the missing element when making their choice.

Language in primates

It would go beyond the scope of this chapter to address the issue of primate language. Since the results of language training in apes have however bearing on the question of the cognitive abilities of non-human primates, some references will be listed here. Ape language studies have been undertaken with chimpanzees (*Pan troglodytes*, e.g. Gardner and Gardner 1969; Rumbaugh 1977; Terrace 1979), orangutans (*Pongo pygmaeus*, Miles 1982) and gorillas (*Gorilla gorilla*, Patterson and Linden 1981). The literature has been extensively reviewed. The reader is referred to the critical evaluation of the subject given by researchers in various disciplines in DeLuce and Wilder (1983) and to the references therein.

Conclusions

These, and a wide variety of other studies, prove the intellectual skills of monkeys to be well developed. When tested in appropriately designed tasks, monkeys rarely failed to perform well even in tests involving conceptual abilities. More important than performance in particular kinds of tests, however, are the underlying cognitive processes. Although they are still poorly understood, the following interpretation of monkey performance can probably be safely given: monkeys are not gifted abstract thinkers. When they can solve a problem associatively, they are likely to do so (e.g. Massel *et al.* 1981). Some performances, however, urge a cognitive explanation. Monkeys appear to learn what to do in order to cope with new situations. Their ability to detect common features underlying perceptually rather diverse experiences appears to be a characteristic property of primate cognitive functioning. Monkeys form concepts that enable them to generalize from familiar to novel objects despite possible gross changes in their physical appearance. Monkeys can, at least in some contexts, use the mental representation of objects or of relations between objects as a guide to immediate actions.

In his paper on the social function of intellect, Humphrey (1976) raised the question of the use of such skills by a monkey in the field. He suggested that the intellectual abilities of non-human primates have evolved and are primarily used in social life. However, when monkeys are tested on learning tasks they perform in a strictly controlled laboratory environment on tasks that involve highly artifical stimuli. The fact that monkeys can show their skills under the unnatural and unfavourable laboratory conditions argues against major constraints in their application to naturally occurring problems in everyday monkey life. While they may have evolved under the selection pressure acting on individuals in a narrow social or non-social context, these intellectual skills appear to have become generalized strategies, potentially available to the individual in a wide range of contexts. We may therefore not expect to find obvious natural counterparts of particular skills. Rather, we may ask on what occasions primates actually *use* their capacities.

2 Cognitive processes: their use in social life

In this section, some aspects of the social behaviour of non-human primates will be examined with the above question in mind. First, however, I will attempt to identify some of the special demands, if any, that social partners impose on an individual monkey.

'Social cognition': theoretical considerations

In humans, social cognition is held by many to be distinct from non-social cognition (for a brief review see Bearison 1982). It is argued that social cognition cannot be reduced to the mere application of cognitive skills to the social context (e.g. Selman 1980). If this really were the case, we might also expect non-human primates with their long-standing adaptation to social life to have evolved specialized cognitive abilities. This assumes that the social environment is defined by unique features whose requirements can best be met by an individual possessing skills that are fundamentally social in nature. The social life of monkeys and apes is characterized by the development of individualized relationships with conspecifics. Partners in such a relationship respond specifically to each other. No such mutual relationship can possibly arise through repeated contacts with any part of the physical environment, nor is it a likely outcome of interactions with members of another species, e.g. predator or prey.

Although individual recognition and personal relationships are not confined to them, primates appear to be outstanding in their capacity to respond to a variety of individual features, to draw fine distinctions between individuals and to develop and maintain simultaneously many different kinds of relationships finely tuned to the individual characteristics of the participants (e.g. Hinde 1983; below). Highly developed learning abilities that enable an individual to perceive, remember, and compare an immense amount of detailed information are certainly a necessary prerequisite for primate social life but such a life does not seem to require abilities differing in kind from those applicable to the non-social environment. There seems to be no reason to assume that social know-how in non-human primates can only be acquired by cognitive processes that are qualitatively different from those involved in learning about the physical world. However, conspecifics have one property that distinguishes them from any other animate or inanimate object, and that, if adequately exploited, can be used to improve one's dealing with them—they function more like oneself than any other organism. Social knowledge could thus be gained by introspection, by using one's own subjective experience as a basis for predicting and understanding the behaviour of others. However, even in apes the degree of self-awareness is still moot (see above), and I know of no evidence that lower primates are capable of any such experience. Perhaps, the ability to reflect upon the self and to differentiate self from other has an evolutionary precursor in the ability to construct a decentred view of the environment, both with respect to the social and non-social world. The concept of decentration was first used by Piaget (1950). It will be used here in the loose sense of an individual being able to perceive objects or events not only in relation to itself and its momentary spatial location or motivational state, but also in relation to other possible locations and states of itself and other individuals. The ability for decentring is needed for taking account of another's viewpoint, e.g. for working out another's visual field and motives and intentions. These are fundamentally social-cognitive skills in that they can effectively be applied only to conspecifics, much less to other living organisms and not to inanimate objects. In human psychology, these and related abilities were among the research topics that initially defined a domain known as social cognition (Serafica 1982). In selecting the literature for this paper, I paid special attention to possible instances of decentration in the social context and typically social-cognitive functions in the sense outlined above.

Coping strategies

Little work has been done on the relation between social interactions and cognitive development in ontogeny. Mason and co-workers have conducted a series of experiments with caged rhesus monkeys in order to establish the role of social experience in the development of cognitive abilities (Mason and Kenney 1974; Mason and Berkson 1975; Mason 1978). In these experiments, the natural mothers of infants were replaced by various surrogates mimicking certain maternal features. Infants raised with mobile surrogates more readily perceived a non-social test situation as a problem requiring a solution than those raised with stationary surrogates (Mason and Berkson 1975). Monkeys raised with dogs were more attentive, more responsive, and more flexible in their reaction to environmental changes than those raised with inanimate surrogates (Mason and Kenney 1974). It proved crucial to the differential influence of inanimate and animate mother substitutes that the latter provided the infants with many opportunities to learn the effect of their own behaviour on objects and events (Mason 1978). The generally good performance of monkeys raised with dogs suggests that they had acquired coping strategies through their interactions with their early attachment figure that allowed them to handle both social and non-social novel situations more effectively than infants lacking 'social' experience. These findings do not exclude, of course, the possibility that other, non-social early experiences could produce a similar effect on cognitive development.

Differences in a generalized approach to problems also occur between species and can result in the superior performance in laboratory tests of problem-solving skills of one species in the absence of a difference in cognitive abilities (Visalberghi and Mason 1983). These findings show the importance of taking account of the general dispositions of species and individuals when interpreting seemingly cognitively demanding achievements.

Knowledge about individuals and relationships

In trying to understand the structure of groups of non-human primates, primatologists have found principles concerned with kinship and dominance status particularly useful. Correlations are reported for different species between dyadic behavioural characteristics and the genetic relatedness and the dominance ranks of the individuals involved.

Monkeys draw fine distinctions between related individuals. Rhesus monkeys spend more time near close kin than distant kin and unrelated individuals (Berman 1983). In a number of Old World monkeys, grooming occurs more frequently among kin than non-kin (e.g. Sade 1972a,b; Dunbar 1980; Silk et al. 1981) and is directed preferentially towards higher ranking animals rather than towards subordinates (Stammbach 1978; Fairbanks 1980; Seyfarth 1980). Rhesus aid close relatives threatened by other individuals more often than distant relatives (Massey 1977; Kaplan 1978; Datta 1983).

What do such findings tell us about the cognitive abilities of the animals and about their understanding of the social structure of their group? In order to answer this question, we obviously need a more detailed description of the behaviour than the summary results cited above. Certain types of agonistic interactions—as alliances—seem particularly promising for revealing cognitive processes in the social context. In an alliance, an individual aggressively intervenes in a dispute between others.

The socially experienced laboratory-raised rhesus monkeys in Anderson and Mason's (1974) experiment apparently knew the dominance relations between two other animals—in a situation where an inferior monkey A was threatened by a dominant animal B, A typically responded with a fear grin to the aggressor and directed threats to a previously uninvolved monkey C, whereupon B now redirected its threat towards C. Irrespective of the dominance relation between A and C, C was subordinate to B in all cases and was not amongst B's preferred partners. Both these attributes of the relationship between B and C were not obviously apparent when A chose its 'victim'. At the very least, A anticipated the behaviour of B from past experiences with B and C. This shows that monkeys do not have to rely on cues immediately available in the behaviour of others when they use each other as 'social tools' (Kummer 1982). Similar agonistic events involving three or more individuals have been observed in other species as well. De Waal (1976) regularly observed the following episode in a captive group of Java monkeys (*Macaca fascicularis*): an older monkey, when fleeing from an infant that enlisted aid from a non-specified group member by 'pointing' at his aggressor, signalled submission to the infant's mother and/or the alpha male. Unfortunately for the present purpose, it is not mentioned whether these potential allies were already paying attention to the situation and therewith showing their intention to intervene on behalf of the infant.

Immature baboons (*Papio cynocephalus ursinus*) formed alliances primarily with high-ranking rather than low-ranking adult females (Cheney 1977). They joined individuals that already controlled the dispute. They were able to choose this 'correct' side although they generally rarely interfered (in 6 per cent of 850 agonistic interactions involving immatures). Few opportunities for learning specific constellations of individuals favour, of course, a cognitive interpretation of such events. It would also be relevant to know whether monkeys participating in alliances join specific individuals rather than predicting who is likely to win or assessing dyadic characteristics. The former is more likely for Cheney's data since most alliances are formed with a few members of high ranking matrilines.

Frequency of interactions does not underlie the differential behaviour towards kin and non-kin in gelada baboons (*Theropithecus gelada*), as Dunbar's studies of the social relationships of adult females show (Dunbar 1980, 1983). Females only rarely interacted with animals belonging to another reproductive unit and within their own unit had only one or two close partners with whom they frequently interacted (Dunbar 1983). However, females supported members of their matriline that did not belong to their preferred partners in disputes with unrelated females. The probability of support did not depend on the frequency of interaction. The author concludes that gelada females remember matriline membership 'in the abstract' (Dunbar 1983).

Do monkeys really recognize aspects of relationships, and do they have insight into the group's structure? A study often cited to support this claim is Bachmann and Kummer's (1980) work with caged hamadryas baboons (*Papio hamadryas*). They found that a male rival was more likely not to challenge the female in possession of another male when the female showed a preference for her present owner, as measured in choice tests. Though suggestive, the study does not justify the conclusion that baboons judge aspects of a relationship because, as Kummer (1982) himself pointed out, the experimental design did not preclude the possibility that the rival recognized the preference of the female only from signals given by one animal of the pair. Also, apparent reactions to relationships may in some cases be adequately explained by a response to a perceived *Paargestalt* (Kummer *et al.* 1974) that does not require higher mental abilities.

Free-ranging vervet monkeys (*Cercopithecus aethiops*), it has been claimed, are able to perceive their group as hierarchically structured into different social units. Cheney and Seyfarth (1980) replayed the screams of juveniles to their mothers when the juveniles were out of sight of the mothers. Mothers responded to the screams by orientating towards the loud speaker with significantly shorter latency and longer duration than did control females nearby. Moreover, the controls, upon hearing the call, looked at the mother, often before she herself showed any noticeable reaction. This is a cleverly conceived experiment, and I am personally convinced that unobtrusive and simple behaviours such as looking and orientating will turn out to tell us more about a non-human primate's knowledge and cognitive competence than dramatic social events. I do not agree, however, with the interpretation of the findings of that particular experiment. In what way does the behaviour of the control females differ from that of a monkey in the laboratory orientating towards the location of an expected stimulus? What closely resembles a conditioned response should not be interpreted as showing the vervets ' "arrange" individuals into higher-order units, apparently on the basis of matrilineal kinship' (Seyfarth 1983). Mothers intervened in 22 per cent of all occasions when their offspring screamed during agonistic interactions and threatened their offspring's opponent. Thus, reactions of mothers are strong and frequent and important enough to attract the attention of other females who may memorize the event. This, and the reaction shown in the playback situation can arise without any understanding whatsoever by the females of the relationship between the individuals.

In summary, I have not found any strong evidence in the literature that non-human primates have insight into the structure of their group that requires some sort of concept of abstract entities such as relationships.

'Social cognition': empirical evidence

In vervets, grooming between unrelated individuals increases the probability that they will subsequently react to the playback of each others' vocal solicitation for aid in alliances by looking towards the speaker. If the groomer had been a close kin of the groomee, however, the latter did not look longer than when it had not been groomed in the last hour (Seyfarth and Cheney 1984). Thus, the effect of what an individual does to another depends on the prior history of the two. Grooming, and plausibly other behaviours as well, assumes a distinct interindividual value or meaning for the interacting partners. This seems to meet Selman's (1980) criterion for a strictly 'social' behaviour in humans—to qualify as social, it must have some interpersonal, psychological meaning for the participants in an interaction.

Beck (1973) reported on a case of cooperative tool use in a sexually bonded pair of hamadryas baboons. The female brought the tool to the male from a cage he could not enter. While the male simply grabbed the tool from other animals, he lip-smacked at 'his' female when she was in the other cage, and elicited her following response when she got the tool by walking away and looking back at her. The male thus seemed to anticipate the female's reaction and used a species-typical signal in an instrumental way in order to get the tool. Even more important, the female's share of the food provided by the male was highest during phases of low cooperation by the female. The author felt that the male left the food on purpose. The first observation suggests the insightful use of a behaviour in an unusual context, the other could tentatively be interpreted as manifesting the male's understanding of the effect of larger food rewards on the female.

Stammbach and Kummer (1982) found that grooming interactions among three or more hamadryas females were 'personal' in that the females specifically responded to the identity of their partners. Each female seemed to have distinct attributes for each of the other females. What these attributes were, is largely unknown. In an experimental study of the development of dyadic relationships between nine adult rhesus males, Sigg (1981) found a linear dominance hierarchy among the males although they had only pairwise interactions with each other. There was no correlation beteen rank and body size or weight. Instead, dominance relationships appeared to be decided on the basis of such personality traits as 'self-assurance'. The last two examples indicate that monkeys perceive and react to individual features of conspecifics that have no obvious correlates in overt behaviour or physical appearance. That these features have so far escaped the attention of primatologists suggests that they are not simple but complex and possibly cognitively demanding.

Some examples given in the previous paragraph have shown that non-human primates can predict how particular individuals will react not only to themselves but also towards others. Whether this is evidence for some level of decentration cannot easily be decided, as the example of the vervets shows. Kummer (1982) discussed the possibility that hamadryas baboons may be capable of spatial perspective taking in the context of hiding from each other. He reports an anecdote suggesting that A knows when B cannot see it though A sees B. Experiments are needed to establish what experience A needs to be able to predict what B sees, and whether A realizes that B may have a visual field differing from what its own would be in B's place because of, e.g. a difference in body size.

Premack and Woodruff (1978) assessed the ability of Sarah, a test-sophisticated, wild-born chimpanzee, to impute mental states to a human actor in a videotaped scene depicting a problem, e.g. an actor trying to reach an inaccessible food item or to escape from a locked cage. Sarah's choice of the still photograph showing the correct solution is evidence that she assigned a purpose to the actor since the possibilities of mere physical matching or other simple associations were precluded (but see a comment by Savage-Rumbaugh et al. 1978).

Cognition and complexity

The characteristics of cognitive events were given in the introduction. The attribute 'complex' is not needed though intuitively we recognize a relation between complexity of content and the cognitive mode. What is meant by complexity? A complex event or structure consists of many different elements connected to each other through a network of pathways or relations. The more elements and connections there are, the more complex is the overall structure. In social structures and events the elements are individuals, the connections between them are manifest in their relationships and interactions. The social structure of many primate species is unmatched in its complexity according to these criteria by any other non-primate species (e.g. Wrangham 1983; Eisenberg 1973). However, complex social structures can be the product of the relatively simple, non-cognitive behaviour of individuals and their propensities to associate with particular other individuals (e.g. Hinde 1974; Vaitl 1978). Some social interactions, especially triadic and polyadic events, are highly complex in that the animals involved relate their behaviour to each other and instantly perceive a number of diverse pieces of information and compare them with memorized comparable episodes (see above). Does this complexity necessarily imply that the monkeys use higher cognitive abilities? Kummer's (1957) observations on the ontogeny of the protected threat in hamadryas baboons suggest that tripartite tactics can be mastered by trial and error learning (Kummer 1982). An understanding of why these tactics work is not necessary, nor any insight into the social relationships of other monkeys. It would be relevant to know to what extent non-human primates extrapolate types of interactions across individuals or classes of individuals and contexts. Do they require information about each possible combination of individuals, either by direct interaction with them or by observing others interacting? Since they have at least a low-level capacity for inferential reasoning, responses to properties of dyadic relationships neither immediately perceivable nor directly experienced in the past is a reasonable possibility. Unfortunately, direct evidence is lacking. Altmann (1981) proposed an experimental scheme that would allow one to test whether non-human primates respond to the dominance relation as an abstract property of a dyad. A similar idea was expressed by Kummer (1982). Dennett's (1983) suggestion that one should tentatively ascribe levels of intentionality to an animal and thus arrive at testable hypotheses about what it knows, believes and wants appears as a suitable method for analysing complex and possibly cognitive events.

Conclusions

Cognition is not an unitary term but subsumes heterogeneous processes. When speaking about an animal behaving cognitively we need to specify in detail what it does and what kinds of cognitive abilities are implied. Primatologists observing social groups in the laboratory and the field are impressed by the great variability and subtleness of non-human primate behaviour. The behaviour underlying interindividual relationships seems particularly sophisticated. It certainly presupposes excellent memory and learning capacities. It is less obvious what higher cognitive abilities over and above these basic capacities are needed to explain the behaviour of non-human primates. I join Seyfarth's belief that monkeys know more than we realize (Seyfarth 1983) but, in my opinion, we still lack the kind of data necessary to infer the internal processes that give rise to the behaviour of interest. I believe that in order to find the most advanced cognitive processes involved in social behaviour, we need experiments that borrow from the rigorous criteria of formal cognitive laboratory tasks, and that we need experimental subjects that are socially sophisticated individuals and live in a rich environment. Such techniques are currently being used by myself to investigate the possibility that monkeys form concepts about interindividual relationships.

Acknowledgements

I thank Prof. H. Kummer, Dr H. Sigg, and Dr A. Stolba for their critical comments on the manuscript and Prof. R. A. Hinde for editorial help.

3

The social and non-social world of non-human primates

DOROTHY CHENEY AND ROBERT SEYFARTH

Why did intelligence evolve? Although the question can be answered teleologically—'intelligence would not have evolved unless it served some biological function'—it is nevertheless more difficult to explain the evolution of intelligence than to explain the evolution of simpler traits such as antler size or fighting ability. Indeed, because scientists have found it easy to speculate about, and even measure, the selective advantage of the latter traits, there has been a tacit assumption that the same general approach can be applied to intelligence. A variety of studies, however, has shown that animal 'intelligence' is not a unitary phenomenon, but differs depending on the particular combination of species, task, and environment. In the case of primates, it has been suggested that intelligence is particularly marked in the social domain, leading to the hypothesis that intelligence may originally have evolved in part as the result of the need to manipulate and compete with other group members.

Captive primates tested with objects often face problems that are logically similar to the social problems confronting primates in the wild. Despite this similarity, however, the performance of primates in these two contexts often seems to differ strikingly. To cite just one example, McGonigle and Chalmers (1977) and Gillan (1981) demonstrated transitive inference in captive squirrel monkeys and chimpanzees, respectively, but were able to do so only after considerable training with paired stimuli. In contrast, field observations suggest that even young monkeys may discriminate among other animals' relative ranks, apparently from their observation of dyadic interactions (Cheney 1978; Seyfarth 1981; Datta 1983). Observations and experiments have also suggested that primates regularly group together individuals on the basis of their relationships with others, and appear to understand enough about the behaviour and motives of others to be capable of deceit and other subtle forms of manipulation (see below). Such observations are both intriguing and frustrating, because they suggest the existence in the wild of striking mental abilities that, with some notable exceptions (Woodruff and Premack 1979; Premack and Premack 1982), have not been documented or duplicated in the laboratory.

Because of the qualitative differences between field and laboratory stimuli, one possible explanation for the animals' differing performance suggests that primate intelligence is relatively domain-specific. This hypothesis argues that group life has exerted strong selective pressure on the ability of primates to form complex associations, make transitive inferences, and predict the behaviour of fellow group members. Thus abilities that seem to emerge only with human training in captivity may readily occur in primates under natural conditions, but primarily in the social domain (Jolly 1966; Chance and Jolly 1970; Humphrey 1976; Kummer 1971, 1982; see also Rozin 1976). Similarly, when captive chimpanzees solve technological problems that require foresight and an understanding of the consequences of past decisions (Döhl 1968), they may be demonstrating abilities for which they have been preadapted as a result of the need to make equally strategic decisions about each other (de Waal 1982).

This domain-specific hypothesis posits that natural selection may have acted to favor complex abilities in the social domain that are, for some reason, less easily extended or generalized to other spheres. However, while the hypothesis may be intuitively satisfying, it has not yet been systematically tested. In this paper, we attempt to investigate this issue in more detail by describing some observations and experiments on the social and non-social knowledge of free-ranging vervet monkeys (*Cercopithecus aethiops*). In the first part of the paper, we examine primates' understanding of the behaviour and social relationships of other group members, focusing in particular on kinship classification. In the second part, we describe a series of field experiments that were designed to present monkeys with logically similar problems using either 'social' or 'non-social' stimuli.

Although our field experiments are less precisely controlled than laboratory tests, they have a number of advantages. First, problems of motivation and human training are circumvented. Second, free-ranging primates daily encounter logically similar social and non-social problems, thus permitting a direct test of performance in the two domains. Third, our subjects regularly deal with objects in the external world that are either relevant or irrelevant to their survival. It is therefore possible to examine whether the monkeys' understanding of non-social objects is a function of their relative importance to them.

At the outset, two points should be emphasized. First, throughout the paper we draw a distinction between the performance of primates in the 'social' and 'non-social' domains. While we believe that this distinction is a heuristically important one, we recognize that the boundary between these spheres is ill-defined. Second, in evaluating observations and experiments, we make no claims about the mechanisms underlying performance. Our experiments measure only the responses that particular stimuli evoke, and not the processes (mental or otherwise) that underlie such responses.

Many of the results we describe could, for example, result either from relatively simple associative learning or from more complex cognitive processes. Our aim is not to argue for one of these alternatives. Instead, we use experiments to determine which of two stimuli is more salient, and to suggest that animals form some sorts of associations more readily than others.

Study site and subjects

Most of the research described in this paper was conducted on three free-ranging groups of vervet monkeys in Amboseli National Park, Kenya. Vervet monkeys live in stable social groups consisting of a number of adult males, adult females, and their juvenile and infant offspring (Cheney *et al.* 1985). As in most Old World monkeys, female vervets remain in their natal group throughout their lives, maintaining close bonds with maternal kin. Males, in contrast, emigrate to neighbouring groups at sexual maturity, often in the company of brothers or natal group peers (Cheney and Seyfarth 1983). Within each group, males and females can be ranked in linear dominance hierarchies that predict the outcome of competitive interactions over food, water, and social partners. Offspring acquire dominance ranks immediately below those of their mothers, such that all members of a family typically share adjacent ranks (Cheney 1983).

Social knowledge

1 The prediction and manipulation of other individuals' behaviour

There are numerous examples in the primate literature suggesting that non-human primates are capable of predicting both the behaviour of others and the consequences of their own actions. Since it is difficult to conduct research on the social perceptions of non-verbal animals, most of these examples are anecdotal. Taken together, though, such anecdotes gain in persuasive power and suggest that primates may have a sophisticated understanding of the relationships, motives, and intentions of others. For example, Kummer (1982) describes a case in which a juvenile female hamadryas baboon repeatedly left her adult male leader to mate with a juvenile male behind a rock where the adult male could not see her. Between matings, she peeked around the rock at the adult male. The female behaved as if she could ascribe motives to other individuals, and that she was aware of the consequences of her own behaviour.

The capacity to deceive suggests that individuals understand enough about the motives and beliefs of others to be able to manipulate them effectively. Although it is impossible to show conclusively that monkeys and apes are capable of the same sort of conscious deceit that we take for granted in humans, there are some intriguing suggestions that non-human primates occasionally attempt to mislead others. Woodruff and Premack (1979) have shown experimentally that a chimpanzee may misinform a particular individual about the location of food when that individual has failed to share food in the past. Similarly, vervet monkeys may selectively withhold information about the presence of danger, depending upon their dominance rank and the presence or absence of close kin (Cheney and Seyfarth 1981, 1985). In the wild, low-ranking individuals, who are excluded from resources by others, alarm-call at lower rates than high-ranking individuals, even though they appear to have the same opportunities to spot predators. Experiments with captive vervets have also demonstrated that individuals alarm-call more in the presence of close kin than when they are with unrelated animals (Cheney and Seyfarth 1985). As with those cited above, these examples do not prove that animals are capable of consciously manipulating others. We can say only that their behaviour *functions* as if to deceive, and use these data as a base from which to design the appropriate experiments.

Chimpanzees also attempt to camouflage their intentions or moods. De Waal (1982) describes a number of cases in which chimpanzees were observed to cover and hide their own fear grimaces when approached by more dominant individuals. Similarly, young chimpanzees in the wild have been observed to cover their mouths and even to throw silent temper tantrums when following others on a raiding incursion into the range of another group, where loud noises and detection could lead to a violent encounter (Goodall *et al.* 1979). These examples again suggest that primates are capable of predicting the consequences of their actions or moods on others, and that they adjust their own behaviour accordingly. Indeed, in describing a case in which a male chimpanzee formed a long-term alliance which resulted in the eventual domination of a third individual, de Waal (1982) argues that chimpanzees may be capable of formulating strategies based on the prediction of events that will occur many months later.

2 Kinship classification

a. Classification of objects and conspecifics

In captivity, it is possible to train both monkeys and apes to group objects into superordinate classes on the basis of either functional or visual similarity. For example, Savage-Rumbaugh and her colleagues have been able to train chimpanzees to sort objects, pictures of objects, and even symbols for objects, into classes (Savage-Rumbaugh *et al.* 1980). Their results suggest that it is possible to teach chimps to classify objects according to some shared abstract feature, and not simply according to association or physical similarity.

There is also evidence that monkeys will learn, after extensive training, to sort objects according to form or colour (Flagg and Medin 1973; Garcia and Ettlinger 1978; Essock and Rumbaugh 1978). They appear to have some difficulty, however, in learning to recognize the constituent parts of compound objects. For example, in an experiment where cebus monkeys were presented with a picture of a triangle inside a square, the animals treated the object as a unified whole, and did not later recognize that it was in fact composed of two different parts (Cox and D'Amato 1982). This suggests an inability to recognize that objects can be associated or grouped into higher order units.

Curiously, however, monkeys do seem to reveal such an ability, and without human intervention, when the stimuli to be associated are other individuals. For example, in an experiment conducted on free-ranging vervet monkeys, we played the tape-recorded scream of a particular juvenile to three adult females, one of whom was the juvenile's mother. In most trials, mothers looked toward the speaker for longer durations than did control females, suggesting that females were able to distinguish among the screams of individual juveniles. More interesting, however, was the finding that in a significant number of trials control females looked at the mothers, without any prior cues from the mothers themselves. This result suggested that females associated particular screams with particular juveniles, and these juveniles with particular adult females (Cheney and Seyfarth 1980, 1982). Vervets therefore appeared to classify individuals on the basis of close association, and perhaps even according to matrilineal kinship.

b. Redirected aggression

Observational data on redirected aggression provide further suggestions that monkeys may be able to recognize the kin relationships of other individuals. Redirected aggression occurs when one monkey is involved in a fight with another and subsequently threatens a third, previously uninvolved, individual. Among rhesus macaques, baboons, and vervets such aggression is frequently directed against a close relative of one of the prior antagonists (Judge 1982; Smuts 1985).

We examined redirected aggression in two groups of vervets where there was a large number of individuals of known kinship. Whenever an animal was involved in an aggressive interaction, we noted the frequency with which it threatened a close relative (mother, offspring, or maternal sibling) of its antagonist within the next two hours. (Both here and below, the resulting measure often underestimated the actual frequency of such kin-directed aggression, since, if the original fight occurred shortly before the termination of the daily observation session, individuals could not always be observed for a full two hours after the original fight.) We then compared this figure with the mean daily frequency with which the animal threatened that individual.

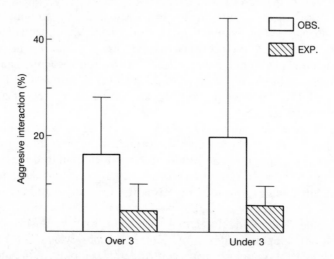

Fig. 3.1 The frequency with which vervets redirected aggression against individuals following aggression with those individuals' relatives (observed), and following a period when no such prior aggression had occurred (expected). Histograms shows means and standard deviations for individuals in two social groups. Two-tailed Wilcoxon tests compare observed and expected probabilities. For animals over three years of age, $N=25$, $T=15.5$, $p<.001$; animals under three years, $N=16$, $T=17$, $p<.01$.

Adult females and juveniles were significantly more likely than expected to threaten a particular individual if they had recently been involved in aggression with that individual's relative (Fig. 3.1). This redirected aggression was not the result of close spatial proximity between kin, since the average amount of time elapsed between the first fight and the subsequent redirected threat was 24 minutes, long after the original participants had separated.

Further analysis of redirected aggression revealed a more subtle recognition of kin relationships. In both groups, vervets were significantly more likely than expected to threaten an individual when that individual's close kin and their own close kin had previously been involved in a fight (Fig. 3.2). Thus, for example, in a group that included two sisters (Carlyle and Shelley) and an adult female (Maginot) and her offspring (Trollope), Carlyle was significantly more likely to threaten Maginot if there had previously been a fight betwen Shelley and Trollope. The monkeys therefore showed evidence of recognizing the relation between their *own* kinship bonds and the kinship bonds of others.

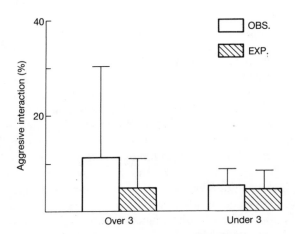

Fig. 3.2 The frequency with which vervets redirected aggression against individuals following a fight between these individuals' close relatives and their *own* relatives (observed), and following a period when no such prior aggression had occurred (expected). Legend as in Fig. 3.1. For animals over three years of age, N=25, T=66, p<.01; animals under three years, N=15, T=47, N.S.

At this point, we should reiterate that these data do not reveal the mechanisms underlying kin recognition. The ability to recognize matrilineal kinship could occur because animals are able to construct hierarchical taxonomies of their fellow group members, or it could be the result of simple associative learning. As Premack (1976, 1983) points out, unambiguous documentation of the more complex hierarchical process could be demonstrated only through examination of the animals' understanding of analogy, which explicitly demands that animals understand the similarity between sets of relations. Experiments conducted on captive chimpanzees have suggested that language training is an essential prerequisite for analogous reasoning (Gillan *et al.* 1981; Premack 1983).

Nevertheless, the examples on redirected aggression suggest that we should not completely rule out the possibility of analogical reasoning in free-ranging primates. In our observations individuals behaved as if they recognized that the kin relations of *others* were analogous to the kin relations of *themselves*. At the very least, this recognition seems to require some knowledge of the relationships of both oneself and other group members. Such knowledge would be non-egocentric, because it appears to demand that individuals attend to interactions in which they themselves are not involved. Furthermore, it is possible that, even in the absence of language training, this understanding may involve the use of some abstract category, such as kinship.

c. The development of kinship classification

It is difficult to specify exactly when the complex associations described above are acquired during development, because even very young juveniles seem to recognize the dominance ranks of others and associate preferentially with kin. Among baboons and rhesus macaques, for example, juvenile females as young as one year interact with their infant siblings more than with other infants and, when interacting with unrelated infants, prefer the infants of high-ranking females to those of low-ranking females (Cheney 1978; Berman 1982). In this respect, juvenile behaviour resembles that of adult females, who also compete to interact with the members of high-ranking families (Seyfarth 1977, 1983). Similarly, juvenile baboons and vervets form alliances with unrelated animals in direct relation to those animals' ranks, such that high-ranking animals receive the most, and low-ranking animals the least, alliances (Cheney 1977, 1983).

However, although juveniles seem to recognize the ranks and associations of others, there is some evidence from data on redirected aggression that they may be less adept than older animals at recognizing complex social relationships, in particular the similarity between their own and other animals' kinship bonds. In the simple case described above, when vervets redirected aggression against a close relative of their prior antagonist, juveniles aged less than three years behaved the same as older animals (Fig. 3.1). There was an age difference, however, in the more complex case. Although older animals often threatened the close kin of individuals who had recently fought with their own relatives, younger juveniles did not do so (Fig. 3.2). Thus the understanding of this complex relation may develop more slowly than the more simple recognition of associations between particular individuals.

Non-social knowledge

The previous section has described the rich understanding that monkeys appear to have of their own and other individuals' social relationships. Is this understanding, however, extended to other objects in the external world? Below, we describe a number of experiments that were designed to address this question. Since the experiments are discussed in detail elsewhere (Cheney and Seyfarth 1985), they are only summarized here.

1 Relevant aspects of other species' behaviour

Vervets in Amboseli are preyed upon by a number of different carnivores, eagles, baboons, and pythons (Cheney and Seyfarth 1981). The monkeys give acoustically distinct alarm calls to different predators, and experiments have shown that each of these alarms evokes qualitatively different escape responses (Seyfarth *et al.* 1980). Calls given to carnivores, for example, cause monkeys to run into trees, while calls given to eagles cause monkeys to look up in the air. The monkeys' alarm calls therefore function to designate different types of danger in the external world.

Vervets are not the only species to give alarm calls to predators, however, and it would seem advantageous for the monkeys to distinguish among the alarm calls given by other species. We investigated the vervets' recognition of other species' alarm calls through playback experiments, using the calls of the superb starling (*Spreo superbus*). Starlings give two acoustically distinct alarm calls to predators, both of which differ substantially from the vervets' own alarms. One starling alarm—a harsh, noisy call—is given to various terrestrial predators (including vervets), all of which prey on starlings or their eggs, but only some of which prey on vervets. The second alarm—a clear, rising tone—is given to many species of hawks and eagles, two of which prey on vervets.

We played tape-recordings of the starling's calls to the monkeys from a concealed loudspeaker, following the same protocol previously used in tests of the vervets' own alarm calls (Seyfarth *et al.* 1980). Three starling calls were used: their ground predator alarm call, their aerial predator alarm call, and, as a control, their song. Results are presented in Fig. 3.3.

Playback of the starling's ground predator alarm caused a significant number of monkeys to run into trees, while playback of the aerial predator alarm caused a significant number of vervets to look up. In contrast, the starling's song elicited little response. In this test, therefore, where the behaviour of another species was relevant to the vervets' survival, the monkeys' recognition of another species' calls was similar to their recognition of their own calls. Research on the acquisition of the meaning of other species' alarm calls is now in progress (Hauser in prep.).

2 Apparently irrelevant aspects of other species' behaviour

The alarm calls of other species represent one end of a continuum of biologically relevant and irrelevant stimuli in the external world. It is perhaps not surprising that vervets discriminate among such alarms, since they are so obviously important to survival. Can similar knowledge be demonstrated, however, for aspects of another species' behaviour that are apparently unrelated to the monkeys' survival? This seems an important

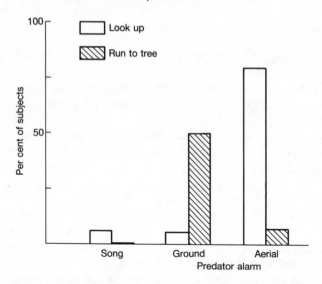

Fig. 3.3 Responses of vervets to playback of three different starling calls. Number of subjects for playback of song, terrestrial predator alarm, and avian predator alarm were 17, 18, and 15, respectively. Ground predator alarms evoked more running to trees than either song ($X^2 = 11.4$, $p < .01$) or avian predator alarms ($X^2 = 7.3$, $p < .05$); avian predator alarms evoked more looking up than either song ($X^2 = 18.2$, $p < .01$) or terrestrial predator alarms ($X^2 = 19.0$, $p < .01$).

question, because one striking feature of human intelligence is our inclination to accumulate information about the world that is not directly relevant to our survival. Can the same be said of vervet monkeys? Are vervets as good naturalists as they are primatologists?

To address this question one must first identify two comparable features of the monkeys' environment, one social and biologically relevant, the other non-social and apparently irrelevant to the monkeys' survival. Then experiments must be designed to compare the monkeys' behaviour in these two domains. As a relevant social test, we asked the monkeys how much they knew about the ranging behaviour of other vervets. As a non-social, apparently irrelevant test we asked the monkeys how much they knew about the ranging behaviour of other species that neither compete nor interact with vervets in any obvious way.

Vervets aggressively defend their group's range against incursions by other groups. Females and juveniles are active participants in inter-group encounters, and give a distinctive vocalization when they spot the members of another group (Cheney and Seyfarth 1982a). In testing the vervets' recognition of other groups' membership and ranges, subjects in one group were played the inter-group call of an animal from a neighbouring group, either from the true range of the vocalizer's group or from the range of another neighbouring group. In these paired trials, subjects responded with significantly more vigilance to calls played from the 'inappropriate' range than to calls played from the 'appropriate' range (Cheney and Seyfarth 1982a).

Subsequent experiments followed the same design, but used as stimuli the calls of other species. Vervets were played the calls of two species that are habitually found in or near water, the hippopotamus and the black-winged stilt (*Himantopus himantopus*). These two species were chosen because neither competes nor interacts with vervets, and both are therefore of little biological importance to the monkeys. Nevertheless, each is a species whose vocalizations are so restricted to wet areas that any indication of calling in another habitat might be regarded, at least by humans, as anomalous.

Subjects were played hippo or stilt calls in paired trials, from either the swamp ('appropriate') or dry woodland ('inappropriate') habitat. The monkeys responded to the playbacks either by looking in the direction of the loudspeaker or by apparently ignoring the call (Fig. 3.4). The monkeys' responses, however, did not vary depending on the habitat from which the calls were played. Vervets responded to both types of calls as if they did not recognize that calls played from a dry habitat were inappropriate.

These negative results, of course, cannot distinguish between the failure to recognize an anomaly and the failure to respond to one. It is entirely possible, for example, that vervets recognize that hippos belong near water, but simply do not respond to the hippo calls emerging from a dry woodland. Negative results *are* of interest, however, when contrasted with similar experiments that do evoke responses. Although vervets fail to respond to hippo or stilt calls coming from an inappropriate area, under comparable conditions they respond strongly to the calls of another vervet. The different performances are particularly striking given that the trials using conspecific calls asked subjects to assess the appropriate location of different *individuals*, whereas the hippo and stilt calls required only a gross recognition of the appropriate location of different *species*.

3 Associations between other species

As discussed earlier, vervets seem capable of forming associations between other group members based on observations of social interactions. A further series of experiments was designed to test the vervets' ability to associate different species with one another.

Vervet monkeys regularly come into contact with Maasai tribesmen, who bring their cattle into the park to graze. Although the Maasai do not

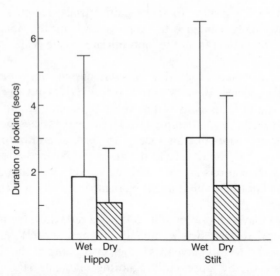

Fig. 3.4 Duration of looking toward speaker (in seconds) after playback of hippopotamus and black-winged stilt calls from wet and dry habitats. Values shown are means and standard deviations. N=10 subjects for hippo calls; N=18 subjects for stilt calls. Duration of responses to calls played from different habitats did not differ significantly (p>.10).

prey on vervets, they occasionally throw sticks or rocks at the monkeys, with the result that their approach causes increased vigilance and flight. Cows themselves pose no danger to the monkeys. Nevertheless, since cows never enter the park without Maasai, a cow alone potentially signals the approach of danger. To test whether monkeys have learned to associate cows with Maasai, we played the lowing vocalizations of either cows or wildebeest (*Connochaetes taurinus*, a common ungulate) to vervets in paired trials.

As Fig. 3.5 indicates, playback of cow vocalizations caused subjects to look toward the speaker for significantly longer durations than did playback of wildebeest calls. This increased vigilance suggests that vervets associated cows with danger, and that they responded to the apparent approach of cows as they would to the approach of Maasai themselves.

4 Secondary cues of danger

When leopards make a kill, they frequently drag their prey into trees, where they can feed without harassment from other predators. This behaviour is peculiar to leopards, and local humans recognize that the sight of a fresh carcass in a tree denotes the proximity of a leopard. We examined whether vervets knew enough about the behaviour of leopards to under-

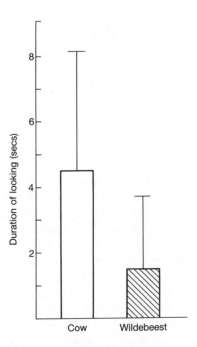

Fig. 3.5 Duration of looking toward speaker after playback of wildebeest and cow vocalizations. Legend as in Fig. 3.4. Duration of responses was longer after play-back of cow vocalizations (Two-tailed Wilcoxon test, N=17, T=20.5, p<.05).

stand that a carcass in a tree in the *absence* of a leopard represented the same potential danger as did a leopard itself.

In conducting the experiment, we placed a limp, stuffed carcass of a Thompson's gazelle, in a tree, before dawn, approximately 50–75 m from the monkeys' sleeping trees. The carcass was positioned in such a way as to mimic its placement by a leopard (indeed, our attempt fooled at least one tour bus driver into thinking that a leopard was in the area). In total, we presented the carcass to one group of baboons and four groups of vervets. One of the vervet groups had seen a leopard in a tree with a carcass only four days earlier, and had uttered alarm calls even when the leopard temporarily left the tree.

Despite the groups' previous observation of leopards with carcasses in trees, neither baboons nor vervets alarm-called at the sight of the carcass alone. Moreover, there was no increased vigilance in the direction of the carcass over that which might have been expected by chance. In all cases, the monkeys behaved as if they did not recognize that a carcass in a tree denoted the proximity of a leopard.

As a further test of monkeys' knowledge of secondary cues of danger, we tested the vervets' recognition of python tracks. Pythons often prey on vervet monkeys (Cheney *et al.* 1981), and when vervets encounter a python, they give alarm calls to it and closely monitor its movements through the area (Seyfarth *et al.* 1980). Pythons lay distinct, wide, straight tracks which cannot be mistaken for those of any other species, and which often denote the presence of a python in a nearby bush. Vervets in the three study groups have often watched and alarm-called at a python as it laid down a track and then disappeared into a bush. Despite their experience with pythons, however, vervets do not seem to recognize that a python track represents potential danger. Over an eight month period of systematic observation, the monkeys ignored both real tracks and artificial tracks that we laid down in the monkeys' foraging route. Indeed, on some occasions, at least one individual subsequently entered a bush where a track led, encountered the hidden snake, and alarm-called at it.

In the preceding experiments, vervets performed well when the secondary cues of danger were auditory stimuli like alarm calls, but poorly when the secondary cues were visual stimuli like carcasses or tracks. There are at least three explanations for these results. First, auditory cues may be more salient than visual ones. This explanation is limited, though, because it fails to explain why, in the first instance, natural selection has favoured different abilities in the visual and auditory domains. It might also be argued that visual stimuli like tracks provide more opportunities for the extinction of a conditioned response than do alarm calls, because tracks will often be old and therefore more irrelevant than alarm calls. Many animal species, however, can be trained to attend to stimuli that are only occasionally reinforced. Given both the rate at which pythons prey on vervets and the frequency with which vervets have been observed to ignore tracks and enter bushes that contained pythons, the apparent inability to attend to tracks seems maladaptive.

We can also hypothesize that the vervets' communication system has evolved primarily to solve social problems, and that this has both shaped and limited their use of signals outside the social domain. Vervets use vocal signals both in the presence and absence of visual contact. If animals are foraging in dense bush, a vocalization can tell them that another group is approaching, or that a snake has been seen nearby, without any supporting visual information (Cheney and Seyfarth 1982b). In contrast, visual signals are limited to occasions when animals are in sight of one another. Vervets do not, for example, make use of each other's tracks when monitoring incursions by neighbouring groups, nor do they visually mark aspects of their physical environment to denote their rank or group membership. As a result, their lack of attentiveness toward the visual cues of predators may be related to their limited use of visual signals as secondary cues in their social interactions.

5 Cooperation and reciprocity

Cooperative alliances among humans are characterized by the exchange of goods or services between individuals. Significantly, such exchange is not limited to any particular domain. Exchange may involve actions (for example, reciprocal support in an aggressive coalition), individuals (the exchange of spouses between two villages), or material goods (the donation of money or food to cement an agreement). In contrast, while non-human primates frequently reciprocate past affinitive acts with future cooperation, the exchange of objects is rare (Chance 1961; Chance and Jolly 1970; Kummer 1971; Reynolds 1981).

In interactions involving both kin and non-kin, monkeys and apes may exchange grooming, alliances, and tolerance at food sites (e.g. Packer 1977; Seyfarth 1977; Chapais and Schulman 1980; de Waal 1977, 1982). Primates remember past interactions and adjust their cooperative acts depending on who has previously behaved affinitively toward them (Seyfarth and Cheney 1984). Such reciprocity, however, rarely involves the use or exchange of objects. Primate tool use, which has received considerable attention because of its relevance to human evolution (e.g. Beck 1974), is striking in part because it is relatively rare. By comparison, observers of primates are continually struck by their ability to form alliances and bonds that allow them to use other individuals as 'social tools' to achieve a particular result (e.g. Kummer 1968; Chance and Jolly 1970; de Waal 1982). Similarly, although parties of baboons and chimpanzees often hunt and kill prey, there is little evidence that such hunts are truly cooperative, or that meat is genuinely shared (Kummer 1968; Wrangham 1975; Busse 1978; Strum 1981; Teleki 1981).

Reciprocity among monkeys and apes therefore appears to occur more commonly in the form of social interactions than in the exchange of material goods. Before we conclude, however, that non-human primates differ from humans in restricting their cooperative acts primarily to the social domain, a number of caveats should be mentioned.

First, the relative rarity of food sharing among non-human primates may result partially from the fact that, with the exception of meat, the food of non-human primates is simply not worth sharing. Primates feed primarily on leaves and fruit that are not easily monopolized or hoarded by one individual. There may therefore be little benefit in acquiring food directly from another. Individuals may derive greater benefit through tolerance at a particular feeding site, and indeed, grooming, copulation, and other affinitive behaviour do occasionally increase the frequency with which subordinate individuals are able to feed near dominant animals (e.g. Weisbard and Goy 1976). Secondly, while non-human primates seldom provision each other or exchange material goods for future beneficial acts, such patterns of exchange do occur in other species. For example, in the courtship displays of many birds and insects, the male offers food to his mate.

Although cooperative behaviour in some animal species is occasionally characterized by the exchange of material goods, we do not know whether such patterns of exchange are at all modifiable. While humans can readily substitute a behavioural act for a material one, such flexibility in the 'currency' of reciprocal acts has seldom been convincingly documented in other animals. More research is clearly needed before cooperation and reciprocity in non-human species are fully understood. For the moment, however, we may hypothesize that, as in other aspects of their behaviour, reciprocity in primates appears to occur more often in the social than in the non-social domain.

Discussion

When interacting with each other, vervet monkeys are apparently able to form complex associations between individuals. Within a local population, vervets can both recognize individuals and associate them with particular groups. Within their own groups, the monkeys appear to recognize other individuals' relative dominance ranks, as well as the close associations that develop among kin. They also seem to remember who has behaved affinitively toward them in the past. Finally, there is an intriguing suggestion that monkeys may understand the relations between their own kinship bonds and the kinship bonds of others.

Curiously, however, vervets seem less able to form similar associations about non-social aspects of their environment, even when to do so would confer a selective advantage. Although the monkeys do recognize and respond to other species' alarm calls, they seem to ignore the visual cues associated with some predators. Similarly, vervets seem disinclined to collect information about their environment when it is not directly relevant to their own survival. This is perhaps not surprising, but it does point out a potential difference between monkeys and humans, who are naturally curious about much of their environment, and who engage in many activities that have little practical survival value. Finally, although vervets and other primates exhibit many forms of cooperation and reciprocity in their social interactions, comparable behaviour using non-social currency (for example, food sharing) is relatively rare.

We believe that these results can help us to understand the intelligence of non-human primates, and to specify more precisely how the minds of monkeys and apes differ from our own. It is now widely agreed that species-specific predispositions affect animal learning (Seligman and Hager 1972; Hinde and Stevenson-Hinde 1973; Johnston 1981). As a result of evolution in different habitats, the behaviour of different species depends not only on the logical structure of the problems they face, but also on the particular stimuli involved. Among primates, evolution appears to have acted with particular force in the social domain. As a result, while monkeys are able to form and make use of complex associations in their social interactions, similar associations are formed less readily when dealing with other species. Within the social group, the behaviour of monkeys suggests an understanding of causality, transitive inference, and reciprocity. Despite frequent opportunity and often strong selective pressure, however, comparable behaviour emerges less readily in dealings with other animal species or with inanimate objects.

Some of our generalizations about domain-specific performance may be less applicable to apes than to monkeys (Premack 1976). For example, unlike monkeys, apes do make occasional use of visual symbols in their social interactions. Free-ranging chimpanzees make sleeping nests each night, and, when the members of one group make incursions into the range of another, they have been observed to make aggressive displays upon encountering their neighbours' empty nests (Goodall *et al*. 1979). Similarly, the captive chimpanzee Vicki was able to sort pictures of animate and inanimate objects into distinct categories without previous training (Hayes and Nissen 1971). Whether or not a monkey would be capable of similar classification is not known, because the relevant experiments have not yet been conducted.

While we have emphasized the importance of social pressures in the evolution of intelligence, we do not mean to imply that ecological pressures are irrelevant. Field data on many species indicate that primates frequently range over large areas and remember the locations and phenological patterns of water and a variety of plant foods (e.g. Rodman 1977; Wrangham 1977; Sigg 1980; Sigg and Stolba 1981). As a result, it has been argued that ecological pressures have played a major role in the evolution of primate intelligence (e.g. Clutton-Brock and Harvey 1980; Milton 1981).

This hypothesis emphasizes that the distinction we have drawn between social and non-social intelligence is not a simple one, and that it is unproductive to oppose one unifactorial ecological argument against an equally unifactorial social one. Primate memory has no doubt evolved as a result of the need to remember both the location of spatially dispersed food resources and previous social encounters. Nevertheless, primates do not appear to manipulate objects in their environment to solve ecological problems with as much sophistication as they manipulate each other to solve social problems. The challenge of exploiting widely dispersed and ephemeral food items may thus have led to increased intelligence not simply because food collection itself becomes more difficult, but because animals must compete with one another to obtain it.

There are some intriguing parallels between the apparent social acuity of non-human primates and the cognitive development of children. In the past, many students of human development believed that infants' knowledge of the social and non-social world developed at similar rates (e.g. Piaget 1963). Recent studies question this view and suggest that an understanding of certain concepts may appear at an earlier age when the stimuli involved are animate (especially other people) than when they are inanimate. For example, Hood and Bloom (1979) examined the development of children's expressions of causality, using two- and three-year-olds as subjects, and found that children readily discussed the intentions and motivations of people in causal terms. They did not, however, talk about causal events involving objects (see also Gelman and Spelke 1981; Hoffman 1981; MacNamara 1982). These results, together with those presented here, suggest that there is, in both human and non-human primates, an evolutionary predisposition which makes it easier for organisms to understand relations among conspecifics than to understand similar relations among things. Compared with humans, non-human primates exhibit this predisposition in an extreme form; they show sophisticated cognitive skills when dealing with each other, but exhibit such skills less readily in their interactions with objects. Among humans the predisposition is more subtle, but nevertheless may appear in the earliest years of childhood, when infants exhibit remarkable social skills while at the same time remaining ignorant of much of the world around them. For a few years, children reveal the results of selection acting on the primate brain; selection that has made them particularly sensitive to the emotions, behaviour, and social relations of their conspecifics.

Acknowledgements

Research was supported by NSF grant BNS 82–15039 and the H. F. Guggenheim Foundation. We thank the Office of the President, Republic of Kenya, for permission to conduct research in Amboseli National Park, and Bob Oguya, the Warden in Amboseli, for help and cooperation. We thank B. Musyoka Nzuma for field assistance, E. Mottram and the staff of the National Museums of Kenya for help in preparing a Thompson's gazelle carcass, and D. and C. Nightingale for help in recording cow vocalizations. We thank D. Dennett, S. Essock-Vitale, D. R. Griffin, H. Kummer, C. Ristau, B. Smuts, and F. de Waal for their comments on earlier drafts. During manuscript preparation the authors were fellows at the Center for Advanced Study in the Behavioral Sciences, where they received support from NSF (BNS 76–22943) and the Sloan Foundation (82–2–10).

Discussion

The two papers in this section were discussed together.

Selective pressures in the evolution of human intelligence

It is often assumed that human cognitive capacities evolved under selection pressures acting through the necessity of coping with the complexities of social life (see pp. 7–8). However, *Vauclair* called attention to the hypotheses of Parker and Gibson (1979), which emphasize the importance of tool-making, tool-using, complex hunting, and food-sharing, rather than purely social issues. In addition *Dasser* noted that the orangutan apparently has considerable cognitive abilities and yet lives a largely solitary life.

In reply, *Cheney* said:
'I discuss this issue to some extent in my paper. No doubt, selective pressures placed on animals to find food and remember where food is located are important to the evolution of intelligence. Intelligence is not a unitary phenomenon, and it is unproductive to oppose a unifactorial ecological argument against an equally unifactorial social one. Nevertheless, primates do not seem to manipulate objects in the environment to solve ecological problems with as much sophistication as they manipulate each other to solve social problems. The challenge of exploiting ephemeral and dispersed food sources may therefore have led to increased intelligence *not* just because food collection itself becomes more difficult, but because individuals must compete with *each other* to obtain it.'

(It must be noted that Parker and Gibson, (1979), like Rozin (see p. 4), emphasized that selection must have operated on *specific* abilities. Just as the ability to grasp probably arose as an adaption for small-branch clinging while hand-catching insects, and was later used in many other contexts, so our several intellectual capacities were presumably initially selected for *specific* functions—*Eds.*).

Social vs. non-social cognition

There were considerable differences of opinion over how best to conceptualize this distinction (see also pp. 4–8). *Chandler* (see also Chandler 1976) argued that the distinction does not require the assumption of two distinct cognitive modes, but only the implication that social events contain features never or rarely present in non-social events. *Tizard* emphasized that the distinction lies between the domains or contexts in which the skills are applied, with individuals differing in the effectiveness with which they apply skills in social and non-social contexts. *Dasser* agreed with this, though implying that the skills used differed between contexts. She emphasized that there was no clear evidence for the distinction in primates, with the possible exception of the ability for perspective taking.

Rutter argued that, while social and cognitive domains are conceptually distinct, 'It is an empirical question whether or not the two domains reflect different systems. However, if that question is to be answered it is crucial that a very clear separation be made between social and cognitive *variables*. In ordinary circumstances, of course, social relationships and cognitive functioning tend to be closely intertwined. This will be so both because social contexts constitute important settings for all forms of learning and because social relationships will be influenced by skills of various kinds. Nevertheless, if we are to understand the nature of functional connections between socialization and cognitive competence we need to pay systematic attention to cases where the usual connection is broken. Such cases exist, as shown by the example of autism which is characterized by a gross failure in social relationships but in which there may be ordinary, even superior, functioning on at least some types of tests of general intelligence.

A further rather different point is that it is likely to be an oversimplification to ask whether socialization and cognition reflect the same or different systems. It is a commonplace in biology for systems to be separate and different but yet for them to be functionally interconnected. For example, the CNS and endocrine systems are separate; they show quite different age trends in development and they are subject to different influences. Nevertheless, the endocrine system is subject to CNS control through the hypothalamic-pituitary connections. With respect to interpersonal relationships and cognitive development I suggest that we need to separate the measurement of the two sets of variables in order to assess the variety of ways in which the two may interact or connect.'

Taking up Rutter's point about autism, *Weiskrantz* argued that a basic cognitive deficit could have very dramatic social consequences. 'For example, it has been reported that autistic children can suffer from a recognition memory deficit (Boucher and Warrington 1976). This would mean that they live in a continuously strange, and hence very frightening, world.'

Bryant queried whether normal human infants really are more precocious in the social as compared with the physical domain (cf. p. 5), for proper experiments can reveal surprising abilities in the latter (Harris 1984). He emphasized that the distinction between (i) being able to make a cognitive move in principle and (ii) recognizing when that move is necessary, is important for understanding human cognitive development. Any difference between social and physical cognition might involve either, and it is important to specify which. Autism could be an extreme example of the importance of context, autistic children functioning better in non-social than in social contexts.

Finally, *Perret-Clermont* gave a certain perspective to the whole discussion by asking whether the capacity to distinguish between a physical and a social event is a matter of social or of physical cognition.

Social vs. physical curiosity

Cheney had suggested that humans tend to be more curious about the non-social environment than other species, even when this is of no adaptive value. *Perret-Clermont* replied that this type of curiosity could be understood as a cultural product of the milleu rather than as a 'natural' characteristic. This would be especially the case for 'those children who live in a social environment that values curiosity *per se* about the physical environment (for instance in the socio-cultural milieu of academic parents, or for pupils in certain science classes)'.

Are primates superior to other species?

Weiskrantz asked Cheney:

'Is there an assumption in your papers that primate cognition is somehow inherently superior to that of other species? The categorization shown by pigeons (Herrnstein 1979) and the spatial learning by rats in the radial maze (Olton 1979) exceeds any capacity demonstrated in monkeys that I have seen.'

   Cheney replied:

'Few cross-species comparisons have demonstrated any inter-specific differences in "intelligence" or problem-solving abilities. Thus although it is tempting to regard primates as more "intelligent" than other animals, to date the necessary experiments that might demonstrate this simply haven't been done. Moreover, it is likely that inter-specific differences will be found not in the ability to solve a problem *per se* but in the *method* chosen to solve the problem (i.e., the *means* by which the problem is solved).

Nevertheless, there is some indication that primates may learn more about the external world than some other species. Whilst vervet monkeys attend to the alarm calls of starlings and impala, playbacks of vervet alarm calls to starlings and impala have shown that these species do *not* attend to vervet alarm calls.'

Laboratory vs. field techniques

Cheney had implied that complex cognitive abilities were more readily revealed in field studies than in the laboratory. However, *Weiskrantz* suggested that:
'Differences in speed of learning or cognitive capacity between field and laboratory could be due to the inadequacy of laboratory *techniques*. For example, for many years cross-modal transfer was thought to be impossible for rhesus monkeys. But with correct techniques it can be demonstrated in the laboratory very easily and quickly (Cowey and Weiskrantz 1975; Weiskrantz and Cowey 1975).'

Cheney replied:
'Your work on cross-modal transfer demonstrates very clearly that laboratory techniques strongly affect performance. Similarly, field experiments also suggest that the choice of stimuli influence performance. The point is not that field experiments are better than laboratory ones, but that some stimuli may be more salient than others. Thus the failure to perform or learn a task in one context may be due more to the particular stimulus or technique chosen than to any inability *per se.*'

(In addition, experiments reveal performance but not necessarily capacity. And performance may depend on factors in addition to cognitive abilities such as 'behavioural style' as revealed by latency to approach, tempo, and vigour of activity—see e.g. Visalberghi and Mason 1983—*Eds.*).

Evidence for cognition

These data raised the question of the criteria necessary for imputing cognition, when many results could be explained at a simpler level, such as associative learning. Therefore, *Weiskrantz* asked:
'Is associative learning cognition? If so, do *Aplysia* demonstrate cognition?'

Cheney replied:
'In studying animal learning, we are limited to studying the responses that particular stimuli evoke. This is especially true of playback experiments, which reveal nothing about the mental processes underlying a response. Thus, a given response may result from simple associative learning, or it may be due to some "higher" cognitive process. We can make no conclusions about the mechanisms that provoke a response. More generally, however, what is cognition? If an animal learns something through association, does this rule out the possibility of some "higher" mental process?'

Olfactory vs. visual cues

Vauclair questioned an assumption behind some of Cheney's experiments:

'In Cheney's situation, the gazelle's carcase would not have an associated smell of a leopard and the python's track would be too old to carry olfactory cues. Therefore any alarm response would be based solely on visual cues. How do we know that normal detection is not initiated also by other cues?'

Cheney replied:

'Olfaction is unlikely to be as important as vision in detecting secondary cues of danger because olfaction in monkeys is poorly developed relative to other mammals. In fact, the visual and olfactory abilities of monkeys and apes are very similar to our own. It is therefore safe to assume that olfactory cues are not relevant in assessing the importance of a carcase in a tree or a python's track.'

Section B

4

Functional aspects of relationships; the dialectics and the role of affect
EDITORIAL

After the preceding two chapters on non-human primates, the remainder of the book is concerned with our own species, with the next chapter forming a bridge between the two. The three chapters in this Section provide overviews relevant to most of the chapters that follow.

We have seen that in non-human primates and in all human societies individuals develop in a gradually expanding network of relationships. In many human societies this involves growing up in an extended family, the relationship with the mother being soon paralleled by relationships with other nurturant adults and caregiving siblings or other relatives. This extended family is usually set within a small community, and the children have frequent contact with individuals of several generations. Such a situation, which was probably common far back into human history, differs from that prevailing at present in many parts of the Western world. Children grow up in nuclear families, more or less isolated from other families and not infrequently with only one parent. Contact with a wider world occurs largely in institutionalized settings—day-care centres, playgroups, preschools, and so on, and with members of only limited age groups—teachers and same age peers. However, in monkeys and in both traditional and western human societies, which relationships are formed is not determined solely by the vicissitudes of chance meetings. Rather, selection amongst possible partners is determined by predispositions (biological and cultural) in both infants and others. This is the theme of Attili's chapter: she emphasizes how the sequence of relationships a child forms, a sequence which we take for granted, can be considered as functional in a biological sense, adapted to the needs of the developing organism.

Her approach involves a perspective not shared by those developmental psychologists who see the developmental task solely as a preparation for adulthood, and forget that it involves also meeting the demands of childhood. To take a biological parallel, a caterpillar is not a miniature butterfly, and many of its characteristics have been produced by natural selection operating to make it successful as a caterpillar (Tinbergen 1959). Similarly children must develop in ways that meet their immediate needs, as well as preparing the way for adulthood. For example, Attili points to the propensity of children fresh in preschool to form a relationship first with a teacher, which will provide them with the security necessary for them to form relationships with peers, which in turn promote socio-cognitive development. Thus in considering the expanding network of relationships we must not limit consideration solely to their consequences for adulthood.

In the following chapter Hartup takes up the issue of the complex interplay between relationships and cognition. He reviews data indicating that the more important contributions to cognitive development come not primarily from isolated interactions with anonymous others, but in the context of long-term relationships. This raises the issue of what sorts of relationship affect what aspects of cognitive functioning and by what processes they do so, and what aspects of cognitive functioning are especially important for dyadic relationships. Hartup's overview paves the way for the more detailed discussion of particular issues in subsequent chapters.

In general, the factors determining the course of development over age are not necessarily the same as those producing individual differences at any one age. However Hartup, and in a later section Radke-Yarrow and Sherman, suggest that developmental changes and individual differences have common sources—Hartup focussing on the child's relationships and the latter authors on an interplay between 'motivational factors and environmental factors which are normally consistent across families, but when discrepant may produce aberrant development'. This links to a further question—why, given an influence discrepant from the norm, does development take the particular divergent course that it does? In some cases, no doubt, the divergence can be seen as a mere consequence of the inadequate or excessive force of the influence in question. In others, however, the developing organism may respond in a way that is adaptive to the circumstances in which it finds itself. The example of the children brought up by somewhat inexpressive and cold mothers is discussed by Attili, and demonstrates that the developing organism must be seen as possessing at each stage a series of alternative strategies—though which is chosen may affect subsequent possibilities.

In the third chapter in this section, Rutter provides a far-reaching review of studies concerned with environmental influences on intelligence and on scholastic achievement. Perhaps more clearly than any other author in this book, Rutter distinguishes between 'social behaviour' or 'socio-emotional and behavioural functioning' on the one hand, and 'cognition' or 'cognitive development' on the other. Aspects of socio-cognitive functioning, as shown for instance by the child's views of himself and others and his styles of interaction with parents, teachers and peers, which may have important cognitive components and are the concern of other chapters in this book, are excluded here (but see Rutter 1984c). However the studies reviewed do show that certain factors, such as family discord or institutionalization, may affect a child's social behaviour or relationships, or his task involvement in school, and thus his school performance.

In general, the data reviewed show clear effects of the social environment on IQ or school performance, though such effects do not account for a large proportion of the variance. It will be apparent that many of the data reviewed come from studies which had the advantage of large sample sizes but were not designed to measure the more subtle social variables discussed in other chapters.

Studies of family influences on cognitive development often necessitate the teazing apart of environmental from genetic influences. It is worthwhile emphasizing that this is by no means an easy task. First, a given environmental factor may have differential effects on individuals who differ genetically; the genetic constitution affects responsiveness to the environment. Second, the child actively selects and creates his or her own environment from that which is provided—how the child does so will be affected by his or her genetic constitution (e.g. Jaspars and Leeuw 1980). This is termed 'active' genotype-environment correlation by Plomin and Leohlin (1977; Plomin and De Fries 1983; Scarr and McCartney 1983). Third, parents may be predisposed genetically to give their children both an environment and genes that are conducive to the development of particular characteristics—so-called 'passive' genotype-environment correlation. For example, parents who are themselves shy might both hand on genes associated with a predisposition to develop a behavioural style that might be labelled 'shy' and also to create an environment in which the children saw few strangers and/or saw their parents behaving with reserve to strangers (Plomin pers. comm.). Finally, parents and others may react differently to children of different genotypes—so-called 'reactive' correlation. This issue is taken up again in the discussion following Rutter's chapter.

5

The extent to which children's early relationships are adapted to promote their social and cognitive development

GRAZIA ATTILI

Introduction

Human social and cognitive abilities must have evolved through the agency of natural selection. However the precise consequences of these abilities through which natural selection acted have recently been the subject of debate. Humphrey (1976) has suggested that most practical problems confronting higher primates (and presumably early men) could be solved without recourse to creative intelligence—a view supported by recent studies of chimpanzees and of contemporary Bushmen (Teleki 1974). Most of the techniques necessary for maintenance could be acquired by relatively simple learning processes or by imitation. Humphrey (1976) thus suggests that the primary function of higher mental functions lay in the complexities of social living.

Knowledge of practical techniques and of the nature of the environment, though crucial to individual survival, are in fact best acquired in the context of a social community. It is the social group 'which provides both a medium for the cultural transmission of information and a protective environment in which individual learning can occur' (Humphrey 1976 p. 307). Furthermore relationships with particular others are crucial for individual survival and reproduction. Higher primates need to be skilled in managing their interpersonal relationships, for social success means survival and biological fitness. The more complex the society in which individuals live, the more intelligent they must be to exploit each others' behaviour and to adjust their own behaviour to others in a continuous calculation of the balance of profit and loss. It thus seems that social skills and cognitive skills must have evolved together.

How is the learning of survival techniques facilitated by interpersonal relationships? A general pattern is found in most higher primates' social groups. Young remain dependent for a long period during which they are free to experiment and explore without having to defend themselves. Furthermore the young remain for a long time in contact with older members of the group and can learn by imitation and transmission of information from them (Humphrey 1976). Ethologists assume that individuals have been adapted to select relationships which promote their social and cognitive development and, by a circular process, through their socio-cognitive skills they can select relationships which better contribute to their biological fitness. That this assumption applies also to man is suggested by the fact that similar principles in the way individuals form relationships during their ontogeny can be traced in non-human primates and in human beings. Studies of non-human species in fact show that animals do not form relationships at random, but have propensities to form relationships with particular categories of others. These studies, and in particular studies of monkeys, have been used to throw light on the dynamics of human relationships (Hinde 1979, 1983).

Studies of non-human primates

a. Mother-infant relationship

Certain principles similar to those that can be applied to human infants can be recognized. The first relationship is with the mother. This differs from later relationships in many ways: it develops through the physiological processes of pregnancy and parturition; it develops and wanes gradually, and it is throughout complementary (i.e. the partners do different but complementary things together, e.g. a mother suckles and an infant is suckled, Hinde 1979, 1983a). The nature of the relationship is of course affected by both participants and the stimulation provided by each partner to the other. is crucial. On the one hand, the mother's behaviour supports and cares for the infant (Lee 1983), and in large measure her behaviour determines the nature of the relationship. But on the other the behaviour of the infant is also crucial: if an infant monkey clings to a non-lactating female she is likely to show maternal responses and may even lactate (Harlow and Harlow 1965).

b. Relationships with individuals other than the mother

Starting from this bond the non-human primate infant forms an expanding network of relationships with other individuals. These relationships differ from that with the mother in that they are not primarily physiologically determined, in that they start mainly through gradual acquaintance, in that they can be disrupted by a sudden event (as when dominance relationships are reversed in a fight), and in that they can be reciprocal or complementary in nature. Furthermore, at least among adults, a fixed sequence of interactions sometimes seems to be crucial to the formation of a relationship. Strange gelada baboons, for example, go through the sequence fighting, presenting, mounting, and grooming before establishing a settled relationship, with fighting being a means for establishing relative dominance (Kummer 1975).

For the young animal, the mother remains the social focus, and the further relationships formed mirror those of the mother. Positive interactions, such as approaching, time spent near, and friendly contact initiations, are distributed amongst other individuals by infants in a manner similar to their mothers'. The group members with whom infant rhesus monkeys initially interact are primarily close kin and their offspring, followed by distant kin or unrelated individuals. Certain consistencies in the distribution of interactions among companions by infants up to 30 weeks of age have been identified: associations occur with close kin more than distant kin, female companions more than male companions of the same age and maternal kinship category, younger immatures of the same sex, and maternal kinship more than older immatures (Berman 1983). These patterns of distribution are maintained even when individuals become more independent from their mothers; the mothers' early influence has long-term consequences on the development of social networks.

Furthermore, in many species a female's dominance status is determined by that of its mother. Individuals whose mother dies not only lose a partner for important interactions such as grooming but may also lose their rank. The pattern of relationships formed by the infant are thus determined in part by the dominance status of the mother and of her lineage. The formation of early relationships with maternal kin is clearly adaptive, for individuals increase their inclusive fitness* by providing protection to close relatives. If an infant is threatened by other group members he/she receives protection not only from the mother but also from other close female kin who thus contribute to the acquisition of rank (Datta 1983).

c. Relationships with peers

As they grow older, infants spend an increasing time with peers, developing relationships different in quality from those of their mother and differentiated according to the age and sex of their companions.

While the early relationships with mother and close female kin provide a secure basis for survival, relationships with peers are such as to augment

* Natural selection acts to ensure that individuals not only maximise their own reproductive success measured in terms of their descendants in subsequent generations, but also that of their relatives (who share a high proportion of their genes) to an extent devalued according to their degree of relatedness. This referred to as maximising 'inclusive fitness'.

the development of social and cognitive skills. Although conclusive evidence is lacking, play must surely contribute a great deal to the development of social relationships, providing opportunities for immatures to interact regularly with individuals inside and even outside their families. The relationships so formed are such as to foster social and cognitive development. Appropriate responsiveness plays a crucial role. Play permits individuals to test each other, to learn about the reactions of others, and to modify their own behaviour accordingly (Bruner 1972).

The choice of particular others in terms of age, sex, rank, and genetic relatedness is probably such as to make them more effective for cognitive and social development. Juvenile and subadult baboons play more with their sibling than with other animals. Females play primarily with infants and with infants of mothers of higher rank than their own; males play mostly with like-sexed partners (Cheney 1978a). This particular distribution of play-partners has short-term benefits as well as adaptive consequences (Cheney 1978a). Thus the fact that immature males play more often and more roughly than females (Hinde and Spencer-Booth 1967; Kummer 1968) has various probable consequences: rough play trains males to fight in their future life as adults (a social skill) and improve their ability to assess the fighting capability of others (a cognitive skill). This is important because, whilst females' ranks are determined by those of their mothers, males' ranks are determined by their own abilities, and affect their long-term inclusive fitness (Cheney 1977).

Preference for play partners of similar age and sex (similarities which extend to weight and social skills) makes play less likely to break down (Suomi and Harlow 1972): individuals can thus obtain maximum profit from playing. They can thus learn about their environment, make social comparisons with individuals at a similar stage of development, and test each other's relative abilities. The preference that immature female baboons have for infants as play partners in play which often involves 'aunt' behaviour (Rowell et al. 1964), may have two effects: it enables the infants to establish bonds with individuals outside their own families, and provides the females with opportunity to practise motherhood (Cheney, 1978a).

d. Reciprocal altruism

Relationships may be established selectively with individuals who not only provide immediate benefits in terms of opportunities to develop social and cognitive skills, but also subsequent benefits through reciprocal altruism. Some types of grooming in primates, for example, are analogous to the giving of social approval (Hinde and Stevenson-Hinde 1976), and monkeys groom selectively those individuals who are likely to bring later benefits.

Thus immature male baboons tend to groom mostly subordinate unrelated adult females; immature females, by contrast, direct their grooming mostly to adult females (and their infants) who rank higher than their mothers (Cheney 1978b). These differences are to be understood in terms of the fact that female baboons remain in their natal troop throughout life, while male baboons leave their troop when sub-adults. It is adaptive for females to form relationships with more dominant females who may later provide aid in aggressive encounters. For males, it is adaptive to form relationships with low rank females because with them they can practise mating without interference from the adult male (Cheney 1978b).

How far non-human primates form relationships that will subsequently be of value to them because they have foreknowledge of the possible advantageous consequences of their actions is not clear. Hinde (1983b) suggests it is safer to assume that non-human primate individuals have been adapted by natural selection to foster this sort of relationship, but that evidence is also accumulating that some primates 'may engage in more complex manipulations of their social environment than they have hitherto been given credit for' (p. 69).

Children's relationships

Are principles similar to those considered above applicable also to human infants? Are human beings also adapted to foster relationships with individuals who provide them with short-term/long-term benefits or do they form relationships at random? Which short-term benefits might bring long-term benefits for human beings, increasing their inclusive fitness? The pattern, considered above, of a network of relationships expanding from one or more blood relatives is ubiquitous in human societies. Most children grow up in a family, interacting daily with parents (nearly always with the mother), with siblings, with relatives, and with friends. The generality of this pattern and the fact that something similar occurs, as we have seen, in many non-human primates justifies the assumption that it was present also, as Bowlby (1969) says, in our 'environment of evolutionary adaptedness' (Alexander 1974; Hinde in press). However the adaptive significance of the relationships formed by the developing child have until recently been little discussed. Children's social bonds have been considered mostly in terms of short-term social functions, neglecting any interpretation in terms of biological functions. One consequence has been an overemphasis on the mother-child bond and a neglect of the relationships children form with figures other than their mothers (such as relatives, siblings, and peers). Furthermore the mother-child bond has itself been perceived too narrowly: psychoanalysts, for example, interpreted it as concerned solely with the satisfaction by the mother of the infant's primary needs.

a. Mother-infant relationship

Bowlby (1969, 1973) has placed the mother-infant bond in an evolutionary framework thereby enlarging previous views of its functions. According to attachment theory (Bowlby, l.c.) the relationship is due to a 'primary' biologically-based need for the infant to be in touch and to maintain proximity with a preferred individual usually conceived as stronger and/or wiser (Bowlby 1977). This relationship is seen as formed by reciprocal expectations and behaviours, and as being both the basis and the origin of the child's subsequent social and emotional development. Natural selection favours individuals who survive and reproduce. The biological function of the mother-child attachment is seen in the security a mother provided for the infant in the evolutionary environment by protection from predatory attack. Behaviours such as following, clinging, crying, and smiling are considered by Bowlby as biologically-based and adaptive, and are the result of selective pressures which have moulded those responses of the infant that lead to his/her survival. It is assumed also that the mother's behaviour is also moulded by natural selection: her sensitivity and her readiness to respond promptly to her child's signals are crucial to the infant's survival; the mother's own inclusive fitness is ensured through the survival of her child.

Security, of course, is not the only benefit provided by the mother. Very early she becomes a social referent for the first learning. She also teaches, plays with, and interacts with the child in many ways from birth on, promoting the development of his/her social competence. Furthermore through very early dialogical structures based on proto-turn-taking she provides the child with facilities for the acquisition of language. These occur in the early synchronized interpersonal exchanges described by Kaye (1977) for sucking, by Schaffer *et al.* (1977) for vocal behaviour, and by Stern (1974) for visual interaction.

Through these early 'pseudo-dialogues', based mainly on the mother's initiative in replying and giving a communicative meaning to her infant's signals, the child starts to acquire the capacity for reciprocity. By the end of the first year the child learns that dialogues are two-sided, based on interchangeable and reciprocal roles. Playing with his/her mother in give-and-take games, for example, a child learns how to achieve cooperative actions through prespeech communicative acts. The mastery of procedures for joint action can be considered the precursor of early forms of grammar (Bruner 1977).

How early an infant realizes that his behaviour has communicative value and can be used intentionally in order to influence the partner's behaviour is still an open issue. It is true that cognitive mechanisms such as differentiation of ends from means, self from others, object permanence, representational skills, and growth of attention span must develop before dialogic ability can appear. But evidence is accumulating that exchanges between mothers and infants based on reciprocal expectations appear much earlier than they were supposed to appear according to earlier Piagetian assessments of the development of symbolic activity and representational level (cf. Bryant p. 246–7). Two-month-olds and their mothers are involved in patterns of exchanges which are mutually generated, in which the interactions of both partners are essential and in which both adjust their acts in order to fit those of the other. The pattern of action is sustained by the mother who complements the infant's acts. But the child makes demands on the mother, looking attentively at her face and smiling. The mother follows the infant and responds to his calling out by smiling and speaking (Trevarthen 1977). If the behaviour of the mother becomes suddenly unfriendly and unresponsive, violating the baby's expectations, the infant shows distress and in this way regulates the subsequent behaviour of the mother (Cohn and Tronick 1982).

Whether these exchanges involve intentionality by the infant is a controversial matter (Dunn 1982). But what is important is that the human baby responds in a particular way to a particular adult behaviour. And even though an infant who smiles in the first weeks of life to his/her mother is not doing it with the intention of communicating with her in order to modify her behaviour (Kaye 1982), the fact that he/she responds sensitively to the parent's behaviour, and that he/she soon comes to behave differently to mother, father, and strangers (Brazelton 1982; Fogel *et al.* 1982), and to be sensitive to aberrant behaviour by the mother (Cohn and Tronick 1982) shows that in the first 3 months of life a baby already has expectations about the patterns of behaviour that the mother or the father are going to show. This implies a certain representational level even though the span of memory is limited. The 'socialization through interaction' which occurs in mother-infant exchanges implies the sharing of a particular history of communication (Newson 1977). So in early mother-infant interactions cognitive and emotional reactions and experience are already closely interwoven (Fogel *et al.* 1982).

At one time, children's cognitive development was studied mainly in non-social contexts, in interaction with objects. But an infant's attention to an object is static and seems almost to paralyse his behaviour. In interaction with the mother the cycle of attention is followed by regular withdrawal, and each pause of one partner can be used by the other for giving or waiting for a reply (Brazelton 1982). The infant's behaviour and attention span when interacting with the mother differs very much from when he/she is 'interacting' with an object, and the different expectations elicited by interacting with an object versus interacting with a person, already postulated by Piaget (1937), emerge early, even at 4 weeks (Trevarthen 1977; Brazelton 1982).

This difference in the kind of responses infants show to persons and to objects suggests that human beings are adapted to interact with human beings. Furthermore signals directed to a particular adult have adaptive value in so far as they attach the adult to the infant and create in the adult an expectancy for a particular type of interaction. What types of interaction contribute to attachment and provide security is an open issue (Bretherton 1980).

Studies investigating the quality of attachment in mother-child dyads stress the importance for the child of being securely attached in order to be able to learn better and to explore, for example, the physical and social environment (Ainsworth 1979; Ainsworth et al. 1978). Children who in Ainsworth's strange situation* have been classed as 'secure' (group B babies) spend more time than 'insecure' babies in exploratory play and are more cooperative with an unfamiliar adult; furthermore they show, later on, higher scores on a variety of problem-solving tests.

This is not to say that optimal cognitive and social development is ensured only when a mother-child relationship reflects the 'ideal' features as assessed by the standard 'strange situation technique'. Interactions which lead to security certainly change with a child's age and may change according to the culture and circumstances (Hinde 1982). 'Conditional strategies' (Maynard Smith 1979) allow the mother's and the child's needs to come to terms with the context of the total situation (Hinde 1982). In other words it is possible to argue that relationships to particular individuals, and in this case to the mother, are adjusted in such a way as to bring maximum benefit to the child according to the current situation.

Children with mothers who are emotionally inexpressive and reject physical contact with the infant show avoidance on reunion with mothers (Main and Weston 1982). This can be considered a conditional strategy which paradoxically permits maintenance of proximity with a threatening partner in a particular situation (Tinbergen 1959). The avoidance can be interpreted as a way for the infant to maintain self-control and behavioural organization (Main and Weston 1982), and is thus an appropriate strategy in the situation. By partially avoiding a mother of this type, the child removes a somewhat threatening structure from view, permitting proximity with a figure who can be seen as a source of security, and maintains a flexibility in behavioural organization by reducing attention to a partner who has power over his/her behaviour patterns (Main and Weston 1982). At this price the child is able to adjust more quickly to a new environment

---

* A technique for categorizing the parent-infant relationship.

and perhaps thereby to augment his/her social and cognitive development as a securely attached baby does.

Another common feature of mother-child dyads concerns the complexity contributed to the relationship by the expectancy created between the participants. Each partner learns about the other and about him/herself within a regulatory system which can be seen as a model for learning (Brazelton 1982). This regulatory system can work properly only in so far as individuals can use a wide range of styles and are able to select the most appropriate one according to circumstances. Indeed a mother-child relationship which produces successful adults in one situation may not do so in another. On the basis of this assumption Hinde (1982) suggests a possible explanation of some otherwise inexplicable results: mother-infant dyads classed on the basis of the 'strange situation' as 'anxiously attached' and 'avoidant' are claimed to be frequent in North Germany (Grossman and Grossman 1981), but not in the States. If this finding is correct, it could be due to the fact that cultural values differ between the two countries. By seeking for obedience rather than compliance North German mothers might prepare their children for an adult life adequate to the social desiderata of their culture. The best mothering style varies with the situation, and an 'avoidant type' relationship could be best for producing individuals able to cope with the society into which they have been born.

Of course it is often necessary for a child to face violations of rules (as happens when a child has a rejecting mother) even in normal face-to-face interaction. Stern et al. (1977), referring to violations such as the double bind messages (e.g. a face posture and gaze contact which signal an intention to play displayed together with a vacant expression) that a caregiver may express in some contexts, label them 'mismatches' and suggest that they facilitate rather than impede the dyadic exchange. Giving the term 'mismatch' a wider meaning, a Piagetian meaning, in the sense of a discrepancy in the external situation which a child has to adjust to and to assimilate, we may now ask which relationships provide a child with the opportunity of having encounters involving cognitive mismatches.

b. *Relationships with related individuals: the siblings*

Are mismatch characteristics, such as differences between the level of cognitive complexity in the behaviour of the partner and that of the child, more effective in parent-child relationships or in peer-peer relationships? It has been argued that a child is most attentive to and influenced by moderately mismatched stimuli, and that this occurs in peer and especially in sibling interactions more than in adult-child exchanges (see Youniss 1980 and Chapters 21 and 22). The moderate difference in age in relationships with an older sibling provide a context in which a child can experience the point of view of a person who has greater maturity but is still an equal. Parent-child (and adult-child, in general) relationships are complementary, characterized by interactions in which the behaviour of each partner differs from, though complementing, that of the other. Relationships between children are reciprocal (Piaget 1965; Sullivan 1953). Child and adult have difficulties in understanding the reasoning and perspective of the other. Relationships with other children, who are at a roughly similar level of maturity, facilitate the development of sensitivity and understanding of the self and of others (Lewis and Rosenblum 1975; Foot *et al.* 1980; Younnis 1980; Hartup 1983). Children, just like non-human primates, spend a great deal of their time inside the family with siblings.

By 1 year, second born children spend as much time in interaction with their sibling as with their mothers (Lawson and Ingleby 1974; Abramovitch *et al.* 1980; Dunn and Kendrick 1982). Familiarity, intimacy, and sharing each other's interests allow sibling relationships to include elements of the direct reciprocity that can be considered the main feature of relationships between peers (Dunn 1983). At the same time siblings provide one another with somewhat different environments; complementary exchanges can be experienced as well. Reciprocal features are clear in both positive and negative interactions: communicative sequences between the children include a high frequency of imitation of older by younger siblings (Pepler *et al.* 1982; Dunn and Kendrick 1982), suggesting that older siblings play an important role in the infant's ability to master the environment (Lamb 1978). But imitation is shown by older siblings too, so that both children are engaged in co-active sequences which highlight the reciprocal features of their relationships. Positive social actions such as cooperation in games, showing physical affection, and attempts to help and comfort are directed not only from first born children to younger, but also from younger to older siblings. Negative behaviour is more frequent from elder to younger, but during the second year of life intentional teasing is shown also by second born siblings (Dunn and Kendrick 1982).

Interestingly, mutuality is not always found. Very often there is an affective mismatch involving friendly behaviour from the younger one, and hostile and aggressive behaviour from the older (Dunn and Kendrick 1982). The effort each child has to make in reducing the incongruence of their dyadic exchanges may well affect and promote cognitive and social skills.

In other respects the complementary features of caregiving and teaching between siblings (mainly from the older to the younger one) resembles the parent-child relationship (Whiting and Pope Edwards 1977). The teaching behaviour involves instruction in physical skills, game procedures, and use of toys (Pepler *et al.* 1982). Attitudes to social rules and the first ideas of justice also often develop first between siblings (Freud and Dann 1951). Older siblings not only teach, but are able to adjust their behaviour to a very young baby—a sort of 'motherese' such as that used by 4-year-old children addressing unrelated 2-year-olds (Shatz and Gelman 1973, 1977), has also been found in 2–3 year olds addressing a baby sibling who is 1 year old (Dunn and Kendrick 1982).

Not only younger children may get benefits from being taught and cared for by older siblings. It is possible to suppose, and this hypothesis has been tested experimentally but not for siblings by Doise and Mugny (1981), that through this sort of exchange elder siblings also improve their intellectual abilities: facing the cognitive difficulties of another child, children reflect upon a problem-solving task in a productive way (Dunn 1983). Furthermore older children, especially females, may within these interactions practise caregiving abilities which will be useful later.

c. Relationships with strange adults and peers

Sibling interactions may help to smooth the passage to relationships with peers. It is still controversial whether adult-child and child-child systems are partially (Vandell 1980) or totally (Lewis et al. 1975) independent, or whether they are closely interdependent (Eckerman et al. 1975). Another question concerns the extent to which relationships formed outside the family, not only with peers but also with strange adults, are affected by the child's 'working models of how attachment figures are likely to behave towards him in any of a variety of situations' (Bowlby 1973), and thus ultimately on the type of relationship he/she experienced with his mother.

Children differ in the extent to which they interact with strange adults and with peers, and in the content and the qualities of these interactions (Attili 1982; Hinde et al. 1983). An interactionist approach seems to be the best one. As discussed above, peer relationships provide a child with benefits not available in adult-child relationships, but what a child gets from peer relationships may be influenced by relationships at home, not only with the mother, but also with the father and the siblings (Hinde et al. 1983).

Regarding the role of peers in child development then, several questions arise. Do relationships with unrelated peers provide security so that a biological protective function can be ascribed to them? If they do, what are the contexts in which peer relationships have this function and through which sort of interactions is it achieved? Are they the same as those where this function can be attributed to an adult-child relationship or are they different? If they do not provide security, what benefits do they bring? Should relationships with peers be considered as adaptive or are they randomly selected? And do interactions among equals augment general cognitive abilities or do they contribute only to the development of socio-cognitive abilities such as interpersonal perception, understanding of self, and sensitivity?

Children adopt conditional strategies not only in accord with their mother's nature, as discussed above, but also in accord to the different partners with whom they interact. The main difference between interactions directed to an adult, even though he/she is a stranger, and to other children concerns, as already discussed, their complementarity versus reciprocity. Not unexpectedly Hinde *et al.* (1983) found that inside a preschool a high frequency of neutral conversation, friendly responses, active, and reactive hostility characterized interactions with children. Interactions with the teacher involved, by contrast, many control and dependency responses. Furthermore, children replied to adults when they received a social opening, but more rarely took the initiative themselves. And even though children disconfirmed and failed to comply with their peers more than with adults, over time (42–50 months) interactions with adults decreased whilst those with children increased and the proportion of children's statements disconfirmed by other children diminished.

Children choose as preferred partners those individuals who, in accord with the situation or environment they are living in, can better provide them with short-term/long-term benefits. I found (Attili 1984) that on arrival in an unfamiliar preschool, children chose the strange adult (the teacher) as a preferred interactant, spending most of their time in social exchanges with her and not with their peers. Behaviour shown to the adult resembled in many ways the attachment behaviours shown towards the mother, including behaviour related to seeking and maintaining proximity, and to displaying affectionate feelings through body contact. Other types of behaviour were relatively infrequent in child-adult relationships. Aspects of play, for example, were very rare. Peer-peer relationships emerged more gradually, but over time children came to direct their behaviour primarily to peers. In preferred relationships with peers also, behaviours such as actual proximity, approaching, contact seeking, looking, and body contact were frequent, but a number of other types of interactions, rare in relationships with adults, played an important part. These involved reciprocity and symmetry in turn taking, as in play—this was true not only for imitative and social play but also for negative or positive/negative social behaviours like the playful provocation of an aggressive act, conflict over possession of objects, or aggression. It seems likely that a relationship formed with the adult (and the adult contributes to its formation) provides a child with a secure basis from which he/she can come to terms with the preschool strange environment.

However once established, peer relationships, if they are preferred relationships, also may provide security. This security may be mediated by the same patterns of behaviour that also characterize the relationship with the teacher and permit these relationships to be labelled as secondary attachments. But peer relationships contain simultaneously other types of behaviour involving a reciprocity in turn-taking, like those used in social play, which promote the development of cognitive processes (Attili 1984).

In social play preschool children are capable of abstracting from the play theme the social interaction's implicit rules, and of distinguishing between reality and pretence (Garvey 1974). Whether representational abilities and cognitive processes underlying young children's peer interactions are better promoted by the presence of objects for the children to play with than by interacting directly with a peer is still debated. As we have seen, the development of cognitive processes in play has usually been studied within a Piagetian perspective in terms of less or more pretend activity with objects. Pretend play with objects has been considered a means for sharing meaning with peers. This process may in fact have a powerful effect on a child's relationships. On this view, the type of activity that can be performed with objects, by toddlers for example, affects the social interaction with peers. Specific activity with objects elicits not only cognitive performance related to objects but also the interpersonal coordination of such activities (Stambak et al. 1979).

On the other hand, even at 18 months children seem to formulate expectations and representations about the peer's behaviour independently from a pretend meaning given to an object. 'Le jeux cache-cache' where children alternate in provoking surprise in each other by appearing and disappearing in and out of a paper box implies not only a strict interconnection between cognitive and affective messages so that each child plans, and in some sense runs, the partner's action inside a shared project (Maisonnet and Stambak 1983), but also a cognitive ability of mentally representing the action of the other, of having expectations about his/her partner's next move, and on the basis of that modifying his/her own behaviour. The socio-cognitive ability of adjusting his own behaviour to the other (an ability which implies expectations and representation of the other) develops as much in social exchanges outside play contexts, and is promoted mainly by the familiarity children acquire with each other. Interactive time, including playing with alternating roles or talking, increases during the school year inside a preschool. By contrast hostile aggression decreases. Specific aggression, i.e. aggression due to a conflict over objects, remains constant (Attili, in press; Attili and Cavallo-Boggi 1983). It could be suggested that if interactions among peers were elicited mainly by interactions with objects, we would have found a simultaneous increase both in interactive time and in specific aggression. Familiarity might thus be considered an operative factor which mediates the development of social and cognitive abilities. In peer interactions, as in adult-child exchanges, cognitive and social abilities are closely related.

To form at first a relationship with the mother, followed by those with other family members (adults or siblings) and later on with strangers—such as the caregiver or the teacher who form a bridge to relationships with strange peers—might be considered an adaptive sequence. The adult and child peers of our study (Attili 1982; Attili *et al.* 1982) met in preschool, for example, and were strangers, encountered in a strange environment. The children behaved adaptively in showing some initial fear of strangers (McGrew 1972). In a strange environment adults are better able to protect a child than are peers, and the children thus behaved appropriately in first orienting their behaviour to the adult. In addition, of course, the adult was aware of the children's needs. Whilst in a strange environment attachment to an adult might be considered more adaptive than attachment to a peer, once the environment becomes familiar the reverse becomes the case. Peers' function consists in providing opportunities for the development of social competence not available in relationships with adults. These results might thus be interpreted as evidence for the independence of the social functions that the two relationships provide: that with the adult (even a stranger) seems to satisfy those needs for security and protection elicited by a non-familiar situation and non-familiar interactants (as happened in the first part of the year). After this, relationships with peers were preferred, perhaps because ultimately they are more useful in the development of interpersonal perception and sensitivity to other's signals.

Amongst peers, children's choice was determined by the sex, age, and level of experience of the other child. At the beginning of the school year, newcomers oriented themselves towards older children who had previously attended the same preschool (perhaps because they could learn from them the rules and habits of the kindergarten) but in the course of the year they usually developed a clear preference for children of the same sex and similar school experience, i.e. peers who had also been newcomers at the beginning of the school year (Attili *et al.* 1982).

The content of the interactions towards newcomers and towards children who were already acquainted one with another also differed. Interactions based on behaviours that implied reciprocity were more common when the preferred partner was another newcomer than when he was an 'older' child: social play and playful aggression were more common when the preferred partner was a newcomer, while there was little difference in imitative play according to whether the partner was a newcomer or older (Attili *et al.* 1982). Rough and tumble play was more common in boys than in girls (Attili 1982).

These results are in harmony with the view that relationships with adults and peers have different functions. The similarities—in the choice of preferred partners and in the content of their interactions—with what we have seen in non-human primates' lives are many. It may be suggested that these choices are determined by propensities moulded by natural selection

(Trivers 1971). However it must be remembered that the environment where the human species evolved was different from the one we now live in. Social groups were not characterized by frequent movements of individuals as happens in modern life. The peer of early childhood was destined to be the companion of adult life. To choose as a preferred partner an individual of the same age could be functional in creating a fruitful long-term alliance. Males benefitted from having a same age male as an ally in their adult life—either in achieving high rank after entering another group or simply in hunting. For females an ally of the same sex could be valuable in agonistic contexts as well as in helping with their own children.

To prefer peers to adults, once the need to be protected is satisfied and the environment has become familiar, has a biological function—friendly relationships with familiar persons at a similar level of maturity and with similar physical characteristics and abilities provide, as we have seen, the optimal basis for social development because they allow for direct experimentation in behaviours, correct or incorrect, that are functional for social competence (Hartup 1983; Lewis and Rosenblum 1975; Foot et al. 1980). And social competence can be seen as crucial to survival and reproduction. In the interaction with adults many behaviours, such as negative behaviours, would be discouraged. But these too promote social development, for example the abilities to attack others, and to control the hostility of others as promoted in rough and tumble play.

Conclusion

Studies of non-human species throw light on the dynamics of human relationships and help us to recognize principles in the expanding network of relationships formed by non-human primates similar to those in our own species. This evidence supports the view that individuals do not form relationships at random, but have propensities to form relationships with particular categories of others, who according to different contexts promote the development of their cognitive and social abilities. Within this general framework, relationships to particular individuals may be adjusted in such a way as to bring maximum benefit to the child. Within an evolutionary perspective the function of intellect must be considered as that of managing interpersonal relationships rather than for problem solving in the physical world. This poses the question of whether persons are more adequate than objects in promoting cognitive abilities.

Discussion

Much of the discussion concerned the nature of the conclusions that can be drawn from observational data, an issue that came up also in other contexts (see Chapter 17 and p. 250).

Vauclair stressed the importance of considering the differences as well as the similarities between mother-infant relationships in human and non-human primates. He referred to a study comparing communicative style in relation to object manipulation between humans and chimpanzees (Bard and Vauclair 1984).

6

Relationships and their significance in cognitive development
WILLARD W. HARTUP

Relationships have been thought to serve three functions in social and cognitive development. First, they are the *contexts* in which basic competencies emerge. These include (but are not limited to) language, impulse regulation, the self-system, a repertoire for coordinating one's actions with those of others, and knowledge about the world—including knowledge about relationships themselves. Processes of learning, remembering, and understanding (or components thereof) may be enhanced through relationships too, but exactly how this happens is unclear.

Second, relationships are *resources*. These entities constitute both emotional and cognitive resources that furnish the individual with the security and skills that it takes to strike out into new territory, meet new people, and tackle new problems. Relationships are resources that buffer or protect the individual from stress; they are also instruments for problem-solving, i.e. relationships are used by individuals for goal attainment. Relationships are thus relevant to competence in both solitary and social settings—in other words, to our successes as both Robinson Crusoes and communal workers.

Third, relationships are *forerunners* or *precursors* of other relationships. Only now are we beginning to learn how this occurs (Sroufe and Fleeson, in press) since new relationships are never carbon copies of old ones. Relationships always reflect one's partners—their motives, skills, and expectations. Nevertheless, old relationships constitute templates or cartoons that we use in the construction of new ones.

The connection between relationships and cognitive development is complex. Sometimes, these complexities are stated so that issues of cause and effect are not meaningfully enunciated. In this essay, we address these questions, not so much to solve the dialectical riddles involved as to highlight their complexity. To this end, we turn first to a consideration of dialectical issues, then to social interaction and its implications for cognitive development and, finally, to relationships themselves and how they may be involved in the cognitive development of the child.

Dialectical issues

Relationships contribute to ontogeny but, simultaneously, ontogeny contributes to relationships. For example, a secure attachment between the toddler and its mother facilitates the child's exploration but, at the same time, the widened world of the child forces renegotiation of responsibilities between the child and the mother. Later, the child's increasing size and capacities for controlling impulses contribute to changes in mutual expectations and social interaction.

Similarly, peer relationships are thought to increase the maturity of the child's notions about the intentions and motives of others through successive induction and resolution of cognitive conflicts about these matters (Piaget 1932). Concomitantly, the acquisition of mature notions about intentions and motive states makes the child more competent in communication with other children (Perret-Clermont 1980). The dialectics between the development of relationships and the development of individuals thus extend from the second to the first as well as from the first to the second.

The nature of developmental change. Some of the developmental changes bearing on relationships are universal and have the thrust of strong physiological or social mechanisms behind them (Maccoby 1984). Examples include the early refinements in depth perception and mental representation that contribute to the onset of specific attachments; the emergence of combinatorial skills that coincides with the emergence of peer relationships in the second year; the increases in physical size and strength that necessitate changes in modes of affectional exchange in middle childhood as well as normative regulation of the child's actions; and the onset of puberty which precipitates the restructuring of child-child as well as adult-child relationships.

Other developmental changes affecting the child's relationships are not universal but produce their effects by constraining the developmental sequence (Maccoby 1984). For example, if the child's early attachment to the mother is 'secure,' development will follow a different course from instances in which the attachment is 'anxiously avoidant' (Matas *et al.* 1978). These different sequences may begin with conditions that are more-or-less independent of experience (e.g. temperament) or with differences in socialization (e.g. sensitive mothering). In either case, however, the developmental differences arise from an interaction between genetically-driven and environmentally-driven factors.

Two individuals, not one. The dialectic between ontogeny and relationships involves two individuals—the mother as well as the child, the brother as well as the sister, and the friend as well as oneself. Ordinarily, our attention is given only to one of these individuals, that is, the child. Thus, we note the concordance between the strongly-determined emergence of object permanence in the child and the onset of a specific attachment to the mother (Schaffer and Emerson 1964) or the concordance between the decline in impulsive behaviour that occurs in middle childhood and the simultaneous decline in power-assertive interactions between parents and their children (Maccoby 1984). But these are oversimplifications; these dialectics are complicated by the fact that two actors are developing within these relationships, not one.

In parent-child relationships, one of these developmental trajectories is rapid (the child's) and the other slow (the parent's). But even though the mother's developmental trajectory may be slower than that of her child's, that trajectory has implications for the relationship that exists between them. The advent of motherhood does not occur at the same time in the lives of all women. Yet neither the cognitive nor the social status of the 12-year-old mother in Western society is the same as the status of the 19-year-old mother in relation to an infant (Furstenberg 1976). And children reach significant transitions (e.g. school entrance or pubescence) at psychologically dissimilar times for different women. Consider that the status of the pre-menopausal mother is not the same as the post-menopausal mother in relation to an adolescent offspring. Other examples are easy to suggest, thus underscoring the importance of adult development as a determinant of children's relationships. The problem becomes more complex, of course, when we recognize that these same relationships exert an influence on the adult as well as the child. New evidence linking 'difficulties' in early mother-child relationships to subsequent maternal depression is a case in point (Wolkind and Desalis 1982).

The dialectics between ontogeny and relationships become exquisitely complicated when the two companions in a relationship are on developmental trajectories which are both moving rapidly. Consider, for example, the relationship between two children observed from the time that each is about 18 months of age until each is about three years old. Observing the social interaction between them, we note that 'interacts' become more frequent as the children grow older, 'contact chains' increase in length, and the incidence of 'coordinations' and reciprocated actions becomes more frequent (Mueller and Brenner 1977). Cooperative interaction in the solution of simple problems also increases (Brownell 1982) and emotional contagion becomes more common (Ross and Goldman 1976). At the same time, individual assessments reveal that syntactical combinations (especially two-word utterances) are increasing (Brown 1973); the imitation of two- and three-unit motor sequences is more and more accurate (McCall *et al.* 1977); the ability to use two or more sources of information in solving simple problems emerges (Reiser *et al.* 1982; Sophian and Sage 1983); and both manipulative and pretend play become more elaborated in terms of unit length (Fenson and Ramsay 1980).

One interpretation of these observations is that, toward the end of the second year, concordant changes in combinatorial skills occur across different domains (i.e. language, imitation, and peer interaction) that reflect changes in a common set of processing components. Consider, for example, that the ability to hold two items in memory and to establish relations between them seems to be newly available to the child at about the age of two. Consider, also, that these processing components may be decontextualized, i.e. not specific to one behavioural domain. Recent research suggests that decontextualization in combinational skills occurs at about this time although it may not be complete (Brownell and Nay 1983).

So interpreted, these developmental changes mean that it is not before the end of the second year that either of the children we have been observing possesses the cognitive capacities for sustaining interaction *with someone who does not have these same capacities*. Sustained and coordinated interaction between the children and their mothers is evident earlier—but undoubtedly because their mothers possess the capacities needed to maintain these coordinations. Consider, then, what we have said: the cognitive demands of a relationship depend as much on the cognitive capabilities of the partner as on the cognitive capacities of the focal child.

Can one carry this argument to its extremes? Is it the cognitive status of the companion that accounts for all of the variance in the social interaction occurring between the child and various others? Clearly not. The evidence shows, for example, that 24-month-old children display more frequent coordinations than do 18-month-olds when interacting with *either* 18 or 24-month-old companions (Brownell 1982). Consequently, the interaction between the children we have been observing (even though they are age-mates) must reflect the cognitive status of both of them.

When two children differ in age, both developmental trajectories are moving rapidly but not in phase: the three-year old and the one-year old, for example, are on different sides of the developmental transition involving combinatorial skills. Consequently, one would expect cross-age relationships to be based in 'asymmetrical' interaction more extensively than same-age relationships. Indeed, the empirical evidence confirms this: in most cultures, 'dependency' is more commonly directed by children to older companions than to same-age or younger ones, and 'succorance' is directed by older to younger children more commonly than the reverse (Whiting and Whiting 1975).

It is difficult to account for these asymmetries without making reference to the more advanced cognitive abilities of the older child as compared to

the younger one, and to the greater knowledge that the older child possesses. And the evidence suggests that, among the relevant abilities, are
both understanding the needs of others and the capacity to apply this
understanding to social interaction. For example, Shatz and Gelman
(1973) discovered that four-year-old children are able to assess the cognitive status of their listeners and to adjust their speech accordingly, i.e. they
talk to two-year olds more simply than to four-year olds or adults. But
these accommodations are not manifested only by the older ones. Three-
year olds are known to use more sophisticated messages when interacting
with five-year olds than when interacting with other three-year olds, even
though these accommodations are less extensive than the similar ones
made by the five-year olds (Lougee *et al*. 1977).

Other evidence comes from studies of the decision-making of 6- and 8-
year olds in mixed-age and same-age groups. Six-year olds remain 'on task'
to a greater extent in mixed-age groups than in same-age groups, while
eight-year olds solicit opinions and organize the decision-making to a
greater extent when six-year olds are present than when they are not
(French *et al*. 1984). The observed interaction thus reflects the cognitive
status of both the younger and the older children, not merely the status of
the older ones. Moreover, these comparisons suggest that age differences
in cognitive regulation as well as age differences in other cognitive abilities
are involved.

Summary. The dialectics between the development of relationships and the
development of cognition consist of an integration of these two domains.
To ask whether one is the cause or the effect is to ask the 'chicken and egg'
question. Nevertheless, the two domains are intimately connected. Considering that relationships involve two individuals, not one, but two
developmental trajectories are reflected in every relationship.

Social interaction and cognitive development

Piaget. According to Piagetian theory, the role of the environment is central in cognitive development. Schemata are created or constructed
through the child's interactions with the surround. Knowledge does not
exist in the absence of a structuring activity and certain relevant activities
are to be found in social interaction. The simultaneous occurrence of
changes in social interaction and changes in cognition is not assumed to
represent a causal connection. What is being described is a mutually-
regulated interactive process through which the child is transformed by
social interaction at the same time that the child changes his or her interpretation of the social interaction and participation in it.

In Piaget's writings (e.g. 1968), the cultural universal that is most relevant to cognitive development is the social exchange—including both the
exchanges that occur between adults and children and those that occur
between children themselves. These social coordinations are different from
the equilibration occurring between the child and the non-social environment. In most social situations, however, ' . . . there is no distinct dividing
line between individual activity and collaboration . . . Among older children, there is progress in two directions: individual concentration when the
subject is working by himself and effective collaboration in the group
(Piaget 1968, p. 39).' Consequently, we are suggesting that, in all societies,
there is a special role in cognitive development for the fact that ' . . . individuals ask questions, exchange information, work together, argue, object,
and so forth (Piaget 1974, p. 302).' When individuals work together, they
take note of different things and these discrepancies give rise to the need to
coordinate perceptions. When working alone, on the other hand, there is
nothing to be coordinated (Doise *et al.* 1975). Inventing solutions, then,
and verifying them are two functions of intelligence emerging in the social
context.

Vygotsky. In the process of developing a psychology based on dialectical
materialism, Vygotsky (1978) also assumed the fundamental importance of
social interaction in cognitive development. The relevant processes, however, were elaborated in relation to the construct of *internalization.* Five
assumptions constitute the core of the theory:
  'First, development occurs in the context of the individual's social interactions with cultural agents . . . (people). Because of this focus, speech
  is the most important mediator of developmental change. Second, for
  social interaction to result in development, it must occur within the *zone
  of proximal development,* defined as the difference between the child's
  actual level of development, as measured by individual problem solving,
  and his/her potential level of development, instantiated by the ability to
  solve problems with the help of others. The function of the social agent is
  to tune the problem solving attempt to a level just beyond the actual
  development of the child so as to foster progress. Third, every skill is
  first acquired at the external (social) level and is then internalized into
  the child's mental repertoire. This comprises the transition from other to
  self-regulation.* Fourth, the child is not a passive actor in the process.
  Rather, he/she is actively transforming the social skill. This leads to
  (internal) abstractions that are possibly more powerful than the original

---

* Helping to explain the connection between the social patterns comprising 'culture' and
the psychological characteristics of its members.

external ones. Fifth, skills are first acquired in the context of specific activities and only later become decontextualized. The decontextualization is mediated by a social agent (Azmitia 1984, pp. 12–13).'

In the West, these processes have been most thoroughly documented by Wertsch and his colleagues (1978, 1980) through analyses of the verbal exchanges between mothers and their young children during puzzle tasks. Indeed, the process conforms to the Vygotskian scenario: a) the child participates in the task, as does the mother; b) the basic learning context appears to be the *dialogue* (which may have nonverbal components as well as verbal ones) in which children and their mothers influence one another, making mutual adjustments in their actions; and c) the puzzle is always completed. More specifically, when faced with a problem, the task begins with the mother serving as the main regulating agent; the child is a participant, but a peripheral one. Next, mother and child begin to connect in their utterances about the task, but responses are sometimes inappropriate and the child's interpretations may be incomplete. The mother does more than the child but not so much more that the child is left out. Gradually, 'self-regulation' emerges; the child now needs only minor supports from the mother. Eventually, the mother's participation is not needed at all. Wertsch's demonstrations have not always included comparisons between problem-solving assessments in the interactive situation and the child's solitary efforts to solve the problem, but generalization effects have been demonstrated with older children to problems falling in the zone of proximal development (Campione *et al*. 1983).

Among the most interesting instances of mutual regulation in cognitive development are those in which *the child* invites the cooperation of the adult in some task that the child has already begun. Indeed, the appropriate use of this 'socially-facilitated problem-solving' may be among the most clear-cut differences between socially competent and incompetent children (Van Lieshout 1975; Sroufe 1979). Once the interaction begins, though, these instances run off much as those initiated by the mother: after soliciting the adult's assistance, the child watches or listens, using the adult's actions as cues concerning the schemes that should be activated and in what sequence. These schemes are eventually evaluated and, finally, the child takes over the monitoring functions 'thus restructuring his internal record (Case 1984, p. 32).' In this analysis, mutual regulation is one of two socially-mediated processes producing transformations in cognitive development, the other being imitation. These augment two other developmental processes—independent problem-solving and exploration.

Hot and cold mediation. One should not conclude that the transition from other-regulation of problem solving to internalization involves only 'cold' cognitive constructs, e.g. concepts, factual knowledge, strategies, and rules. Such schemes occur in instructional interaction, of course, but effective 'teachers' use accuracy checks, simplifications, and repetitions, too. In addition, they encourage children, try to maintain their attention, tell them when they are right and when they are wrong, and celebrate their successes (Brown *et al*. 1983).

This mixture of hot and cold elements occurs in peer-mediated as well as in adult-mediated instruction. School children, for example, demonstrate their knowledge that good teaching includes: a) involving the tutee as an active participant; b) estimating the region of the tutee's sensitivity to instruction (as evidenced by adjustments that tutors make to the ages of their tutees); and c) the inclusion of both hot and cold elements in the tutor's repertoire. These elements have been observed in 'informal' peer instruction as well as in situations in which an adult asks one child to teach something to another (Cooper *et al*. 1982; Ludeke and Hartup 1983).

What develops? Developmental psychologists are still arguing about 'what develops' in the cognitive domain. Sternberg (1984) argues that these developments include: a) *metacomponents*—recognizing the problem, selecting basic components for task performance, selecting strategies for combining these components, selecting mental representations or schemes, allocating cognitive resources, monitoring, understanding feedback, deciding how to act on this feedback, and acting on it; b) *performance components* ranging from encoding processes to the occurrence of the response; and c) *knowledge-acquisition components*—selective encoding, selective combination, and selective comparison—used in acquiring new information. Other writers conceiving the individual as an information processing system have formulated similar lists, including ones that apply to social interaction (Dodge, in press).

One caution emerges from this list-making: one must be specific about what one wants to know about cognition before elaborating how social interaction may be relevant to it. For example, social interaction may not be relevant to an analysis of some ways that three-year-old memory differs from the memory of five-year olds. On the other hand, if one wants to know how the inefficient three-year old memorizer becomes the competent adult memorizer, an analysis of social interaction is probably relevant (Azmitia 1984; Bearison 1983). Similarly, social interaction may not be necessary for the emergence of some intelligent behaviours (e.g. special attitudes toward novel tasks) but a *sine qua non* for others (e.g. organizing resources for problem-solving utilizing the other people in one's surround—Goodnow 1984). To date, no one has constructed a theory that differentiates those cognitive components that originate in social contexts from those that originate in individual activity. Knowing them, however, would reduce some of the chaos in research dealing with social mechanisms in cognitive development.

Summary. Several themes emerge from these comments: ' . . . effective mediation involves procedures that enable children to experience a sense of mastery, that let them see that they have some control over learning situations, and that systematic analysis can lead to successful performance . . . Successful mediation involves much more than the act of dispensing pearls of cognitive wisdom (Brown *et al*. 1983, p. 149).' It is perhaps too much to say that these elements constitute a refined theory of the socialization of cognition. Nevertheless, these ideas are embedded centrally in the implicit theories of cognitive development among both parents (Goodnow *et al*. in press) and children (Ludeke and Hartup 1983).

Relationships and cognitive development

Social interaction is not synonymous with social relationships. Robert Hinde (1976) has distinguished three levels in social activity: a) *interaction* (meaningful encounters between individuals); b) *relationships* (aggregations of interactions between individuals that persist over time and that involve distinctive expectations, affects, and configurations); and c) *social structures*. (see also p. xv) One can readily see that the notions about socialization and cognition discussed above are based more on considerations about social interaction than considerations about social relationships. To be sure, our theories differentiate between adult-child and child-child exchanges in terms of their significance in cognitive development, recognizing the implications of the 'control' that marks adult-child relationships and the 'co-operation' that marks peer relationships (Damon 1977; Youniss 1980). But the significance of the attachment between the mother and child, as well as the significance of variations in these attachments, are not mentioned in these accounts. The mother is considered to be an important cognitive agent owing more to her wisdom and her ubiquity than to the special 'tie' or 'bond' that she has with the child. Similarly, children's friendships have little significance in current cognitive theory except as concentrations of interactions with other children. One looks in vain for hypotheses in most developmental theories that attribute special significance to relationships in the cognitive development of children.

One exception exists: theories of ego development. Although the term 'ego' usually refers to a broad psychological 'organization' or 'synthesis', the designated functions always include coping with reference to 'contradictions' or 'tasks.' Most writers distinguish between ego development and the development of intelligence but there is general agreement that cognitive complexity is one dimension of ego strength (Loevinger 1976).

Modern ego psychologists believe that certain perceptual and cognitive functions may arise on a 'conflict-free' basis (Hartmann 1939) but most hold that socially-induced conflicts which are centred on early psychosexual needs form the foundations of ego development. Ego strength is believed to stem from commerce between the child and a firm but empathic caretaker—in addition, an individual with whom a trusting relationship has been formed (Erikson 1950). The qualities in early relationships that are believed to be relevant to effective ego development are thus relatively well-spelled out and form the basis of several ongoing research studies (Block and Block 1980). What I am trying to say here is that at least one theoretical tradition encompasses the dialectic between early relationships and cognitive development even though no claim is made within it that relationships are significant for every cognitive function.

Psychoanalytic theory and cognitive-developmental theory are both interactionist in their views of the organism and the environment. Piaget's interactionism, however, assumed an 'average expectable environment' (Shantz 1983), or so it seems, since organismic considerations drove most of the relevant studies and individual differences in cognitive abilities were usually explained as differences between 'advanced' and 'delayed' development. Psychoanalytic interactionism, on the other hand, is a bit different. Environments are assumed to vary in the extent to which they support the child in meeting developmental challenge; in turn, these experiential variations have developmental implications. This theoretical tradition, therefore, provides some basis for generating notions both about the manner in which relationships are involved in normative cognitive development and in the emergence of individual differences. Even so, neither theory very explicitly delineates the dialectics between relationships and cognition.

Nevertheless, I would like to propose the general hypothesis that *relationships account for both normative change in certain cognitive functions and the generation of individual differences in certain cognitive skills*. Not every normative change may be traceable to relationships, and many individual variations in cognitive functioning may be independent of social experience—even ones that are closely tied to social interaction (e.g. perspective taking). But I believe that the evidence merits the hypotheses that, as a cognitive mediator, the mother is not just 'any old adult' and that a friend is not merely 'one of the gang'; rather, these are special individuals playing special roles in cognitive development.

Cognitive skill and social relationships should connect because the meshed interaction sequences that constitute relationships necessitate that . . . 'one knows one's own needs and goals, and how the other's behaviour may affect the achievement of these; know the other's needs and

goals and how oneself impinges on them; know the responses that the other is likely to exhibit in reaction to one's own behaviour, and then, possess the capability of performing the responses necessary to bring about the desired effect. This last also requires the ability to interpret instantly feedback from the other . . . to make the necessary corrections to keep the interaction moving in the desired direction (Berscheid in press, p. 17).'
Consider that mutual friends, as compared to 'unilateral' friends, are more accurate in assessing the characteristics that the two have in common as well as being more aware of their differences (Ladd and Emerson 1984). In addition, the commitment built into close relationships should make them better-suited than other social contexts to the dialogues necessary for cognitive change.

Normative change. I would like to suggest that *the cognitive functions most closely linked to social relationships are the 'executive regulators'—the planning, monitoring, and outcome-checking skills involved in problem-solving.* Different writers decompose cognitive regulation in different ways (see Flavell and Markman 1983; Sternberg 1984) but the following can be mentioned: a) predicting one's capacity limitations; b) being aware of the repertoire that one has available for problem-solving; c) identifying the problem; d) planning with respect to strategy; e) monitoring the routines one uses; f) evaluating outcomes; and g) using them to make adjustments in one's activities.

Why should these be derivatives of close relationships? Because the shift from 'other-regulation' to 'self-regulation' that is so well-documented as a mechanism in their development seems to depend on finely-tuned dialogues with an effective mediator that centres on them. Rarely are component skills the centre of attention in these dialogues; more commonly, these exchanges deal with how to assemble various actions in some kind of order, how to make sense of the task, and how to monitor one's behaviour. There may be other skills that are socially mediated, but it is difficult to argue that these are not. The dialogues of mothers and their children are centred on them (Ninio and Bruner 1978; DeLoache 1983), and so is peer tutoring (Cooper *et al.* 1982; Forman and Kraker in press). On this basis, then, I suggest that we concentrate our efforts on cognitive regulation in studying the dialectics between relationships and cognitive development.

In doing this, we must be sure to examine cognitive regulation in collaborative or social situations as well as individual situations. Most writers believe that these functions are largely decontextualized, and that there is relatively little difference in collaborative and non-collaborative problem-solving. But, to the extent that decontextualization may not be the case, it is in social problem-solving that we should see the clearest manifestations of relationships—assuming, of course, that such manifestations exist. The meagre evidence available concerning relationships and cognitive development (as opposed to the influence of social interaction on cognitive development) almost exclusively concerns collaborations in problem-solving (Matas *et al.* 1978) and these collaborations are increasingly recognized as key elements in everyday intelligent behaviour (Goodnow in press).

What now? We do not need to document further the mother's centrality in the external regulation of the young child's cognitive activities. Even though other cognitive agents may supplement or supplant the mother in some cultures, systematic surveys are not required to document the mother's salience as a cognitive agent. Rather, we need evidence confirming that mothers employ the same dialogic strategies in day-to-day interactions with their children that they use in more academic tasks (e.g. those used in research).

Even more interesting, I believe, is whether mother-child relationships constitute a basis for cognitive socialization that is different from: a) the relationships that others have with children, and b) the interactions between the child and other people. Overall, the literature tells us very little about the mother as a cognitive mediator in comparison with other cognitive agents—only that she interacts more frequently with the child than they do. One would guess that mothers are able to define the zone of proximal development more accurately than other mediators might. But mothers seem to tune their dialogues with their children broadly to their children's ages rather than to their children's developmental status (Gleitman *et al.* 1984). Mothers and strangers differ in the extent to which their presence supports cognitive maturity in the child's play—a difference thought to be mediated by emotional reactions in the child (Ainsworth and Wittig 1969; Maccoby and Feldman 1972). Comparative studies, however, are needed to tell us whether conditions prevailing in 'relationships' have any bearing on these developments.

The same arguments can be made about peer relationships and cognitive development. We know that social interaction with peers has a variety of outcomes relating to cognitive coordinations (Doise *et al.* 1975). We also know that children spend more time with their friends than with other children (Medrich *et al.* 1982). But we know very little about friends, compared to nonfriends, as cognitive mediators. Among school children, social collaborations between friends and nonfriends differ. First, friends prefer to work under cooperative conditions more than nonfriends (Philp 1940). Second, they talk more and engage in more frequent social contacts, and the affective atmosphere is more felicitous. Third, conversations reveal a greater attention to individual viewpoints and, simultaneously, stronger pressure to coordinate them. Greater attention to equity rules and the more frequent use of mutual directives ('Let's do it this way' as opposed to 'Put your block over there') are observed between friends than between nonfriends (Newcomb *et al.* 1979). Fourth, friendships support more extensive exploration of the test materials and acquisition of information about them (Newcomb and Brady 1982). Fifth, small groups of friends more effectively utilize environmental resources than do similar groups of nonfriends (Charlesworth and LaFrenière 1983).

In situations involving cognitive conflict, friends and nonfriends also differ in their negotiations. Friends give more explanations of their views to one another than nonfriends, and also criticize their partners more freely. Under the pressure of disagreement, friends move to more mature solutions through discussions than nonfriends, although the frequency of change in views does not differ (Nelson and Aboud in press). We have an inkling then, that friendships support social problem-solving and cognitive change—especially in regulative activities—in ways that interactions between nonfriends do not. No information exists, however, about the long-term consequences of friendship interaction in cognitive regulation. Having many friends may make one socially salient (Hartup 1983), but we do not know whether having many friends makes a difference in cognitive development (but see Krappmann this volume).

In summary, we hypothesize that close relationships should be marked by more effective cognitive dialogues than interaction occurring outside relationships. Within relationships, the child and its companions know one another well, share commitments to one another, and share a commitment to continuity in the relationship—conditions that should make the individuals sensitively-tuned to one another's cognitive needs. If this is so, the exchanges occurring in close relationships, as contrasted with other exchanges, should more effectively mediate the internalization of cognitive regulatory functions.

Individual differences. Pair-to-pair variations in relationships can be described in many ways. Hinde (1979) has suggested some eight categories of dimensions that can be used to do this: *content, diversity, qualities, patterns of interaction, reciprocity, intimacy, interpersonal perception*, and *commitment*. Only now are these dimensions beginning to be used together to establish 'profiles' or 'categories' of relationships (cf. Pancake, in preparation).

The best-known classification work in children's relationships has been centred on the 'qualities' of mother-child interaction using the classification scheme developed by Ainsworth and Wittig (1969) for assessing the security of the attachment between the mother and the 10- to 18-month-old child (see also pp. 57–62). While many differences between 'secure,' 'anxious,' and 'avoidant' attachments need not concern us, the reader needs to be reminded that secure attachments were originally described as well-functioning 'goal-corrected partnerships.' (Bowlby 1969) Indeed, mother and child in secure relationships are known to be sensitively tuned to one another in terms of proximity-maintenance and also in terms of their emotions and communication. These attachments then, should be especially good contexts for mediating the regulation of the child's cognitive activities. That is, socially-coordinated task engagement—both between the child and the mother and between the child and other individuals—should be better facilitated through secure attachments than insecure ones as a consequence of the more effective dialogues occurring between these mothers and their children. One would not expect differences to emerge between securely-attached and insecurely-attached children in global measures of intelligence—such as IQ—that reflect strong developmental forces. But the cognitive executive functions, the patterning of specific abilities, and the emotional accompaniments of cognitive performance can be expected to differ.

Indeed, the evidence concerning the relation between attachment security and measured intelligence is not consistent (Campos *et al*. 1983) but that with task engagement is relatively clear-cut. For example, at 18 months of age, securely attached children, as compared with insecurely attached children, more often engage in symbolic play, are more enthusiastic and compliant in problem-solving tasks, exhibit fewer frustration-related behaviours, and are more sophisticated in social tool-use (i.e. in initiating and utilizing social coordinations involving the mother) as well as nonsocial tool-use (Matas *et al*. 1978; Bretherton *et al*. 1979). These children are *not* more likely to achieve ultimate success with the task than are insecurely attached children, but rather, their engagement with the task is different. Security in early attachment relationships thus seems to be associated with patterns of cognitive functioning that suggest effective internalization of regulative activities rather than other functions. And indeed, follow-up studies show that these cognitive advantages remain characteristic of securely-attached children into the fourth year in: a) exploratory tasks requiring social coordination with the mother (Arend 1984); and b) more 'academic' tasks in which the mother teaches the child a simple skill (Rahe 1984).

Child-rearing variations (as opposed to parent-child relationships) have been studied in relation to the social and cognitive functioning of older children by many investigators (see Maccoby and Martin 1983). These correlations are instructive because they demonstrate, for example, that authoritative child-rearing (consisting of high parental demands and high parental responsiveness) is associated with cognitive competence in children (Baumrind 1967; 1971). Similarly, parental involvement is associated with measures of 'ego control', including cognitive assessments (Block and

Block 1980). Other studies link disturbances in parent-child relationships to low achievement motivation, academic difficulties, and a variety of other problems with cognitive agency (see Rutter this volume). These studies do not provide direct evidence concerning relationship variations and cognitive development, but the clues contained in these correlations are tantalizing—especially since they are consistent with the notion that the main consequences of relationship variations in cognitive development are individual differences in cognitive regulation.

Friendship variations have not been examined in relation to cognitive development. In fact, no one has ever tried to assess pair-to-pair variations in children's friendships. One study of sibling relationships contains some interesting tidbits: Dunn and Dale (1984) discovered that children whose sibling relationships could be considered 'friendly' engaged more extensively in symbolic play than did children whose relationships with their siblings were 'neutral' or 'unfriendly'. Fragmentary as these data are, they suggest that all child-child relationships should not be regarded as alike, and that pair-to-pair variations need exploration in relation to cognitive development.

In summary, the significance of relationship variations for cognitive development are understudied. The evidence suggests, however, that well-functioning relationships are marked by effective cognitive dialogues which are centred on the child's cognitive agency. Relationships may not account for significant variance in every cognitive capacity, but cognitive regulation may be an exception.

Dialectical issues: a reprise

My comments on 'social interaction and cognitive development' and 'relationships and cognitive development' seem to have been written as though the section on 'dialectical issues' did not exist. That is, little consideration was given in these discussions to the effects of cognitive change on children's relationships; my attention was focussed almost exclusively the other way round. Moreover, little attention was given to the implications of relationships for the cognitive development of the child's companions. And yet I argued earlier that these considerations must be taken into account in any comprehensive research attack in this area.

I admit to these omissions. Moreover, I want to be certain that the reader is aware of them. The omissions themselves occurred for two reasons: a) the relevant studies are scarce and inconclusive; and b) space considerations would not permit more. Hopefully, however, these dialectics will not be ignored as investigators begin to work on the various issues covered in these remarks.

Conclusion

Three arguments were reviewed in this essay: a) that both relationships and individuals need to be viewed in developmental perspective; b) that social interaction, rather than social relationships, has been the level of analysis used most commonly to examine 'socialization effects on cognitive development' or 'social experience and intelligence'; and c) that close relationships serve uniquely to foster the development of the executive functions in cognitive development. The existing literature, thin as it is, is quite consistent in showing that relationships have a bearing on the child's efforts to marshall resources for problem-solving and to apply them, as well as to monitor and modify these efforts.

I feel relatively certain that, if relationships were important to more 'basic' components of perception, remembering, and thinking, we would probably know it already; the documentation would be extensive. Thus, I doubt that further investigation will uncover either memorial or combinatorial functions in intelligence that derive from these sources. Cognitive change, of course, may be revealed in the transformations that occur within relationships, but this does not alter my basic contention: that the significance of relationships in cognitive development consists mainly of their contributions to the executive functions.

I have brushed against several knotty questions in this essay: whether cognitive functions can be separated into those that are 'social' and those that are 'nonsocial'; whether interpersonal relationships are themselves components of intelligence; whether the dialectics involving relationships and cognitive development can be conceived in causal terms. To address these questions directly would take me far afield, even though it may be necessary, eventually, to come to grips with them in deciding whether close relationships have anything to do with cognitive development. Meanwhile, I am content with simpler hypotheses: namely, that relationships are important resources utilized by the child in cognitive development, and that this contribution consists of optimizing the child's efforts to apply and monitor cognitive activities in everyday life.

Acknowledgement

Preparation of this chapter was supported by Grant No. 5 P01 HD 05027 from the National Institute of Child Health and Human Development (USA).

Discussion

Relationships, context, and performance

Cognitive development implies a change in capacity: what we observe is merely performance. *Tizard* made the point that:

'One must be cautious about the step from observing the facilitating effect of a good social relationship on performance to assuming that cognitive development is facilitated. A child observed performing poorly in a negative or distant social relationship may perform better elsewhere. That is, the relationship may affect performance in its own particular *context* rather than cognitive development.'

Stambak argued that, in any case, a relationship need not be close in order to improve cognitive development. For example, two children from different backgrounds and with different approaches to a task might accelerate each other's development. Confrontation may help in acquiring knowledge. (See also her chapter, and Perret-Clermont's discussion of this issue p. 313. It must also be noted that 'confrontation' in an interaction may take place in the context of generally affectionate or of cold relationships, and could be more effective in promoting cognitive growth in the former—*Eds.*).

Shure referred to the reciprocal influences of cognitive development on relationships and vice versa, and asked whether intervention involving cognitive training and intervention involving making friends might not be equally successful.

7

Family and school influences on cognitive development*
MICHAEL RUTTER

Introduction

Few topics have given rise to such prolonged controversy as the relative importance of genetic and environmental influences on intelligence. One might suppose that the immense literature on this issue would provide a wealth of data on the ways in which psychosocial factors serve to influence cognitive development. Unfortunately it does not. That is because most of the arguments have concerned estimates of the relative strength of genetic and environmental forces rather than *how* the latter operate.

The findings on the heritability of intelligence have been reviewed extensively (see, e.g. Jencks *et al.* 1973; Rutter and Madge 1976; Scarr 1981; Vernon 1979). Although there have been extreme claims that there are only minor environmental effects (Munsinger 1975) or near-zero genetic effects (Kamin 1974), the evidence shows that both genetic and non-genetic factors are influential. There is little point in debating the precise level of heritability both because heritability estimates are necessarily population-specific (being affected by the extent of genetic and environmental variations) and because such estimates refer to population *variance* rather than population *level*: a move from a severely disadvantaged to a more normal environment might well increase IQ level whilst hardly affecting genetic contributions to within-population variance (Scarr 1981).

A further important consideration concerns the effect of genetic-environment covariation and the likelihood that to an important extent genotypic differences create different environments (Scarr and McCartney 1983). In other words, parents or teachers treat children differently both because the children themselves are different and because of the adult's own genetically determined propensities. Not only is the environment not statistically independent of genetic influences, but also people tend to make their own environments. This creates two important methodological

---

* Based on a paper of the same title published in the *Journal of Child Psychology and Psychiatry*, 1985.

issues. First, the analysis of psychosocial influences must take into account the possibility that the supposed psychosocial effects were in fact genetically determined. Second, it is necessary to differentiate between psychosocial influences that impinge similarly on all children in the same family and those that affect each child differently (i.e. shared and non-shared environmental influences—Rowe and Plomin 1981).

Most studies of genetic and environmental effects on cognition have been restricted to IQ scores. In contrast, this paper is concerned with cognitive performance more broadly defined to include scholastic and other attainments. The reasons for this are three-fold. First, IQ tests tap performance just as do tests of reading or mathematics; they differ only in that they assess a much wider range of cognitive skills that are less likely to have been taught directly at school (Vernon 1970). Second, many psychosocial factors influence general and specific aspects of cognitive performance in broadly comparable ways. Third, neither intelligence nor scholastic attainment are unitary skills, so that it is necessary to search for possibly different types of influence on their different elements. Not all psychosocial influences operate in the same way. In particular, with both IQ and scholastic attainment tests it is necessary to differentiate between effects on cognitive capacity (i.e. the level of *competence*) and effects on cognitive performance (i.e. the *use* of skills).

It may be that environmental factors influence variance in scholastic attainment more than in IQ (Jensen 1973), although that is not certain (Scarr and Kidd 1983), but that possibility will not be examined here. Also, almost by definition, direct teaching will affect accomplishments in school-taught subjects more than IQ, but such teaching effects will not be discussed. Rather, the concern is for the more general aspects of cognitive growth and performance, with special attention to developmental issues. The main focus is on *which* aspects of the psychosocial environment have the strongest effects on cognitive growth and performance; *how* the effects are mediated, that is what processes are involved; and to what *extent* the effects are seen across the range of ordinary environments and to what extent are they restricted to extremes of disadvantage.

In general the environmental influences that have the greatest impact on social behaviour are very different from those with the most effect on cognition (Rutter 1981, 1984a; Rutter and Madge 1976). Thus, family discord is a potent factor in the development of conduct disorders but of less relevance for cognitive development. Similarly, even a good institutional upbringing predisposes to disruptive and maladaptive social behaviour but has less impact on intelligence. Bereavement predisposes to emotional disturbance but does not impede cognitive development to the same degree. Nevertheless, as emphasized by the very topic of this conference, it would be absurd completely to separate social relationships and cognitive development. Accordingly, some attention will be paid to possible links between the two with respect to the influence of psychosocial variables.

Family influences

Family influences are conveniently considered in terms of the effects of variations in the ordinary environment, of planned interventions, of abnormal environments, and of extreme environmental conditions.

Variations within the ordinary environment

Most studies of psychosocial effects on cognitive development consider the performance of socially disadvantaged groups (Birch and Gussow 1970), correlations between IQ and social class (e.g. Douglas 1964), or correlations between IQ and child-rearing variables within samples of children reared by their biological parents (Wachs and Gruen 1982). However, all three approaches confound genetic and environmental influences. To study the latter it is necessary to control the former or to choose circumstances in which they are unlikely to operate. Three main strategies may be employed:

(i) the study of ordinal position effects;
(ii) the study of family influences after statistically partialling out the effects of parental IQ; and
(iii) the study of parental effects within an adopted sample.

Ordinal position

The examination of ordinal position effects involves several problems, but eldest or only children tend to show slightly higher scholastic achievement and verbal intelligence scores than later born ones (see Rutter and Madge 1976). The difference is small but in within-family comparisons it involves a $3\frac{1}{2}$ point difference between first born and fifth born children (Record *et al.* 1969). How this effect is brought about is not known with certainty, but parents tend to interact differently with their first born than with subsequent children. In general, they do more things with, and are more pressuring with, the first child (Gottfried 1984; Rutter 1981). Thus, Davie *et al.* (1984) found that parents interacted or talked more with their first borns; conversely, youngest children spent more time playing and talking with other children but, compared with eldest children, they were more likely to be followers than initiators. It is plausible, although not demonstrated, that these interactional differences may influence the children's cognitive development. In passing, it should be noted that the ordinal position effects for socio-emotional behaviour are rather different. Whereas eldest children may be at a slight advantage intellectually they are at a slight disadvantage socio-emotionally. Eldest children tend to be more aggressive, less cheerful (Davie *et al.* 1984), more socially assertive (Snow *et al.* 1981) and more liable to show emotional disorder (Rutter *et al.* 1970).

Parent-child interaction, controlling for parental IQ

Ideally, in order to control for genetic factors when searching for environmentally mediated family influences, the effects of both paternal and maternal IQ should be partialled out. No study has achieved this but several investigations have taken into account maternal IQ or education or, less satisfactorily, socio-economic status (Gottfried 1984). In all cases the family environment-child IQ correlations were reduced after controlling for IQ, but to an extent varying greatly across studies. In some (Longstreth *et al.* 1981) the introduction of maternal IQ reduced the home environment effect to insignificant levels whereas in others the reduction was only minor (see chapters in Gottfried 1984). However, there is general agreement that, whatever the extent of the reduction in home environment effects, it does not seem to alter the overall pattern of variables associated with children's cognitive performance.

Because of variations in sampling and measures across studies, firm conclusions about the specifics of family influences are elusive. However, certain inferences may be drawn. First, the variance accounted for by overall home environment variables is typically low—with corrected correlations generally in the 0.15 to 0.35 range. Second, the effects tend to be greater with IQ at 3 to 5 years than with developmental quotients in the first two years. Third, family measures at 2 years and older tend to have greater effects than similar measures in infancy, especially after correction for parental IQ (Yeates *et al.* 1983). In other words, the very early family environment seems less influential than that experienced at the toddler stage and later whether correlations are with contemporaneous or later cognitive measures.

Fourth, whereas no one type of parent-child interaction stands out as of critical importance, the relevant features include: absence of noise-confusion (meaning a background of high intensity sound that is either not discriminable or not meaningful to the child), the provision of a variety of activities and experiences, ample parent-child play and conversation, responsivity to the child, parental nurturance, teaching of specific skills, and opportunities for the child to explore and try out new skills and activities (Gottfried 1984; Wachs and Gruen 1982). Clearly the traditional notion that it is 'stimulation' that promotes cognitive growth is seriously misleading. Although varied and interesting parent-child interactions are beneficial, a high level of noise is not. Moreover, the crucial feature is not so

much the parental 'input' in a stimulus sense as the reciprocity of the interactions, the variety and meaningfulness of their content, and the active role
taken by the child. Nevertheless, this does *not* amount to a non-involved
passive parental style in which the children are left to discover everything
for themselves; both a degree of structure and considerable direct teaching
seem desirable.

Fifth, although differences in the children's characteristics may well
make them responsive to different sorts of home environment, the matter
has been so little explored (with the exception of sex differences) that no
conclusions are possible. However, there is some weak indication that children's needs may change with age. For example, Wachs and Gruen (1982)
argued that physical contact was important only before 6 months of age,
that conversation and responsiveness to the child's non-verbal signals
become important during the second year of life, that responsiveness to the
child's talk increases in relevance in the third year and that the variety of
contacts with a range of other adults (and an avoidance of noise-confusion)
increase in relevance after age three. Bradley and Caldwell's (1984) longitudinal study suggested that by the time children reach school, maternal
responsivity is less important but parental encouragement and the availability of a range of play materials and of experiences remain salient.

Finally, the effects on verbal intelligence may possibly be greater than
those on non-verbal intelligence, and effects on scholastic attainment may
be greater than those on IQ. However, the evidence on these points is
inadequate.

Adopted children

Early studies of families with both biological and adopted children (see
Bouchard and McGue 1981; Rutter and Madge 1976) showed that the correlations for IQ between biological children and their parents were about
twice as high as those between adopted children and their adopting parents
(circa 0.34 versus 0.19). Recent studies have added substantially to this evidence. In a study of transracial adoptions Scarr and Weinberg (1983) found
correlations of 0.43 versus 0.29, and in a study of adolescents an even
greater disparity with an adoptive mid-parent-child correlation of only
0.14. Horn's (1983) findings from the Texas Adoption Project were similar
(0.28 versus 0.15). Both studies also found that adopted children's IQ
scores correlated more highly with those of their biological parents who did
not raise them than with those of their adoptive parents who raised them
but did not share their genes. Clearly part of the parental effects in children's intelligence is genetic rather than environmental; however, some
environmental effect is also suggested. Plomin and De Fries (1983) in the
Colorado Adoption Project assessed environmental effects more directly

using the Caldwell and Bradley HOME interview-cum-observation assessment of the home environment. In line with the biological family findings discussed above, the HOME scores showed little correlation with Bayley developmental scores at age 1 year, but there were significant correlations at 2 years: 0.22 for the adoptive families and 0.39 for the control families (the variance on the HOME inventory being similar in the two groups). Interestingly, the adoptive and control groups did not differ in the level of correlation between the HOME scores at 1 and the Bayley scores at 2, suggesting that genetic factors may not be involved in the predictive ability of HOME, at least before 2 years.

Thus, three different types of study of variations in the ordinary home environment are in fairly good agreement, and hence provide reasonable guidelines regarding family influences that may be influential for cognitive development. However, the family measures account for a rather small proportion of population variance and most of the results apply to very young children. The ordinal position data stand out in showing comparable effects throughout the course of development; however, they reflect within, rather than between, family differences. The findings on age differences in home environment effects are inconclusive. Horn's (1983) data showed a possibly reduced genetic effect for children over 10 years but no comparable increase of family environment effects. Scarr and Weinberg's (1983) data on adolescent sibling correlations suggested that shared familial influences decreased with age. Possibly shared family environment effects are most marked in early childhood, with both non-shared family effects and extra-familial influences greater in adolescence. However, that hypothesis is speculative and requires systematic testing.

Family intervention studies

The findings here can be considered most conveniently in terms of the degree of change entailed.

Adoption

The most complete family change is that brought about by adoption. Family effects *within* adopted groups have been considered already. However, because of deliberate policies in the selection of adoptive parents the range of home environments within such groups is relatively narrow and relatively favoured. A rather different use of the adopted child research strategy is to compare the mean *level* of adopted children's intelligence and scholastic attainment with that expected from the characteristics of their biological parents. Skodak and Skeels (1949) showed that children born to mothers of below average IQ when adopted in homes somewhat above the

average had a mean IQ of 106 compared with the mean of 86 for their biological mothers and 90 to 95 expected if they had been reared by their biological parents. Important criticisms can be made of Skeels' later research
(Longstreth 1981) which do not invalidate this observation, but further
major problems stem from the fact that the early versions of the Binet had
both means and standard deviations that varied greatly with age (Clarke
1982).

More recently, Scarr and Weinberg's (1976) study of socially classified
black children, mostly of mixed race adopted by white families of high
socioeconomic status provided better data showing a similar effect. The
adopted black children had a mean IQ of 106, somewhat above the white
general population mean (although a little below the mean of 111 for white
adopted children), and more than a standard deviation above that
expected from their biological parentage. The adopted black children's
scholastic attainments were also slightly above population norms.

Horn's (1983) data from the Texas adoption project were similar.
Although the children's IQs *correlated* more strongly with those of their
biological mothers than those of their adoptive mothers, the mean IQ of
the children (111.5) was actually *closer* to that of their adoptive mothers
(112.4) than to that of their biological mothers (108.4) (Walker and Emory
1985 discuss this point).

Lastly there are the important findings from the Schiff *et al.* (1982) study.
Thirty-two children born to biological parents both of whom were
unskilled workers were abandoned at birth and adopted by families span-
ning the top 13% of the socio-professional scale. The effects of this change
in social situation were assessed by comparing the children's IQs (mean
109) with those of their biological half-sibs (n= 20; mean 95) reared by the
biological parents and with those of general population studies of children
of unskilled workers (95) and of upper middle class parents (110).
Although the sample was small, in many ways these data provide the
strongest evidence of an environmental enhancement of IQ through the
adoption into socially favoured homes of children from a socially disadvantaged background.

Inevitably there are limitations on these data. Ideally there should be
measures of the biological parents' IQs (these were available for mothers
in the Horn project but other investigators have had to rely on years of
education); and a comparison between the adopted children and their biological siblings reared by their biological parents (partially available in the
Schiff study) would be even more informative. Nevertheless, the findings
indicate a clear effect on intelligence and scholastic attainment of a *major*
change of environment from below to above average social circumstances.
The rather small effect of environmental variations within the adopted
sample may indicate that the removal of adversity matters more than

normal variations within the range of acceptable environments; however this could also reflect the lack of sufficiently discriminating measures within the normal range.

It has been suggested that environmental effects on intellectual development are greater during the preschool years than in later childhood (see Clarke and Clarke 1976; Clarke 1984; Rutter 1981). Evidence in support of this view comes mainly from slightly lower IQ scores of later adopted children compared with children adopted in infancy (see Dennis 1973). The finding is interesting but there are three main reasons why it does not point unequivocally to a greater effect of the early family environment. First, it is difficult entirely to rule out the possibility of biases due to selective placement. Second, the durations of disadvantageous environmental circumstances are not comparable; by definition the late adopted children have experienced more prolonged disadvantage. Thirdly, the amount and character of parent-child interaction in middle childhood is very different from that in infancy. The lesser intellectual gains of older late-adopted children simply could reflect their lesser opportunities for intensive interaction with their adoptive parents. In any case the differences in cognitive outcome between early and late adopted children are small, and late adopted children perform relatively well educationally (Triseliotis and Russell 1984).

Milwaukee project

During the last two decades there have been many varied preschool interventions. With the exception of the Parent-Child Development Centres (see below) most have emphasized mainly some form of out of the home education. However, beginning with Gray's pioneering Early Training project (Gray *et al.* 1982; Klaus and Gray 1968), a number included work with parents as an integral part of the programme. In a few cases this feature, combined with the pervasiveness of the interventions, suggests that they should be considered as family interventions. In the Milwaukee project (Heber *et al.* 1972; Garber and Heber 1977; 1982) the mothers were given home management and job training together with remedial education (but little direct help with parenting); and from infancy up to 6 years the children participated in a structured educational programme with an emphasis on problem-solving and language skills. The families were socially disadvantaged and the mothers had an IQ of 75 or less; 20 were allocated to the programme and 20 served as untreated controls. Both groups have been followed up to 4th grade level. The WISC IQ results at age 10 showed an 18 point between-group difference (104 vs. 86); a significant difference but a smaller one than that when the children were younger. The experimental group also showed scholastic attainments superior to those of the control group but the differences were less marked than those for IQ, both groups declined steadily through the first 4 years of schooling, the between-group difference progressively lessened, and in absolute terms the experimental groups were performing far below national norms by fourth grade. Within the experimental group there was substantial individual variation in outcome, with the poorest outcomes for those in the most disadvantaged homes. The findings, albeit based on small numbers and with some problems in project design and presentation (Page and Grandon 1981), show that a prolonged and wide-ranging intervention programme during the preschool years can produce important cognitive gains. However, the follow-up has also shown that, after intervention ceased, the benefits diminished markedly with time.

Carolina Abecedarian project

The Abecedarian project was similar to the Milwaukee programme in starting in infancy, providing year-round daily care throughout the pre-school years, a focus on language and adaptive social behaviour and a structured approach to teaching (Ramey and Campbell 1981, 1984; Ramey *et al.* 1982). Hence, it is most conveniently considered here; however the project did not include direct work with parents and parental participation was limited and sporadic. Most parents were black, all were seriously socially disadvantaged, the mothers had a mean IQ of 84, and three-quarters of the families were female-headed. There was random assignment to experimental and control groups, with outcome data up to 54 months for 49 experimental and 46 control group children and up to 5 years of age for 27 and 23 children respectively. At 60 months the mean WPPSI IQ (98) for the experimental group was 8 points above that for the controls and far fewer (11 per cent vs. 39 per cent) had an IQ of 85 or below— figures comparable to the findings at 49 and 54 months for the total group. It should be noted that the cognitive differences did not begin to appear until about 18 months of age. The experimental group children also adapted better to new situations, responded more appropriately to task demands, and made better use of language than the controls. Fewer of the experimental group mothers were out of work at follow-up and more had obtained further education during the course of the intervention. The experimental programme was accompanied by an attenuation of the correlation between maternal and child IQ scores, but in both groups the home environment and the children's temperamental qualities (those with 'easy' temperaments had higher IQ scores) continued to predict the children's performance. Although these results are more modest than those of the Milwaukee study, the pattern of findings is broadly similar. Intervention produced limited but worthwhile cognitive benefits; the effects were not manifest in early infancy, better cognitive performance was associated with better social adaption, and there were considerable individual differences in response.

Parental programmes

A few programmes have concentrated on work with parents in order to foster the cognitive development of children from low income families. Thus, Parent Child Development Centres (PCDC) in the USA provided a wide ranging curriculum for mothers including information on child rearing, teaching on home management, and family support combined with pre-school education for the children: the families started attending when the children were infants and stopped when they were 3 (Andrews *et al.* 1982). There was random assignment to experimental and control groups, and multiple methods of outcome evaluation. The results have varied somewhat from centre to centre but all have shown benefits for families receiving the intervention programmes. Mothers in the experimental groups tended to communicate better with their children, were more sensitive and emotionally responsive, used more encouragement, and provided more information in their talk with their children. The children showed more social behaviour and more positive interactions with their mothers. At age 3 years the experimental group children scored 4 to 8 points higher on Stanford-Binet IQ assessments. Preliminary follow-up data 1 year later showed reduced between-group differences.

Tizard and Hewison Reading Study

The last family intervention project to be discussed was quite different both in its timing (primary school rather than infancy or preschool) and in the minimal nature of the intervention provided. A general population study of children in inner city London schools had shown that those who read to their parents at home had markedly higher reading attainments than those who had not received this kind of help from their parents (Hewison and Tizard 1980). The finding could not be accounted for by the children's level of intelligence or other aspects of their upbringing. An experimental study was undertaken to assess the effects of parental involvement in the teaching of reading (Tizard *et al.* 1982). Infant school classes were randomly assigned to non-intervention, extra group reading tuition at school, and parental help ('home collaboration') with reading, the books being provided by the school.

At the end of the 2 year intervention period the children in the home collaboration group showed reading skills superior to their controls (a 12 point difference in one school and an 8 point difference in the other), the groups having been comparable before the intervention. The extra teacher help groups did not differ in outcome from controls. One year after the intervention ended the benefits of parental involvement had markedly diminished in one school (to a $3\frac{1}{2}$ point difference) but had been maintained in the other (11 point difference). Nevertheless, in both schools the parental collaboration resulted in a substantial reduction in the proportion of children reading below age level (79–84 per cent in controls; 46–54 per cent in experimental groups).

There are limitations to the study: randomization was by class rather than by individual; the effects varied by school; there was not a crossover, and it was not possible to identify the effective elements of the intervention. Thus, for example, it is not clear whether the benefits stemmed from increased reading experience, from one-to-one reading opportunities, or from effects on esteem or motivation. Nevertheless, the results are impressive in showing that an intervention as simple as getting (often poorly educated) inner city parents to listen to their children read was associated with important gains in reading skills. It is particularly notable that the benefits followed interventions during middle childhood rather than infancy.

Abnormal environments

Institutional rearing

Various studies in the 1940s and 1950s reported marked intellectual impairment in institution-reared children (see Tizard 1970). However, other investigations showed cognitive *gains* in children transferred from a poor institution to a better one (e.g. Garvin and Sacks 1963; Skeels and Dye 1939), suggesting that the damage did not come from institutional care *per se*. More recent studies of better quality institutions have confirmed that children reared in them have around average intelligence. For example, Tizard and Hodges (1978) found that eight-year olds who had been in institutions from infancy had a mean WISC full scale IQ of 99 despite marked discontinuities in caretaking (some 50–80 parent surrogates), but had gross abnormalities in social behaviour. The mean IQ of 99 was below that (110) in the general population control group but this could be due to the more disadvantaged backgrounds of the institutional children. Similarly, Dixon (1980) found that 5 to 8 year olds reared in institutions from infancy had a mean IQ of 108 compared to a mean IQ of 106 in the comparable group of family-fostered children from similarly disadvantaged backgrounds. Again, the institution-reared children differed markedly in behaviour but not in measured intelligence.

The lack of effects on general intelligence is striking in view of the marked effects of institutional upbringing on socio-emotional behaviour (Quinton *et al*. 1984; Rutter 1981). Nevertheless, although not reflected in IQ scores, there were group differences in classroom behaviour of a kind likely to affect learning. Thus, the institution-reared children in both these studies showed poor task involvement and inattentive overactive behaviour (Roy 1983). How far these behavioural features in fact impeded either cognitive development or cognitive performance is unknown, but clearly the negative potential was there. Other studies have shown that institution-reared children are less likely than controls to continue in education (Quinton and Rutter, unpublished data) and that behavioural disturbance has indirect effects on scholastic attainment as a result of truancy and school drop-out (Maughan *et al*. 1985).

Family pathology

Numerous studies have shown the increased risk of socio-emotional and behavioural disturbance in the children of parents with some form of chronic psychiatric disorder (Rutter and Quinton 1984). Less is known about the cognitive sequelae for children reared by mentally ill parents. However, such data as are available show small and inconsistent effects. For example, Winters *et al*. (1981) found that the children of depressed and of schizophrenic parents had mean IQ scores about a fifth of a standard deviation below controls. On the other hand, problems in school work are often included among social and behavioural difficulties (Billings and Moos 1983). Similarly, there are no consistent effects of parental divorce and death on IQ (Rutter 1981), but there are often temporary decrements in scholastic performance that persist for a year or so (Hetherington *et al*. 1982; van Eerdewegh *et al*. 1982).

Several clinical reports have commented on the poor language development of children subjected to physical abuse by parents. However, most studies have failed to control for other family characteristics and it appears that language and cognitive problems are more a function of serious social disadvantage than of abuse *per se* (Elmer 1977). Moreover, one study showed that children who experienced parental neglect had more impaired verbal abilities than those who suffered abuse (Allen and Oliver 1982).

The available data are not satisfactory but consistently show that patterns of upbringing involving serious discord, discontinuities in parenting, and/or parental deviance carry a high risk that the children will show socio-emotional or behavioural problems; however the risk for intellectual impairment is quite slight although that for difficulties in scholastic attainment is rather greater.

Extreme environmental conditions

Lastly with respect to family influences there are children reared in gross social isolation and severe physical confinement. Skuse (1984) has summarized the findings on nine well documented cases and Clarke (1984) has added another. The findings are striking in at least four crucial respects. Firstly, these grossly abnormal patterns of upbringing were associated with severe cognitive deficits at the time the children were discovered at ages ranging from $2\frac{1}{2}$ to $13\frac{1}{2}$ years (Skuse 1984); all were virtually without spoken language and most showed severe intellectual retardation. Secondly, most showed substantial cognitive improvement following rescue and seven of the 10 achieved normal levels of intelligence and language; two more gained a normal performance IQ but remained seriously impaired in language. Thirdly, improvements in cognitive functioning were usually evident within months after removal from the extreme environmental conditions, although normal levels were achieved more slowly. Fourthly, the published reports contain remarkably few systematic data on possible effects on the children's interpersonal relationships.

It may be inferred not only that severe privation led to severe cognitive impairment, but also that a complete change of environment led to marked cognitive recovery.

Conclusions on family influences

Although there are major lacunae in the evidence, the findings on family influences point to several reasonably firm conclusions. Firstly, children's experiences during their upbringing do produce important effects. Secondly, these are most striking during the later preschool years with less impact during the infancy period; however major environmental changes continue to have major effects on cognitive development throughout childhood. In part, the limited effects seen in infancy probably stem from the difficulties in measuring cognitive performance so early in life, but also early development may show a greater degree of 'canalization', with maturational influences more important than either environmental features or polygenic factors (McCall 1981). Thirdly, there is no marked critical period of cognitive development. The benefits of good experiences in the early years are largely lost if subsequent experiences are bad; conversely there may be substantial recovery if early bad experiences are followed by good ones in middle childhood. Fourthly, the most striking effects on cognitive development and performance are seen with gross deficits in rearing conditions; however, both the ordinal position data and the findings on home environment-child IQ correlations show also smaller effects within the normal range. Fifthly, although the features of the home environment conducive to optimal cognitive development cannot be specified precisely, they clearly include: parental responsiveness to children's signals; varied and positive patterns of reciprocal parent-child interaction and communication; and a range of interesting, varied and meaningful experiences and activities with both parents and other people. Family discord and disturbance affect scholastic performance more by influencing socio-emotional and behavioural functioning than cognition.

School influences

Variations within the ordinary environment

Studies in both developing and industrialized nations have shown that schooling is associated with superior cognitive development (see Rutter 1981). In countries where schooling is available in only some areas, cognitive skills are greater in those children able to attend school. Forced closure of schools as a result of war or political activities has been followed by IQ decrements of 5 points or so, and continued schooling during later adolescence is associated with a mean IQ gain of 5 to 7 points. It is usually supposed that IQ predicts scholastic attainment but it is evident also that schooling and improved educational accomplishments may themselves lead to IQ gains.

The importance of school influences may vary with family circumstances. Relatively speaking, pupils from a socially disadvantaged background tend to lose ground in cognitive performance during the long summer vacation when they are not at school, but this does not occur with those from more favoured families (Heyns 1978).

In recent years increasing attention has been paid to the extent to which ordinary schools vary in their effects on children's cognitive performance (see Rutter 1983c). These have mainly relied on measures of scholastic attainment rather than IQ and it is not known how far variations in the quality of school experiences influence IQ. Nevertheless, the evidence shows important school effects on children's educational performance, after taking account of differences between schools in the social and intellectual distributions of the children at the time of school entry. Differences in mean levels of attainment between schools at the extremes of the ranges studied have been marked (about 1 standard deviation), indicating that school effects can be substantial. However, many schools vary little in their effects so that it is not possible to derive any meaningful overall measure of the general strength of school influences.

The qualities of schools associated with good scholastic performance have been studied by comparing schools performing better with those performing worse than expected on the basis of their intakes. The findings show a variety of school features to be important. First, several might be considered to have direct effects on learning. Thus, children's scholastic performance is better in schools with a clear focus on academic goals and a translation of that attitude into practice through the regular setting and marking of homework, a high proportion of time devoted to active teaching, group planning of the curriculum, and checks to ensure that teachers follow the intended practices. Secondly, there are features designed to foster efficient learning through good classroom management—as by lessons beginning and ending on time, clear and unambiguous feedback to pupils on their performance, minimum disciplinary interruptions, and effective classroom teaching techniques. Thirdly, there are school practices that may operate by aiding high morale—such as adequate but discriminating use of praise and encouragement, good pupil conditions, good models of teacher behaviour, and good care of the school buildings. Fourthly, there are features that serve to set school norms—such as an adequate intellectual balance in the composition of the intake and appropriately high teacher expectations of the pupils. Fifthly, there are actions that may foster pupils' commitment to educational goals; for example, opportunities for most pupils to take responsibility and to participate in the running of their school lives and out-of-school activities shared between staff and pupils.

It is noteworthy that there are substantial positive correlations between different measures of pupil performance. On the whole, schools with good scholastic attainment also tend to be those with high levels of attendance, generally good behaviour, and a high proportion of pupils staying on at school beyond the period of compulsory schooling (Rutter *et al.* 1979). Similar associations are seen at an individual level; children who attend poorly tend to do less well in exams, those with disturbed behaviour tend to leave school early without sitting exams. Correlational data of this type cannot delineate mechanisms with any precision. Nevertheless, the strong inference is that many of the cognitive effects stem from influences on self-esteem, patterns of behaviour and task performance, educational motivation, and attitudes to learning. Presumably pedagogic skills are important in determining both how much is taught and how well it is taught. But children's *responses* to that learning environment will be influenced by whether or not they are present (as determined by school attendance and when they leave school), their involvement in the learning task (as determined by features such as attention to instructions and task performance style), their aspirations (as shaped by the expectations of those about them and their own self-concept), and their interest and commitment to learning (as reflected in what they do outside as well as inside school). Whether or not schools foster effective learning in their pupils will be determined by their qualities as a social organization (Minuchin and Shapiro 1983), as well as by their qualities in relation to the transmission of knowledge as such.

Interventions

Although planned interventions in elementary and in secondary schools have been studied, most research has concerned the effects on IQ and scholastic performance of children experiencing day care (Zigler and Gordon 1982) or some form of preschool education (Clarke-Stewart and Fein 1983). As the literature has been extensively reviewed several times, only the key findings will be summarized here.

Preschool education

Numerous studies in the 1960s and early 1970s showed that preschool interventions led to immediate IQ gains but that the benefits did not persist for long after the start of regular schooling. The loss of advantage from the preschool intervention resulted in part from the IQ gains in control groups associated with entry into formal education, and in part from decrements in IQ in the experimental groups. These largely negative findings promoted acceptance that compensatory education was not worthwhile. Undoubtedly, the original expectation that relatively brief educational experiences in the preschool years would lead to lasting beneficial effects on general intelligence was both naive and unwarranted even by the empirical evidence available at the time (Clarke and Clarke 1976). Nevertheless, recent reevaluations of findings have suggested reappraisal of the view that there were *no* lasting benefits (Berrueta-Clement *et al*. 1984; Lazar and Darlington 1982; Ramey *et al*. 1984; Schweinhart and Weikart 1980).

Lazar and Darlington (1982) reported findings from a pooling of the data from 11 of the best designed investigations with appropriate control groups, in which there was a systematic follow-up when the children were aged 9 to 19 years. High quality preschool programmes were associated with IQ gains during the first 2 years after the programmes finished, there was a lesser gain for up to 3 or 4 years but no detectable effect at the final follow-up. The findings regarding effects on scholastic achievement were less clearcut but there were significant benefits during the early years of schooling, particularly in maths. In some projects the effects on achievement seemed to diminish with time but in the particularly well planned Schweinhart and Weikart (1980) study they increased. In many respects, the most striking lasting benefit of the preschool intervention was the reduced proportion of children who had to repeat a grade or who were placed in special education classes. In addition, the intervention was associated with an increased tendency for the children to report themselves proud of their achievements and for the mothers to be satisfied with their children's school performance and to have higher aspirations for them. In some of the projects teachers tended to rate the children from the programmes more positively on attitudes to learning and behaviour. It was concluded that early education brought benefits through two routes. First, there were direct effects on cognitive performance (as mediated through the learning of specific skills, improved task orientation, and better persistence). Second, there were benefits through non-cognitive effects on children's self-esteem and self-efficacy and on their attitudes to learning; on parents' hopes and aspirations; and on teachers' expectations of and responses to the children. It was hypothesized that the early education experience may change children from passive to active learners who take the initiative in seeking information, help and interaction with others. When this increased motivation to learn is met by a positive response at home and school, long-term cognitive gains can result.

The most thorough analysis of the effects of high quality preschool education is that deriving from the Perry Preschool programme (Berrueta-Clement *et al*. 1984). They have data from five cohorts of low SES 3–4 year old children with IQs in the 60 to 90 range randomly assigned to an experimental or a control group. The experimental children received an organized educational programme directed at both intellectual and social development; in addition teachers made a once per week $1\frac{1}{2}$ hour home visit to see each mother and child. Both groups have been followed to age 19 years. IQ differences between the groups did not persist beyond the early years of schooling but other benefits lasted throughout the follow-up period; the experimental group showed improved scholastic achievement, a higher rate of high school graduation, a lower rate of delinquency, and a higher rate of employment. A causal path analysis suggested that the initial effect of preschool on intellectual performance generated long-term effects through its intermediate direct effects on scholastic achievement and its indirect effects on social maturity and commitment to education. Case studies comparing young people who did and did not respond well to the experimental intervention suggested that long-term success depended on parental and family support for education, the presence of positive role models (particularly those who demonstrated the value of schooling), a sense of responsibility that extends beyond oneself, and an active goal-oriented approach to life. The findings emphasize the probable operation of transactional indirect effects in the path leading to long-term benefits, the interaction between cognitive and social processes, and the importance of an effective mutually supportive 'mesh' between home and school.

Similar processes were postulated by Pedersen *et al*. (1978) in their investigation of the effects on children of one outstanding first grade teacher. The findings showed that the teacher had a major effect on children's academic achievement, work effort, and initiative during the following year, but that there were negligible *direct* effects thereafter. On the other hand, her *indirect* effects were substantial because the children had acquired styles of behaviour that brought success, which in turn reinforced their efforts. Similarly, their behaviour in class made them more rewarding students for the teachers of later classes who then responded to them in ways that facilitated their continuing success. Although these ideas have not been rigorously tested the evidence certainly suggests that preschool education has both cognitive and social effects. The latter may be as important as the former in processes leading to persistence of benefits. However, it is likely that the durability of cognitive gains depends not only on whether the children continue in beneficial environments, but also on how parents and teachers respond to the initial changes in the children's behaviour and performance. The particular patterns of adult-child and of child-child interaction that ensue probably contribute to the processes that determine whether the initial gains are built upon or undermined by subsequent experiences.

Day care and nurseries

These preschool programmes were explicitly designed to foster the cognitive development of children from a disadvantaged background. It remains to consider other forms of day care in which the main aim is to provide good quality caregiving, as well as the general run of nursery schools and play groups that may be seen as providing an early start to general education rather than any form of compensatory special education (see review by Clarke-Stewart and Fein 1983; also Zigler and Gordon 1982).

Although the findings vary between studies, *good quality* day care has no negative effects on intellectual development and may produce some temporary cognitive enhancement. Whether or not there are cognitive gains seems to depend largely on the extent to which there are explicit prescribed educational activities together with opportunities for social interaction and exploration of materials. The findings regarding effects on socio-emotional development are both more meagre and less consistent (Rutter 1981). It seems clear that good quality day care, at least after the age of 2 or 3 years, does not lead to emotional disturbance and may enhance children's social competence. However, again much depends on the quality of care provided, with maladjustment most likely in those who began day care in infancy in centres with little adult-child interaction and conversation (McCartney et al. 1982).

Sometimes it is assumed that nursery schools have much to offer in terms of their provision of a 'stimulating' language environment that can serve to compensate for an 'unstimulating' disadvantaged family environment. The studies by Tizard and her colleagues (1980, 1983a, b; Tizard and Hughes 1984), as well as those of other investigators (see Clarke-Stewart and Fein 1983), suggest that this may be misleading. They found that there was more, and more varied, conversational interchange between parents and their children in upper working class homes than the same children experienced at nursery schools. Also, working class children seemed inhibited at school in their language usage. In that good quality nursery schools *can* aid children's cognitive performance, it may be inferred that the reason they do so is *not* because they provide greater verbal 'stimulation'. Teachers do, however, differ from parents in using language that is more often complex, in asking more questions, and in providing more direct teaching (see Tizard, this volume).

It should be emphasized that the Tizard and Hughes (1984) data refer to relatively advantaged working class girls from small two-parent families (thus, they had a mean IQ of 106, which was some 20 points or more above that of the seriously disadvantaged children in most of the preschool intervention programmes), who were studied in relatively stress-free social circumstances (usually with only the mother and one child at home). Accordingly, it cannot be assumed that the conversational interchanges between parents and children would be as rich in socially disadvantaged homes as in the homes studied by Tizard and Hughes. Nevertheless, three findings suggest that the benefits from preschool education do not largely derive from verbal enrichment. First, the data are convincing in demonstrating that nursery schools do not encourage conversations that explore the development of ideas (because Tizard and Hughes found that the *same* children who appeared verbally inhibited at school were much freer in their use of language at home). Second, preschool education appears to benefit children from average homes as much as those from a very poor background (Clarke-Stewart and Fein 1983)—although the data on this point are limited. Thirdly, children in home based day care tend to talk more with their caregivers than do those in special centres, but the intellectual benefits from the latter are greater (Clarke-Stewart and Fein 1983). It may be tentatively suggested that language experiences at school and at home may both be beneficial, but for different reasons. Perhaps systematic teaching together with questions and answer type interchanges benefit cognitive growth through direct learning, whereas the less structured more inductive conversations at home are helpful through encouraging problem solving and sensitivity to other people. Clearly, children's conversations play a part in their language development and in their intellectual development more generally but we remain surprisingly ignorant of just which aspects of those conversations are most likely to promote cognitive growth.

Nature of influences and effects

In conclusion, it is necessary briefly to consider the possible ways in which psychosocial influences may act to facilitate cognitive development and performance. The evidence clearly demonstrates the reality of such environmental effects, although the effects on IQ are relatively modest within the normal range of environments. The effects on other aspects of cognitive performance are probably greater and the effects of markedly disadvantageous circumstances are very substantial with respect to all aspects of cognitive functioning.

The types of environments that have the greatest effect on socioemotional and behavioural functioning are different from those with the greatest impact on cognitive development. In particular, discord and discontinuity in relationships seem most important for the former whereas opportunities for active learning seem most influential for the latter. Nevertheless, there is overlap. Probably (Rutter 1984a) what is needed for optimal cognitive development is a combination of active learning experiences that promote cognitive competence together with a social context in which the style of interaction and relationships promotes self-confidence and an active interest in seeking to learn independently of formal instruction. The social context considerations necessarily entail environmental conditions likely to foster social as well as cognitive development. However, the provision of active learning experiences also does so. It is clear that these are *not* most usefully conceptualized in terms of 'stimulation'. Rather they involve the child's active, rather than passive, participation; the provision of varied, interesting, and meaningful experiences rather than high level sensory input; direct teaching of specific skills and knowledge; a sensitivity and responsiveness in reacting to children's approaches and questions; and a reciprocity in pattern of interaction. It is obvious that these features bear more than a passing resemblance to the style of authoritative-reciprocal parenting patterns, high in bidirectional communication, thought to be most likely to lead to social responsibility, control of aggression, self-confidence and high self-esteem (see Maccoby and Martin 1983).

The next question is *how* these influences operate. Traditionally, it would be supposed that the direct cognitive effects operate on some form of cognitive 'structure' that is part of the basis of personality. However, the longitudinal data suggest that this may not be the most helpful way of conceptualizing the process; indeed it may be doubted that there *is* an underlying personality structure laid down in the early years of childhood (Rutter 1984b). Of course, even very young children have substantial cognitive and interpersonal skills which show some coherence and organization. However, these skills may not be most helpfully conceptualized in terms of a 'structure' that is either present or absent and which, once present, is fixed and indestructible. Perhaps the key data in that connection are the relative impermanence of any effects independent of later environmental circumstances, and the relatively rapid recovery in middle childhood following removal from extremely damaging conditions. Also it may be relevant that both temporary and lasting environmental effects seem to be *least* evident in infancy. For the same reason, it seems dubious whether the main effects are on brain structure and chemistry. It is not that there is any doubt that marked environmental variations *can* lead to structural effects (see Greenough and Schwark 1984); the scepticism is over these being the main process involved in the continuation of cognitive sequelae following good or bad experiences. In the first place, the physical changes are more evident in the case of either gross sensory deprivation or the restriction of specific sensory inputs (as in vision), rather than in impairments in the *quality* of life experiences as discussed here. Secondly, the modifiability of environmental effects and the relative speed of cognitive changes following radical alterations in life circumstances do not suggest that the processes require any reorganization of brain structure or chemistry.

Perhaps, the direct cognitive effects mainly concern the acquisition of knowledge and skills. However, it is most unlikely that these mainly comprise the gaining of information or specific scholastic attainments. It is striking how very limited are the benefits stemming from interventions that are confined to the school environment. Not only may the experiences at home provide something not readily available in school but also it seems that the skills involved apply as much to processes of attention, perseverance, task performance, and work organization as to particular areas of knowledge. Learning how to learn may be as important as the specifics of what is learned.

In addition to direct cognitive effects, there are equally important non-cognitive effects that influence later cognitive functioning. These include children's concepts of themselves, their aspirations and attitudes to learning, their self-esteem, their commitments to education and their styles of interaction with parents, teachers, and others in the environment. The various ways in which environmental effects may lead to continuities and discontinuities in socio-emotional functioning have been considered elsewhere (Rutter 1984c). However, once again it seems that the processes are not best conceptualized in terms of personality structure; on the other hand, it would be equally misleading to view the effects simply in terms of immediate situational effects. Continuities arise because acquired skills and habits open up or close down opportunities; because acquired styles of interaction influence children's responses to new environments; and because children's self-concepts influence how they perceive new circumstances and how they respond to new challenges. It may be that this transduction of experiences through incorporation into the child's self system partially explains why experiences in early infancy tend to be less influential (because infants lack some crucial components of self-awareness) and why modest changes in the environment during later childhood have such a minor impact on cognitive development (because the habits, concepts, and styles of interaction have had so long to become well established).

Acknowledgement

I am most grateful to Ann Clarke for her detailed and helpful criticisms of an earlier draft of this chapter.

Discussion

IQ and cognition

Perret-Clermont and Chandler felt Rutter used 'cognition' in too narrow a sense.
Perret-Clermont commented:
'You distinguished in your analysis "cognitive" and "non-cognitive" (e.g. self-concept, self-esteem, attitudes to learning, commitment to education, etc.) effects of differing environments. It seems to me that this relies on a too narrow definition of what is meant by "cognition", which is merely equated with "educational success". I wonder if the results of the studies that you reported would not be understood differently (and have different implications for intervention programmes) if *cognition* were defined as *the capacity to make sense of what is going on in the surrounding (physical and social) environment.*'
   Chandler added:
'Certain contemporary critiques of the notion of IQ maintain that the concept artificially restricts attention to more or less formal, person-independent context-free aspects of cognition. In your talk you generally used cognition and IQ interchangeably and reached the conclusion that social interactions have relatively little impact on cognition (read IQ). Might this conclusion not be an artefact of the restricted definition of cognition?'
   Rutter replied:
'I agree with you that my conclusion (that the domains of interpersonal relationships and cognition are subject to rather different sets of family and school influences) stems from my focus on intellectual and school achievement measures of cognitive functioning. Obviously the conclusions would have to be rather different if cognition referred to social cognition. I am *not* arguing that social cognitive skills are irrelevant for the development of interpersonal relationships. Indeed in my own research into autism (Rutter 1983a) I have argued that the serious impairment in social relationships in that condition may well be due to a deficit in social cognition. Nevertheless, my conclusion is not a trivial one based on a particular semantic convention. What the evidence suggests is that to a substantial extent (but by no means completely) the influences on IQ and scholastic achievement differ from those on interpersonal relationships. That partially negative conclusion suggests the importance of differentiating *within* the cognitive domain between social skills and those intellectual skills that relate to academic achievement. That differentiation raises the as yet unanswered question of whether the family and school influences on social cognition differ from those that foster general intelligence.'

Gene/environment correlations

Hinde referred to the different types of correlations—active, passive, and reactive—that could be found in trying to teaze apart genetic and environmental influences on behaviour (see the Introduction to this section). He asked how this conception of correlations related to the meaning of the percent of variance explained by genetic and environmental influences.

Rutter replied:

'Several different issues are involved in your question. First, estimates of the proportion of the variance accounted for by gene-environment *interactions* are necessarily affected by the level of *correlation* between genes and environment. This is important because the study of gene-environment interactions is frequently undertaken through the investigation of adoptees just because the lack of gene-environment correlations allows the better separation of gene and environment effects. The consequence is well illustrated by Bohman and Cloninger's studies of criminality and alcoholism in adoptees (Bohman *et al.* 1981; Cloninger *et al.* 1981). It is evident that the risks increase sharply when there is an overlap between gene and environment vulnerabilities, but because such overlap was infrequent in adoptees the proportion of the variance accounted for by gene-environment interactions was low. The effects of such interactions are likely to be greater in the general population where it is usual for there to be an overlap between genes and environment.

The second issue is that proportion of the variance estimates are not necessarily the best way of assessing the size of environmental effects. As I tried to bring out in my paper, there may be substantial effects on *level* without there being much effect on variance.

The third issue is different yet again. It is clear that children's qualities (temperamental and intellectual) influence their responses to the environments they encounter (Porter and Collins 1982). Such qualities are in part genetically determined. It follows that children's responses in circumstances where there is a correlation between genes and environment (i.e. in ordinary families with biological parents) may not be the same as in circumstances where there is no such correlation (as in adoptive homes or in extrafamilial environments). So far as I am aware, there has been very little investigation of this matter and hence it is not possible to draw any conclusions on the extent or nature of the effect. Nevertheless, as you suggest, it is likely that the sources and level of gene-environment correlations will have an effect on both the strength (and maybe even the direction) of genetic and environmental influences.'

Weiskrantz then asked if anyone had looked at social skills of children in biological vs. adoptive families. Rutter replied not as far as he knew. (Some material on very young children is available in Plomin and De Fries, in press—*Eds.*).

Critical or sensitive periods

Kremin asked Rutter to substantiate the view that there is no critical period for cognitive development in the light of the critical period for language acquisition and the evidence from the case of Genie, a child brought up in social isolation (Curtiss 1977).

Rutter replied:

'My argument was, there was "no *marked* critical period" for cognitive development. That is to say, although children vary in their sensitivities to particular experiences according to their phase of development, the differences in sensitivities are relative, not absolute; the phases are long and without clear limits; and the sensitivities are not necessarily most marked in infancy (see Rutter 1981). The case of Genie is not particularly helpful for several different reasons.

Firstly, the case is unique in terms of both the duration of social isolation and the child's age at rescue; the other cases do not show any critical period effect. Accordingly, even if the evidence from this one isolated case is accepted, the critical period must be very long. Secondly, in order to show a critical period effect it would be necessary to demonstrate a lack of effect for a similar degree and duration of isolation at other ages; such evidence is not available. Thirdly, it is not known whether Genie had any biological handicap.

The last point to make is that age variations in vulnerability to environmental effects are not necessarily synonymous with age differences in response to brain damage. However, even the latter do not show an easily interpretable critical period effect outside the domain of unilateral effects of cerebral damage on language (see Rutter 1983b). This is because there are *several* different kinds of age effects which do not all operate in the same way; some increase and some decrease vulnerability to trauma.'

Hinde noted that, for comparable reasons, the term 'critical period' has long been abandoned by most ethologists in favour of 'sensitive period' (Fabricius 1951; Hinde *et al.* 1956; Bateson 1979).

Kremin commented further as follows:
'The model proposed by Lenneberg (1967)—according to which the critical period for maturation of areas responsible for language development extended to puberty—has indeed been contradicted by some recent studies of acquired aphasia in childhood (see for example, Krashen 1973; Hecaen 1983). Even Genie's case may be cited as a counter example: although she was deprived of linguistic stimulation from age 2 to puberty, Genie acquired some language. However, although Genie *has* language, her linguistic knowledge was far from normal after four and a half years of environmental stimulation (see Curtiss 1977). Moreover, there is some experimental evidence that only her right hemisphere treats auditory verbal information. The problem of a critical period for maturation and hemispheric specialization with regard to language acquisition may thus be posed in a different manner: if the specific cortical area stays without adequate stimulation during maturation this zone cannot acquire its functional capacity. Cerebral plasticity may indeed guarantee some sort of compensation. Studies of hemidecortication in infancy show however that the solitary right hemisphere shows a pattern of linguistic performance which is different and quite restricted with regard to syntax in comparison with right hemidecorticate subjects (see Dennis and Whitaker 1976).'

Mediating mechanisms

Wertsch asked why more importance was not placed on mechanisms which might mediate cognitive development, such as 'learning to learn'.

Rutter replied:
'My inference regarding the probable importance (for continuities in cognitive development) of variables such as 'learning to learn' and non-cognitive features such as self-esteem is just that—an inference, and not a demonstrated fact. There has been regrettably little research on such possible mediating mechanisms in rela-

tion to the effects of either family or school influences on intellectual functioning. Nevertheless, in my view, there is a need to invoke such mechanisms to account for the findings on the long-term benefits of preschool and school experiences (when such benefits do occur). It does not seem that the benefits stem from direct effects on cognition as such. Rather it appears that such continuities as exist derive either from effects on children's responses to learning tasks or from effects on children's styles of interaction with their teachers. After all, in the long run, the goal must be to enable children to *continue* learning when they are outside the direct influence of parents and teachers. What remains uncertain is just what skills and qualities are required for that to happen; even less is known about what features in the environment serve to foster those skills and qualities. Yet it is useful, I think, that we are pressed to pose that question and to design research that might provide an answer.'

Negative environment and social deficits

Sherman asked:
'If the data may be fairly summarized by the following example, what may one conclude about the necessary or sufficient influence of social experience on cognitive development?

1. An enduring positive social environment may promote cognitive development (e.g. IQ improvement due to adoption—Scarr and Weinberg 1976).

2. Institutional rearing produces enduring social problems (especially with peer relationships) and behavioural problems, yet no negative effects on IQ outcome.'
Rutter replied:
'It would not be correct to conclude that an enduring positive social environment enhances cognitive development, but an enduring negative one is without effects on cognition even though it leads to social deficits. There are plenty of examples of adverse environments which do indeed retard intellectual development. Rather the conclusion is that the environments that influence intellectual development tend to differ from those that influence socio-emotional development. Nevertheless, as you point out, there is a paradox. Rearing in an institution seems to lead to impaired attention to, and involvement in, learning tasks in the classroom but yet general intelligence is relatively unaffected. Why do the attention deficits *not* lead to poor cognitive performance? There are, I think, two possible answers to that question. First, it is quite likely that the attention deficit is not an all-encompassing one. It *is* present in the classroom situation, but it may not be present in other contexts. That possibility suggests the second answer—namely, that although the IQ is relatively spared there may be greater effects on scholastic attainments that reflect classroom teaching. Tizard's further follow-up of her institution-reared children will be informative in that connection.'

Rutter asked Tizard if, in her study of children who spent their first few years in residential nurseries, she found (as did Roy 1983) that there seemed to be few effects on IQ although there were marked effects on behaviour, and especially, on task involvement in the classroom:

'One might postulate that the latter would lead to impaired scholastic attainment in spite of the approximately normal IQ at age 8. Does the recent follow-up at 16 years show any effects on attainment? If so, are they accompanied by any changes in IQ? Also, are there differences in later academic attainment according to whether the children were adopted, remained in the institution, or returned to their biological parents?'

Tizard replied that the analysis of the follow-up data on the children at age 16 years would be available in 1985.

Abnormal environments and alternative strategies

Perret-Clermont asked:

'In "normal" environments it is sufficient for the child to learn the "normal" accounts of what is going on to give sense to the environment and develop his skills within it. In "abnormal" environments (for instance formal institutions or pathological families) could it be that children who fail to develop certain cognitions are in fact, by doing so, somehow *adapting* to environments that could not cope with other attitudes and self-images on their behalf? Have you met in your studies children who develop alternative means (e.g. social skills, specialized interests) to make sense of the environment and to deal with it or to abstract from it?'

Rutter replied:

'You raise the possibility that successful adaptions to an abnormal environment (such as an institution) may involve a rather different set of skills from those involved in responses to a normal environment. I agree. As Hinde (1982) has argued elsewhere, adaptation cannot be conceptualized as an absolute quality. Certainly it is possible that some of the behaviour seen in institution-reared children may represent attempts to deal with the environment they experience. For example, over-friendly, attention-seeking, disinhibited behaviour, perhaps represents an attempt to elicit personal interactions in a setting that tends to lack intense personal relationships. Nevertheless, what is striking is that this behaviour tends to persist even when the children move to environments in which manifestly it is maladaptive.

As to whether in our own studies we have encountered children who have developed unusual ways of making sense of their environment, I do not know. I doubt whether our measures would have picked up the unusual quality of such skills. However, what *is* clearly evident in our follow-up of institution-reared girls is that good functioning in adult life was associated with good experiences in school (Quinton and Rutter 1984). Such experiences tended to involve some sort of success, but the successes ranged from social to sporting to academic. The pattern of findings suggested that the 'good' experiences led to an enhanced sense of self-esteem and self-efficacy that enable the individuals to cope better with the life demands and challenges that they faced. The results illustrate, I think, the interplay between coping skills (perhaps a facet of cognition) and successful social adaptation.'

Section C

8

Adults and peers in relation to cognitive development
EDITORIAL

As the previous chapter indicated (p. 95), children reared in social isolation show severe cognitive deficits. On exposure to a more normal social life, they often show rapid improvement in intellectual functioning. Although the increase in social interchange is not the only change they encounter when released from isolation, it is difficult not to regard it as a major factor in their improvement. This in itself does not necessarily mean that variations in the social environment affect cognitive development under more normal conditions, but it strongly suggests that they may do so. The chapters in this section present primarily descriptive data concerned with the influence of adults and peers on cognitive development. Whilst the authors acknowledge the two-way influences between relationships and cognitive abilities, their primary concern here is the influence of the former on the latter.

It is often argued that the infant does not distinguish between the self and the outer world. Discussion of this issue almost necessarily seems to involve a language of metaphor—we use words like 'distinguish' or 'self', words like 'know' and 'believe', that have a real meaning to us but may be quite inappropriate for the infant. The formulation itself can thus easily lead to the creation of non-existent problems. But there are reasons for thinking more than that—that the formulation is actually erroneous. In all organisms, even invertebrates, once an elementary nervous system has been formed the consequences of self-induced movement differ from those of movements imposed by the environment (see von Holst and Mittelstaedt 1950). The human foetus already has a complex motor repertoire, so that in that respect differentiation between self and other, in so far as it is a real issue, can reasonably be thought of as occurring before birth (see also Lewis and Brooks-Gunn 1979). This of course in no way denies data indicating that an infant or young child may have difficulty in taking another's perspective—though these two issues (and others) are often confused in the concept of egocentrism.

The further development of the self-system, and especially the emergence of the self as a subject, agent, and knower, and the emergence of the self as an object, including the definition of self in terms of appropriate categories with respect to the external world, surely depends on interactions with other individuals (Mead 1934; Sullivan 1953a,b; see a recent review by Harter 1983). Some aspects of the relationships of young children likely to be important in this way were discussed in Chapters 5–7. Here we select two further problems to exemplify some of the issues.

The first has been chosen to illustrate the subtlety of the ways in which a relationship with an adult may contribute to the development of the self. It concerns how an individual comes to discover that mental states can be shared between individuals. The interpersonal qualities of the interactions of even young infants have been described in some detail (e.g. Trevarthen 1979), but the processes involved in the infant's mental development are still far from clear.

Given that the infant distinguishes between self and other, he soon develops means of influencing the behaviour of the other, but how does he learn that the other can appreciate what he is feeling? Whence does true 'intersubjectivity' arise? Stern *et al.* (in press) have recently described a process of 'affect attunement' which may have this function. They noted that mothers often respond to their infant's expressive movements by a movement which differs from that of the infant in modality and/or form but resembles it in intensity. Some examples follow:

Example 1. A nine-month-old girl becomes very excited about a toy and reaches for it. As she grabs it, she lets out an exuberant 'aaaah!' and looks at her mother. Her mother looks back, squinches up her shoulders and performs a terrific shimmy of her upper body—like a go-go dancer. The shimmy lasts only about a long as her daughter's 'aaaah!' but is equally excited, joyful, and intense.

Example 2. A nine-month-old boy bangs his hand on a soft toy, at first in some anger, but gradually with pleasure, exuberance, and humour. He sets up a steady rhythm. Mother falls into his rhythm and says 'kaaaaa-bam, kaaaaa-bam', the 'bam' falling on the stroke and the 'kaaaaa' riding with the preparatory upswing and the suspenseful holding of his arm aloft before it falls.

Example 3. An eight-and-one-half-month old boy reaches for a toy, just beyond reach. Silently, he stretches toward it, leaning and extending arms and fingers fully out. Still short of the toy, he tenses his body to squeeze out the needed extra inch of reach. At that moment, his mother says 'uuuuuh . . . uuuuuh!' with a crescendo of vocal effort, the expiration of air pushing against her tensed vocal cords. The mother's accelerating vocal-respiratory effort matched the infant's accelerating physical effort.

Example 4. A ten-month-old-girl accomplished an amusing routine between herself and mother, and upon doing so looks at mother. The girl 'opens up' her face (mouth opens, eyes widen, eyebrows raise, and then closes it back in a smooth arch). Mother responds with saying 'yeah'. However, she intones it so that it's pitch line is 'y h'. The mother's prosodic contour has matched the child's facial-kinetic contour.

Example 5. A nine-month-old-boy is sitting facing his mother. He has a rattle in his hand and is shaking it up and down with a display of interest and mild amusement. As mother watches she begins to nod her head up and down keeping tight beat with her son's arm motions.

In each case the mother's response involves not imitation but a sort of cross-modal or inter-modal matching of the internal state, and is largely automatic and unconscious. (Piaget (1951) recorded comparable cross-modal matching of the caregiver's behaviour by infants of about the same age, though he regarded it as 'intelligent confusion'). Stern and colleagues (l.c.) regard cross-modal matching by the mother as 'peaks' of a process of interpersonal communion which is probably much more continuous. If the mother responds too much or too little the baby turns round and looks at her, as though to enquire, 'What is going on?' Stern stresses that the mother could not convey a relation between her inner experiences and those of the infant by simply imitating the infant's movements, but that this intensity matching of the infant's movements in another modality could achieve just that function. Such a process could lead to the infant 'discovering' that there are other minds which are different from, but can yet respond to his own, and thus contribute to the acquisition of his sense of self. This 'affect attunement' seems to be widespread if not ubiquitous amongst mothers, who are often not conscious of their actions. Whilst it must be admitted that there is no proof that 'affect attunement' has the function suggested, it is difficult to believe that it has no function and a better suggestion is yet to be made.

We may now turn to the second issue. Are all relationships equally valuable in providing opportunities for cognitive development? Some, and especially those with a psychoanalytic orientation, regard the mother-child relationship as the prime determiner of all future relationships (e.g. Bowlby 1969). And some developmental psychologists see cognitive development as depending in large measure on the internalization of skills first actualized in interaction with a more competent other (e.g. Vygotsky 1978; see Chapters 19, 20, and 22), or depending upon the imitation of a competent model (Rosenthal and Zimmermann 1978). However the parent-child relationship is inevitably complementary. Sullivan (1953a,b) stresses that adults teach, provide rules, and give the order to the world that children seek. Children thus conform. But the radically different perspectives of parent and child pose difficulties for the child in acquiring real understanding from interaction with the parent. Peers, by contrast, interact with a sense of equality—each is free to behave in the same way as or differently from the other. Mutual understanding is reached by a process of mental interchange, and the child can achieve a sense of his own nature by a comparison of the self with other *compar-*

able individuals (see also pp. 58–65). Cognitive development arises especially from differences from other comparable individuals in opinion or approach to problems (e.g. Piaget 1976a; Doise *et al.* 1974, 1975; Damon 1983a; Youniss 1980; Ruble 1983). In addition, just because parents can be so sensitive to children's abilities, they may provide poor stimulation to the child to overcome communicational misunderstandings and cognitive shortcomings. Of course this does not mean that the parent-child relationship is unimportant for the acquisition of cognitive skills: it is crucial early in development, and on its nature depend characteristics of the child that will in turn determine what the child can get out of peer relationships (Hinde and Tamplin 1983). Clearly relationships of different types are important in different ways and at different stages in development.

Given the importance of interactions with one or more sensitive others for development in the early years, one must ask whether extra-familial relationships can substitute for the interchanges that take place within most families. In recent years the provision of preschool education has become increasingly widespread in Europe and North America, and Rutter in a previous chapter has reviewed some recent studies indicating that preschool experience can have positive effects on cognitive functioning and other aspects of personality. Tizard's study (Chapter 9), however, points out that the interaction between preschool children and their teachers is usually impoverished in comparison with the interactions which many children have with their mothers, and suggests that it may not be so conducive to cognitive growth. Whether or not the latter is the case, her findings certainly do not provide an argument against preschool education, and Tizard would be the first to agree that it may convey other benefits—for instance in interaction with other children. And of course there is no implication that some teachers may not provide interactions more conducive to cognitive growth than some mothers, or that preschools may not provide much of value in addition to peer interactions, or for that matter that cognitive growth is all that matters. But Tizard's study does have important implications for the design of preschool education, and for the benefits we should hope to obtain from it. As Emiliani and Zani (1984) have elegantly shown, what goes on in educational institutions is critically related to the goals and values of the staff, and Tizard's data underline the importance of clarifying the aims of the preschool.

The aim of the research carried out by Stambak, Ballion, Breaute, and Rayna is that children should be provided with an environment (social and physical) where they can realize their abilities. They emphasize that a child who performs poorly in school may do much better if the conditions are changed. A setting suitable for peer interaction can reveal complex socio-cognitive functioning even in two-year olds. Although their work involves primarily the analysis of detailed observations of peer interaction, they go to the heart of the dialectic between social inter-actions and cognitive development. The behaviour of children in creating puppet shows must depend on a surprisingly sophisticated level of social awareness—a level considerably beyond that which would be expected from their interactions with adults, and which seems to go beyond that found in the stories generated by children of a comparable age (reviews Mandler 1983; Shatz 1983). Whilst Stambak and her col-leagues argue that these puppet performances must depend on a shared and deep-seated logical framework, at the same time they speculate that the challenges and provocations of social and puppet play, including the discrepancies between the real and the pretend world and the conflicts between the participants, promote cognitive growth—a view entirely in keeping with Doise's experiments in a later chapter.

Krappmann similarly emphasizes the importance of relationships in cognitive development and especially of reciprocal relationships. Empiri-cally, Krappmann demonstrates a close correspondence between the friendship patterns of young children and their school performance. Whilst relations between IQ or school performance and sociometric status have been established previously (e.g. Roff *et al.* 1972), Krappmann has used a more sophisticated approach for categorizing the children in terms of their relationships with their peers. As discussed more fully later (pp. 239–45), a correlational study cannot demonstrate the direction of causation between school achievement and social status. However Krappmann considers various hypotheses including the existence of indirect links between the two, and the evidence strongly suggests 'intense social relationships between children of the same age generate experiences which influence cognitive development.'

9

Social relationships between adults and young children, and their impact on intellectual functioning

BARBARA TIZARD

'One of the most crucial ways in which culture provides aid in intellectual growth is through a dialogue between the more experienced and the less experienced.' J. S. Bruner

I want to start by explaining why I think that this dictum is especially true in the preschool years. I shall go on to argue that parents are generally much more likely to be effective in giving such aid to young children than nursery school teachers, in large measure because of differences in social relationships.

I shall support my arguments with evidence from a study I carried out with Martin Hughes, Gill Pinkerton, and Helen Carmichael (Tizard 1984; Tizard and Hughes 1984; Tizard *et al.* 1982, 1983 a,b). In this study we recorded with radiomicrophones the conversations of four-year-old girls at morning nursery school, and then in their homes in the afternoon. The girls were within three months of their fourth birthday.

The sample, which is decribed more fully elsewhere, was made up of 30 girls, half of them middle class and half working class, from nine different part-time State nursery classes. They were randonly chosen from those who met our criteria, and no schools or mothers declined to take part. There were no children from single parent or large families in the sample, but otherwise the girls seemed representative of those attending schools in southern England. Because our interest was in the role of the adult in learning, only adult-child conversations were transcribed.

The study addressed a number of issues, including the contexts in which learning occurs at home, the evidence for social class differences in mother-child and teacher-child interactions, and a comparison of the educational contexts of homes and nursery schools. In this chapter I shall focus only on one aspect of cognitive behaviour, that is, the way in which the children seemed to be attempting to make sense of their world through conversation with adults.

Let us return to Bruner's dictum, quoted above. Traditional nursery education lays little stress on children learning from adults. The twin cornerstones of its pedagogy are learning through play and learning through socialization with other children. The value of learning through play was first argued by Froebel 150 years ago. Today, the nursery school world is mainly given over to play. The physical environment largely consists of a great variety of play materials, and staff interactions with children generally occur within the context of play. (Although I am describing my experience in British nursery schools and playgroups, a very similar situation exists in many Western countries.)

The epitome of this pedagogy can be found in 'open plan' nursery schools, a fairly recent trend in Britain. Here up to a hundred children, instead of being cared for in separate classrooms, are encouraged to wander through the whole building and the shared playground, sampling the vast array of toys and play materials made possible by this arrangement. The theoretical basis for this kind of educational practice is Piagetian. It rests on the doctrine that the child's active perceptual and motor exploration is an essential precursor to abstract conceptual learning. By providing the child with bottles and jugs of different sizes and a tray of water, she will develop concepts of volume—the use of building beakers of different sizes will develop concepts of balance, and also of seriation.

Equally important in nursery school practice is the stress on learning from other children. Both the theory and practice of this branch of pedagogy are weak. Exactly what children learn from each other as opposed to what they learn from adults is rarely defined, nor is there agreement on what the adult's role, if any, in this process should be. It is assumed that just as, if you put a child together with a pile of building bricks she will learn about balance, so, if you put her in a group of other children she will acquire useful social knowledge.

The pedagogy I have described obviously 'fits' with important aspects of young children's functioning—they do love to play, and they love to be with other children. Moreover, there are certainly some kinds of knowledge that children learn from play (although I would dispute that this requires a lot of play materials). There are other kinds of knowledge that they learn from other children (although I would argue that we know very little at a scientific level that might guide us in planning this socialization. It is, for example, by no means obvious that setting up a large peer group is the best way to help children develop social skills or knowledge). This pedagogy, however, pays scant regard to what children can learn from adults. In the past twenty years nursery school teachers have attempted to repair this omission by deliberate efforts to involve children in adult-led conversation, or in structured verbal exercises.

These verbal encounters are, of course, dialogues, but they differ strikingly from spontaneous conversations in the home. In the first place, teachers aim at developing verbal skills and simple attribute concepts, whilst parents are primarily concerned with the *content* of the talk, that is, with transmitting a great range of human culture. Secondly, teacher-child conversations are dominated by the adult, both in terms of the sheer amount of talk, and in its direction. In this they differ from conversation at home, which tends to be evenly balanced between parent and child. Because teachers' conversations are based on questioning the child, they fail to capitalize on an important aspect of young children's cognitive repertoire—their enormous intellectual curiosity. As every parent knows, children of this age are constantly asking questions. In our observations we found that on average children of both social classes asked their mothers 26 questions an hour. We know from normative studies that question-asking starts to decline in frequency soon after this age. It seems unlikely that a cognitive behaviour would develop, become dominant, and then decline unless it plays a significant role in the child's development at a particular age. A study of our transcripts suggested to us that question-asking may be particularly prominent around the age of four because of two co-existing factors:

(1) the child's enormous ignorance, and lack of an adequate conceptual framework in which to organize her experiences; and

(2) her awareness of the misunderstandings, failures of communication, and confusion that this gives rise to.

At an earlier age, the child is even more ignorant, but she does not have an equal awareness; at a later age, she has acquired an adequate enough conceptual framework to cope with her experiences.

The ignorance of the four-year old is so vast that a great imaginative effort is required to apprehend it. We have recently begun to get some understanding of how the world must seem to babies. Because four-year olds seem 'like us' in a way that babies do not, we forget that much that we take for granted they have yet to learn. Occasionally, as parents, we gain an insight into this ignorance. I remember one of my children looking aghast when I referred to someone walking along 'with their eyes on the ground'. But a study of our home transcripts showed how basic the gaps in their understanding can be. It is not simply that they lack a specific item of information, such as the example just cited of a particular usage of language, or the name of a capital city. More fundamentally, they lack the underlying conceptual framework to which the information can be integrated. For example, they cannot grasp the concept of a capital city, because they lack a conceptual map of the world as made up of different countries, cities, and towns.

This does not, of course, mean that their minds are a blank. Because they are such active thinkers, young children are constantly constructing their own theories to fill the gaps, but these theories often prove unsound when tested against experience. Amongst the false theories that we detected in our manscripts were the following: that age and size are associated, so that a child larger than another must be older and will be an adult sooner; that adults' grow, like children do; that adults were not necessarily babies in the past; that tradespeople pay their customers, not the other way round; that parties occur only on birthdays, hence if someone holds a party it must be their birthday; that parents switch on street lights from within their own house; that clock numbers tell the day of the week; and that people with the same first name have the same second name.

It is because these theories are constantly coming unstuck, and because of their need to gather the necessary information to understand events and test their theories, that in our opinion young children are constantly asking questions. (Of course, sometimes they ask them to gain attention, or sustain a conversation, but in studying our transcripts we were rarely convinced that these were the sole motives.)

Their task is made more difficult by the frequent failure of their parents to detect their misunderstandings, or if they do detect them, to give adequate explanations. In the following conversation, Lynne's mother failed to detect that in referring to 'Nanny's Daddy' Lynne was talking about her grandfather, not her great-grandfather. Lynne was drawing a picture of her grandmother ('Nanny'), when she asked:

Child: Where's Nanny's Daddy?
Mother: Nanny's Daddy? He's up in heaven.
Child: Oh, he's not up, not really up . . . ?
Mother: Yes, he died.
Child: No, not Nanny's . . . He's not . . .
Mother: He is, he got very, very old, and he just died.
Child: Is Nanny up in heaven?
Mother: No, she came up here the other day, didn't she?

Lynne broke the tip of her felt tip pen at this point, and the topic of conversation changed.

This short conversation illustrates very nicely the bewilderment a four-year old can easily fall into. Children of this age have difficulty with the relative nature of terms such as 'father' or 'husband', and this can often create misunderstandings. The confusion is confounded by Lynne's inexperience. An older child would probably have realized that she would have been told if her grandfather had died in the past few days, and his death would not have been referred to so casually. Lynne lacked the experience that might have helped her realize that there was a misunderstanding, and instead tries to make sense of what her mother is saying. In her last question, 'Is Nanny up in heaven?', she seems to be making an inference in an attempt to clarify the situation; her reasoning seems to be that if her grandfather who was alive and well a few days ago, is now up in heaven, then maybe her grandmother is there too. Lynne's mother puts her right on this one, but fails to spot the underlying misunderstanding.

When questions and misunderstandings were centred round complex topics, the children often failed to absorb their parents' explanations straight away. Instead, there was evidence of a lengthy process of struggling to understand. I will give one example from our transcripts which illustrates the way in which the child's confusions about the relationship between money, work, and consumer goods were exposed, and to some extent clarified. The conversation started while Rosy and her mother were having lunch, and was triggered off by the appearance of the window-cleaner in the garden. Rosy's mother went off to the kitchen to get him some water, and called out to her neighbour, Pamela.

Child: What did Pamela say?
Mother: She's having to pay everybody else's bills for the window cleaner, 'cause they're all out.
Child: Why they all out?
Mother: 'Cause they're working or something.
Child: Aren't they silly!
Mother: Well, you have to work to earn money, don't you?
Child: Yeah . . . If they know what day the window cleaner come they should stay here.
Mother: They should stay at home? Well, I don't know, they can't always . . .

At this point the window cleaner appeared at the dining room window, and cleaned the window while Rosy and her mother carried on with lunch. The conversation switched to what they might have for pudding, and what they might do that afternoon. Rosy, however, was still thinking about the window cleaner . . .

Child: Mummy?
Mother: Mmm.
Child: Umm . . . she can't pay everybody's, er . . . , all the bills to the window cleaner, can she?
Mother: No, she can't pay everybody's bills . . . she sometimes pays mine if I'm out.
Child: 'Cause it's fair'.
Mother: Mm, it is.
Child: Umm, but where does she leave the money?
Mother: She doesn't leave it anywhere, she hands it to the window cleaner, after he's finished.
Child: And then she gives it to us?
Mother: No, no, she doesn't have to pay us.
Child: Then the window cleaner gives it to us?

Mother: No, we give the window cleaner money, he does work for us, and we have to give him money.

Child: Why?

Mother: Well, because he's been working for us cleaning our windows. He doesn't do it for nothing.

Child: Why do you have money if you have . . . if people clean your windows?

Mother: Well the window cleaner needs money, doesn't he?

Child: Why?

Mother: To buy clothes for his children and food for them to eat.

Child: Well sometimes window cleaners don't have children.

Mother: Quite often they do.

Child: And something on his own to eat, and for curtains?

Mother: And for paying his gas bills and electricity bill. And for paying for his petrol for his car. All sorts of things you have to pay for you see. You have to earn money somehow, and he earns it by cleaning other people's windows, and big shop windows and things.

Child: And then the person who got the money gives it to people. . . .

It seems that until the middle of this second conversation ('then the window cleaner gives it to us?') Rosy was under the impression that the window cleaner pays the housewives, and not the other way round. In the course of the conversation the relationship between work, money, and goods is slowly outlined for her, but it is still unclear from her last remark whether she has really grasped all that has been said. The conversation in fact continues later on, after Rosy has watched her mother actually hand over the money to the window cleaner:

Mother: I expect the window cleaner's going to have his lunch now.

Child: He would have all *that* much lunch (stretches arms out wide) because he's been working all the time.

Mother: Mm . . . I expect he gets very hungry, doesn't he? I expect he goes to the pub and has some beer and sandwiches.

Child: He has to pay for that.

Mother: Yes, he does.

Child: Not always, though.

Mother: Mm, always.

Child: Why not?

Mother: They won't give him any beer and sandwiches if he doesn't have any money.

At this point Rosy clearly wonders why he cannot do without money:

Child: But why doesn't he use his own food?

Mother: Well he might do, I don't know, perhaps he brings his own sandwiches, do you think?

Child: He go to a pub and he has his lunch some *and* he has it at his
                home.
Mother: Oh, he wouldn't do both, no.
Child: He would do all of those a few times. But he usually go to a
                pub.

At this point Rosy's mother changed the subject of the conversation, tiring
of it, as was often the case, before the child herself had lost interest.

Rosy's remarks in this third conversation (especially 'not always,
though') suggest that she has only hazily grasped what she has been told,
and her understanding of money transactions still seems shaky. This is not
because she lacked the intellectual capacity, nor because her mother's
explanations were too complex. On the contrary, the reader will note how
closely Rosy was following her mother's explanations and exposing their
weak points. Rather it seems likely that this conversation reveals something which is characteristic of the slow and gradual way in which a child's
understanding of an abstract or complex topic is built up. It may take a
considerable time, as well as several more conversations like the one
above, before Rosy has grasped the complexities of the relationships
involved. She may have to return to the same topic again and again before
she achieves full understanding.

It is significant that Rosy keeps asking questions on this topic. Clearly,
her initial questions were asked out of curiosity aroused by the unusual
event in the daily routine—the arrival of the window cleaner and the subsequent conversation between her mother and the neighbour. But her later
return to the topic suggests something beyond this initial curiosity. Indeed
it suggests, as I have argued above, that Rosy is at some level aware that
she has not grasped the relationships involved, and that her questions are
motivated by her desire to clarify her misconceptions. Why she should be
aware of her own lack of understanding is not clear—she could perfectly
well have continued with her original idea that the neighbour received
money from the window cleaner, and got on with her lunch. The fact that
she did not do so suggests that she was at some level dissatisfied with her
own grasp of the situation, perhaps because it didn't fit with other facts that
she knew—and wanted it sorted out.

We came across a number of conversations of this kind, which we called
'passages of intellectual search', in which the child persistently worked on a
topic which she misunderstood, or about which she was confused. They
were more frequent in middle class than in working class homes. Sometimes the child returned to the same topic several times during the afternoon. In the course of these conversations she would tackle her confusion
from different angles, reflect on her mother's answers, and test them out by
inferences. Because parents often gave explanations that were less than
full, or failed to meet the central point of the children's questions,

advances in children's understanding seemed to depend very much on their own intellectual efforts.

At school, a very different educational situation obtained. There were no 'passages of intellectual search', and the children asked very few questions. Whilst on average they asked their mothers 26 questions an hour, they only asked two questions an hour of their teachers. Moreover, of those questions that *were* asked at school, a large proportion were not asked out of curiosity, but were of the 'business' type—asking for information needed to carry out an activity, for example, 'Where is the glue?'.

It was not the case that the teachers discouraged questions. Their answers to 'curiosity' questions tended, in fact, to be fuller than the mothers', and they less often ignored questions. The problem was that they were very rarely asked them. Instead, it was the teachers who asked the questions. Their questions tended to be of the traditional pedagogic type, that is, the teacher knew the answers, and was using questions to assess the child's knowledge and to develop her skills. Here is one example. June approaches her teacher with a piece of paper and says, 'Can you cut that in half?'

Teacher: Where do you want me to cut it?
Child: There.
Teacher: Show me again, 'cause I don't quite know where the cut's got to go (child shows teacher where she wants paper cut). Down there? (child nods; teacher cuts child's piece of paper in half). How many have you got now?
Child: No reply.
Teacher: How many have you got?
Child: No reply.
Teacher: How many pieces of paper have you got?
Child: Two.
Teacher: Two. What have I done if I've cut it down the middle?
Child: Two pieces.
Teacher: I've cut it in. . . . ? (teacher wants child to say 'half').
Child: No reply.
Teacher: What have I done?
Child: No reply.
Teacher: Do you know? (child shakes head)
Other Child: Two.
Teacher: Yes I've cut it in two. But . . . I wonder, can you think?
Child: In the middle.
Teacher: I've cut it in the middle. I've cut it in *half*! There you are, now you've got two.

The teacher's intentions are clear enough. She does not simply want to comply with June's request, but tries to turn the situation into one of educational value by introducing a simple mathematical idea: namely, that if you cut a piece of paper into two equal pieces then you have cut it in 'half'. Nursery teachers are frequently encouraged to introduce mathematical vocabulary in this way: indeed many educational advisers would congratulate June's teacher on noticing the educational potential of this situation. The problem, however, lies in the way that she attempts to introduce this idea.

At first she asks June how may pieces of paper there are, and then tries to elicit the word 'half' from her. June appears confused by this. She gives what she may feel to be an adequate, if somewhat condensed, reply to the teacher's request: if she has cut it down the middle she has produced 'two pieces'. This however does not seem to be the answer the teacher wants, for she keeps on questioning. Subsequently the teacher rejects two more answers, one suggested by the other child ('two') and one by June herself ('in the middle'). Finally the teacher gives them the correct answer, 'I've cut it in half'!

There are a number of problems with this type of approach. The first is that the teacher may be so focused on the answer she wants to hear that she rejects other perfectly adequate replies. As a result, the child—as happens here—may lose confidence and start to doubt the knowledge that she does have. The second criticism is that the child may in fact know the answer, but be so confused or inhibited by the questioning that she is unable to produce it when required. This also, can be seen happening here. If we go back to the very first turn of the conversation, we see that June's initial request to the teacher was that she should 'cut it in half'. In other words, June started the conversation by spontaneously using the very concept that she was unable to produce when being questioned. (Incidentally, that afternoon we watched June, a working class girl of average IQ, play a fast game of knock-out whist with her mother.)

Another difficulty with this type of dialogue is that children often fail to answer the teacher's questions—we found that over a third of them went unanswered. This was irrespective of the level of difficulty, and seemed to reflect the child's lack of interest in the questions, as well as her uncertainty about whether she was giving the 'right' answer. The teacher's well intentioned attempts to turn a simple request for help with an activity into an educational exercise may seem to the child to be imposing a tedious and unnecessary brake on her plans. In contrast, the child's own questions indicate what the child herself is interested in, and they provide the adult with an opportunity to feed in information at the most appropriate time.

Socratic dialogues of this kind were almost the only kinds of dialogue we saw in our study between teacher and child, other than brief, routine, exchanges. If, as we believe, asking questions and clarifying confusions plays an important role in the child's intellectual growth, and in the process of inducting her into her culture, an educational setting that does not capitalize on this tendency seems handicapped from the start.

Why did the children ask so few questions at school? Part of the answer probably lies in the teacher's reliance on asking questions herself, a technique which David Wood and his colleagues have shown to inhibit question-asking in children (Wood, McMahon, and Cranston 1980). But an important part of the answer lies in the different relationship between mother and child, and teacher and child. The child moving from home to school enters a very different world. Whilst at home her life is centred on a close relationship with her parents, at school adults become relatively unimportant to her, and most of her experiences are with other children, or involve playing on her own. Her interactions with the staff tend to be almost incidental. We found that on average the children had ten conversations an hour with the school staff, compared with twenty-seven at home with their mothers. Many fewer of the conversations at school were on a one-to-one basis, since the teacher was often addressing a small group. Moreover, most of the conversations with staff were very brief; two-thirds were of six turns or less.

These characteristics to some extent follow from the free play environment of the British nursery school, where it is perfectly acceptable, even desirable, for a child to occupy herself well away from the aegis of the staff. But other characteristics of the staff-child relationship are more universal. One of these is the rather low emotional involvement of staff and child. This has several important consequences for learning. In our study, the children did not seem to feel the same urge to communicate significant experiences and feelings to the teachers as to their mothers, and equally the staff were less interested in what the children had to say. This can be illustrated by comparing the conversations which Carol, a working class child, had with her teacher and with her mother about an incident at school. During the morning, a strong wind came up and blew the buckets in the sandpit about. Carol was clearly excited by this event. When she went inside the school she did not describe it to her teacher, but she did confide her intention to tell her mother about it:

Child: I'm telling Mummy that the buckets rolled away.
Teacher: Pardon?
Child: I'm telling Mummy that the buckets rolled away when we were not looking.
Teacher: Are you?
Child: Yeah.
Teacher: That's nice.
Child: A-and the sand went a-all in my eyes.

Teacher: In your eyes? Were they sore? Are they still sore now? Oh, you poor old thing. Do you think if you had a piece of apple, it would make them feel better?

Child: It was in there.

Teacher: Okay, well let's wash it out. (Teacher and child go to bathroom).

In fact, Carol did not tell her mother about the incident, until her mother brought up the topic herself:

Mother: It's very windy today, isn't it? Was the wind blowing the sand at your nursery?

Child: Yeah.

Mother: Hm. The sand is all dry now and when it's windy . . .

Child: And it went right in my eyes, and Mummy I want to tell you something. And it's funny, um, the buckets rolled away and . . . and, we wasn't looking there . . . and we said . . . and we weren't able to catch the buckets.

Mother: Weren't you?

Child: No.

Mother: And what was making the buckets run away? Because they haven't got legs, have they?

Child: No.

Mother: Then what was making them run away?

Child: Rolling.

Mother: Were they rolling?

Child: Yeah. Yeah. They were rolling . . . see, they were standing up, and we was not looking, 'cos, we was making sand castles.

Mother: Mm.

Child: And then it tumbled over, the buckets, and then it went roll, roll, roll, roll. 'Cos the wind blow huff.

Mother: Oh gosh. Very strong the wind, isn't it?

Child: Yeah.

The mother's contributions to this conversation are in a somewhat 'teacherly' questioning style, but nonetheless there is a note of urgency and excitement in Carol's answers—she wants her mother to share her feelings. The depth and intensity of conversations like these is almost entirely absent in the children's talk with their teachers.

The closeness of the parent-child relationship has other consequences for learning. On the part of the child, it keeps her near her mother, and thus able to learn from her. Moreover, what the mother says has salience for the child. She is also able to ask her mother questions and express uncertainties freely. As anyone who has attended a postgraduate seminar knows, confusion and misunderstandings are not likely to be voiced except in a relationship of trust. Within this trusting relationship, the child can also express the anxieties and fears that she is most unlikely to disclose to her teachers.

On her part, the mother's concern for her child means that she has definite ideas of what she wants her to learn. This will vary from one mother to another, but whether the mother emphasizes the value of acquiring general knowledge, musical skills, domestic skills, obedience, or personal survival, it is usually of immense personal concern to her that her child should acquire these skills or knowledge. In contrast, the teacher's educational commitment to a particular child is inevitably held and pursued much less seriously. It is true that the emotional intensity of the parent-child relationship may at times interfere with the child's learning, but on balance, it leads to a great potential educational advantage for the home.

The rather obvious fact, that the staff do not know the children very well, has other important consequences for learning. The mother's greater knowledge of her child makes communication between them easier. In the preschool period, communication often breaks down either because the child does not realize what another person needs to be told in order to understand her, or because she has an inadequate or faulty grasp of the concepts she is using. Because of the mother's intimate knowledge of her child, she is often able to tune in very sensitively to the meaning of her child's remarks. Of course, children have to learn to express their meaning to strangers, but when their skills are not well advanced, their attempts may lead to mutual bafflement rather than communication.

The mother's knowledge of her child also enables her to make a much better match of the educational task she presents her with than is possible for the teacher. This often leads to children performing at a higher level at home than at school. We saw above June's teacher asking her to count two pieces of paper, and laboriously trying to extract the concept 'half' from her. At home, that afternoon, in the course of playing knock–out whist with her mother, June recognized number symbols, counted tricks, and knew which numbers were bigger and smaller than others.

The mother's advantage also lies in the fact that she and her child share a common life, stretching back into the past, and forward into the future. This shared experience not only makes it easier for the mother to understand what her child is intending to say, but also allows her to help in a task that is important for intellectual growth—relating present to past experiences. Paradoxically, since one function of schooling is to extend the child's intellectual horizons, it is at school that conversation is focused on the 'here and now', and at home where links are made between the present, the past, and the future, and the child's experiences are integrated into the world beyond the home. (This latter finding is in part, of course, because the home is the centre for many links with the outside world, whilst schools tend to be closed communities).

It is possible for the school to overcome some of the limitations I have described. Given the motivation, and sufficiently small classes, the social relationship between teacher and child can become much less distant. Teachers can get to know the children out of school, develop a relationship of trust with them, and learn more about their family and their out-of-school activities. But even in these circumstances the parent-child relationship retains great educational advantages; the parent has a much better possibility of integrating the child's present experiences into her past, and a much greater educational commitment. For these reasons I consider it inappropriate, in the present state of knowledge, for professionals to offer parents advice on educational aspects of child-rearing. There is little scientific basis for most of this advice—for example, there is no real evidence that parental time spent playing with children, rather than say, answering their questions, aids their intellectual growth. Moreover, the evidence of this study suggests that the haphazard, unplanned encounters occurring between parent and child in the course of living together are better attuned to young children's intellectual needs than the 'educational' conversations at nursery school.

It may nevertheless be the case that the rather bleak staff-child conversations we have described play an important function in socializing the child into the role of pupil. It may also be argued that the child's perceptual and motor skills will be enhanced by nursery school activities. However, all the children in our study had access to paper, pencils, puzzles, games, and books at home. Nursery school has, of course, other very important functions for both parents and children, especially in terms of socializing the child into a wider world than the family, liberating the parents, and providing opportunities to run and climb not available in many urban homes. What does seem clear is that nursery school does not facilitate the child's own attempts to understand a great range of aspects of human culture. For this reason, I would argue that if children are to spend long hours from a young age in a nursery setting, attempts should be made to reproduce within the nursery some of the features of the home that seem to facilitate question-asking and intellectual search.

Discussion

Influences on intellectual growth

Rutter queried the claim that parental influences on children's intellectual growth are superior to teacher influences. 'Your findings show that the same children who engaged in rich conversations at home did not do so at school. Moreover, this "inhibiting" effect of school tended to be greater in children from working class homes. These findings are of great theoretical and practical importance, but do not show that interactions at home are more likely than those at school to foster intellectual growth. They do not do so both because you did not have any measure of children's intellectual development and because your home-school comparisons alter if instead of taking absolute measures per unit time you take ratio measures (i.e. types of interchange per unit conversation). May that not be relevant?

Furthermore, if your data are combined with the data from other studies they raise a rather different set of questions. Let me note three sets of findings. First, children from socially disadvantaged homes tend as a group to show below average intellectual skills—this is so on IQ tests, on tests of scholastic attainment, on measures of verbal skills, and indeed on virtually every type of cognitive test used. Second, children from such disadvantaged homes tend to show an intellectual increment on starting school; similarly there tend to be benefits rather than deficits (albeit often not persistent) from pre-school education. Thirdly, children from disadvantaged homes, unlike middle class children, tend to lose intellectual ground during long summer vacations. All these data point to the conclusion that school influences enhance intellectual development (when added to home influences).

If these findings are combined with yours do they not suggest that:

(i) the types of conversations that we think are most likely to enhance intellectual growth do not do so; *or*

(ii) the benefits from school do not stem from the types of conversational interchanges that you measured; *or*

(iii) that optimal intellectual growth requires a variety of experiences some of which are most readily provided at home and some at school; *or*

(iv) that your sample did not include the kinds of disadvantaged children included in other studies.

The last possibility, of course, raises the further hypothesis (complementary to your own finding on context effects) that disadvantaged mothers are *capable* of providing the types of conversational interchange needed for intellectual growth, but that in practice they may not do so because of the situational constraints provided by large family size, one parent status, overcrowding, and other features of disadvantaged homes.'

Tizard replied:

'So far as children's development is concerned, it must be the rate per hour of adult input that is important, not the ratio measures. The latter, as I indicated above, reflect the style of talk, but not its frequency.

You are right in pointing out, as did Dr. Yarrow, that we have no evidence to support our belief that the mother-child talk is educationally more effective than the teacher-child talk. However, we did have a measure of intellectual development—the children were all tested with the Stanford Binet scale. The mean IQ of the working class groups was 106, that of the middle class group, 122. The working class children were not 'disadvantaged'—there were no children from single parent, overcrowded, or very large families. We considered them representative of most working class children in southern England, who live in council housing, in families of three or less children.

There is no evidence that attendance at an ordinary British nursery school leads to IQ gains. In fact, a minor reason for undertaking the study was to seek an explanation for the apparent lack of any such impact (cf. Woodhead 1976). The studies which have reported undeniable IQ gains have been those, mainly American, where the traditional regime has been altered, to include structured programmes in which children are withdrawn from the main group to perform IQ-like tasks and games with staff.

A further issue, however, which has been amazingly little addressed, is that of the meaning of these IQ gains. Do they represent real intellectual advances, or simply a familiarization of the child with this type of situation? No one appears to have supported the former claim by studying the child's cognitive behaviour in non-test situations. The fact that the IQ gains are very rapid (they seem to occur within the first six months, and then plateau (Karnes 1973; Woodhead 1976) suggests that the second interpretation may be correct.

It may be the case that conversations in very disadvantaged families differ from those we recorded. If so, it seems unlikely that the amount and quality of staff-child conversation at a traditional nursery school would compensate for any lack at home.'

Peers, as well as teachers, at school

Hinde asked if Tizard meant to cast doubt on what is learned from *peers* at school. *Tizard* replied that we do not know enough about what is learned from peers, especially in the school context. This is certainly an important area for investigation (see Chapters 8, 10, 11, 21, and 22).

Teachers can learn

Shure pointed out that teachers, like mothers, tend to focus on what *they*, versus the children, are interested in. However, both teachers and mothers are happy to learn, as she has demonstrated with her training programme for teaching social skills (see Chapter 14).

Tizard replied:

'My argument is that the qualities of conversation depend very much on the social relationship between the participants. Therefore, I doubt whether changes in teacher training would affect conversations, unless they led to a change in relationships. If teacher training encouraged teachers to get to know children outside of school, for example, and to spend time sitting down and listening to them, the quality of their conversations with children would probably change.'

10

Pretend play and interaction in young children

MIRA STAMBAK, MONIQUE BALLION, MONIQUE
BREAUTE, AND SYLVIE RAYNA

Children who are non-participant, passive, and unresponsive in one setting, may be curious, inventive, active, and communicative in another (CRESAS 1978, 1981). Even children labelled 'mentally defective' by psychologists can, in certain settings, prove to have learning and reasoning abilities similar to those of good students. 'Setting' is, of course, to be interpreted widely here, and includes not only the material environment and educational methods but also, and indeed mainly, the inter-individual relationships between all those present, including children, teachers, and other staff. CRESAS (Centre de Recherche de l'Education Spécialisée et de l'Adaptation Scolaire) research, therefore, focuses on improving the educational setting rather than on efforts to determine the sources of behavioural problems in individual children.

This ecological approach is prominent, for instance, in our studies of young children's peer relationships. In France, early peer relationships are thought to be rare, short-lived, and often aggressive. As a result, in crèches* adult-child relationships are stressed and staff are encouraged to follow a mother-infant model in dealing with young children. However, many staff members find this difficult, and the attempt creates feelings of unease. Because of this policy, peer relationships have been little studied and are disregarded in the organization of crèches.

Contrary to this view, we found, early in our work, that children under two-years old spontaneously form small harmonious groups of two or three, and can often communicate successfully. Although earlier studies of peer relationships used a variety of methods and often led to contradictory interpretations on a number of issues, most also showed a wealth of social interactions among young children (Blurton-Jones 1972; Garvey 1974; Lamb 1977; Mueller-Lucas 1975; Mueller et al. 1977; Nicolich 1977; Flament 1977).

---

* Public institutions in which children from three months to three years of age spend the day. Crèches are open weekdays from 7 a.m. to 7 p.m.

Our first research concerned interactions amongst 12–24 month old children (Stambak *et al.* 1983; Stambak and Verba, in press). Previous research (Sinclair, Stambak, *et al.* 1982) had shown that, with familiar small objects and toys, their activities were often imitative, developing into pretend or symbolic play, and that collections of identical objects led to pre-logical activities, such as collecting, seriating, and establishing one-to-one correspondences. By contrast, several unfamiliar small objects induced exploration of their physical properties and the construction of new objects.

Accordingly, for the purpose of investigating social interactions, two of the above-mentioned situations were used: in the one, the children were given small familiar objects, and in the other unfamiliar objects. In a third situation, the usual objects were removed from the playroom and replaced by large cardboard boxes and cylinders. Groups of 4 to 8 children were observed in these three situations. In all three situations, we saw the children, without adult intervention, being attentive to one another and wanting to do things together. Sometimes cheerful, sometimes serious, they succeeded in carrying out various games together and in solving the problems that sometimes arose. In all our observations, all children took part, at some time or other, in a shared activity; not one of them remained isolated throughout the whole observation period.

With the cardboard containers, the children engaged in various forms of social play: peek-a-boo games, object exchanges, etc. In these games, each participant needs to respond to the proposals of the other(s) if the game is to continue. Their actions are organized as in a dialogue, one child responding to the invitations of another and each child playing a particular role.

With the familiar objects, the children engaged mainly in pretend play, which allowed us to observe their knowledge of many events that recur in daily life and their capacity to detach these events from their real life context. The children's capacity to distinguish fiction and reality was clearly demonstrated by the way they managed to make their intentions clear to their partner(s), e.g. throwing their head well back while pretending to drink from an empty cup, making exaggerated noises of satisfaction, and looking intently at their partner when they wanted to engage in a pretending-to-drink-and-eat situation. Such emphatic procedures (postures, gestures, mimicking) were often the main indication that allowed the observers to conclude that the children's activities were to be interpreted as pretend play. At the ages observed, words were rarely used and then only a few at a time.

In both the container situation and the familiar-object situation, the children constructed extended sequences of interactions, making up 'little stories'; two or sometimes three children combined their initiatives and succeeded in creating a system of shared meanings. The ways and means of getting an episode going showed great similarity. After a first contact between partners, one of them proposes an idea whose content is made more or less explicit. The partner understands totally or partially and acts in consequence. This response elicits a response of the other child and so on. A first meaning is shared and a theme is elaborated. Each partner adjusts and modifies his/her activities according to the responses of the other child.

The variety and efficacy of the means invented by these very young children to make their intentions and expectations clear to their partners were truly surprising. When calling for somebody to join them, when appealing to a partner to do something, in moments of agreement or of conflict, the children interacted with one another in well-adapted and often subtle ways. All these interactions took place spontaneously, without any adult intervention. These very young children manifestly desired to act together, to engage in social exchanges with other children; in our observations, shared activities were far more frequent than solitary ones.

Our first studies thus confirmed the importance and the variety of early social interactions, which concurrently were also being increasingly discussed in the psychological literature. Like other researchers, we began to ask questions about the interactive process: how do children come to organize their social interactions? How does their playing together become gradually more organized and elaborate? What is the importance of social interactions for child development in general?

In the research field of CRESAS, we had to deal with institutional problems concerning children between the ages of two and four. In France, the organization of educational institutions for children of these ages raises many problems, certain aspects of which are studied by CRESAS. We, therefore, decided to extend our work on peer relationships to children between two and four.

Many researchers consider that, at these ages, pretend play is an important factor in psychological development. Authors such as Forbes *et al.* (in press), Garvey (1974, 1979, 1982), Garvey and Berndt (1977), Goffman (1974), Stockinger and McCune-Nicolich (1983), Shugar (1979), and Schwartzman (1976, 1978) see the most important function of pretend play as the interpersonal negotiation of shared meaning systems within which the fantasy action, whatever its specific content, may be understood by the players. In our opinion, the interactive processes in pretend play do not promote only the development of social interaction skills, but also cognitive development. We therefore started our new research project with the aim of specifying the cognitive mechanisms which lead to the successful construction of pretend play sequences.

In the course of our research in a *jardin d'enfants** in Paris, we were impressed by the fact that, in this institution, the directress and her collaborators not only presented frequent puppet shows to the children (either to all the children together, about 60, or to smaller groups of about 10 children), but also encouraged the children to present shows themselves, providing them with a puppet theatre and handheld puppets of a size suitable for young children. Two children go behind the puppet stand (they are visible to the audience from about their middle) and improvise together a show for the other children and for the adult leader of the group (the latter does not intervene in the show but reacts simply as another member of the audience).

This particular situation enabled us to undertake a systematic study of the shows created by two-to-four-year old children. Fifty-two couples were filmed; the shows they presented were of unequal length, but mostly comprised several scenes in succession. In this paper, we present a detailed analysis of the productions of three couples of three-year-old children.

First, a few general remarks: all the adults (researchers, educators, parents) who happened to attend any of the shows produced by the children were amazed by their theatrical capacities. The scenes were well constructed, the themes varied and many theatrical devices were used, so that the shows were indeed entertaining for the audience. The quality of the children's productions was all the more remarkable since they had to improvise together: in fact, they functioned simultaneously as scriptwriters and puppet-actors. In comparison to free pretend play, the puppet situation is highly complex and incorporates many constraints. The children have to accomplish a conventionalized task in a traditional framework with a social aim, the amusement of the audience.

In our analysis, we took into account the nature of the interactive processes in each of the sequences and the linking of the sequences into the show as a whole. We oriented our analysis towards answering the following questions:

1. Do the children show mastery of the rules of puppet shows? Can they use theatrical procedures that differentiate free pretend play from a show for an audience? An analysis of these points should throw light on their capacities for communication in a highly specialized situation: in free pretend play, much can be taken for granted by the players which, in a show for spectators, will need specifying and explaining.

2. Do the children show that they have interiorized certain interpersonal relationships (as they have experienced them themselves or observed them in others in their daily life)? Are they capable of sketching a character and

---

* Institution in which children between two and four are cared for during the daytime. Our research was carried out with the collaboration of the principal and her colleagues.

of acting and coordinating different roles? Analysis of such points should tell us something about the children's conceptualization of social relationships and about awareness of social rules.

3. How do the two children construct the sequences? How do they share their task? How do they reach agreement about the theme and its development? Are the relationships between the two partners symmetrical?

4. What cognitive processes lead to building up these coherent scenes? What are the peculiarities of the reasoning process when it is applied to an enterprise effected by *two* children together? From our constructivist point of view, this cognitive analysis appears especially interesting.

Use of theatrical procedures

a. The construction of the scenario

Right through their totally improvised shows, it is perfectly clear that the children have mastered the *rules of constructing* a puppet show. A set of interiorized rules allows them to build scenarios that comprise both *coherent sequences* with greater or lesser dramatic content, taken from ordinary events in their daily life or derived from imaginary situations, and also *interludes* consisting of dancing and singing which often appear to function as a form of appeasement following the tension of a strongly emotional scene. We observed also certain *short passages* that *punctuate* the action, either to fill a small gap or to create a link between two events.

The children show a constant *awareness of the spectators* and of the need to amuse them, using the following procedures:

1. *Introductory rituals*. At the beginning of a show, or when a new sequence starts, or at the appearance of an as yet unused puppet, certain theatrical formulae are employed with such consistency and frequency, that we have regarded them as 'rituals'. These rituals introduce and characterize a personage, establish contact with the spectators and function as openings:

'Hello, I am called a Man* who has glasses'.

'The Clown, he comes' followed by a chant, 'The Clown, the Clown'.

2. *Different forms of appeals to the audience* taking place at any time during the show:

'Well, children?'

'Knock, knock, who's there?'

At other times, they incite the public to contribute to the action: they call the spectators to witness a particular event:

---

* In all examples, the translation from the French was made to reflect certain forms that are ungrammatical by adult French standards. Here and subsequently, puppet characters are given an initial capital.

'Flopper did poo-poo in his pants' (towards the audience).
Or they may explain something, or ask for approval:
   'We got to take water, all right children?'

b. The acting performance

In the acting of the scenes, the children also show considerable mastery. The children use the visible part of their body to emphasize and supplement the movements of the puppets.

During the show, the children respect the fact that the puppets are the actors: the children do not speak in their own name and do not forget that it is the puppet that has to carry out the action:

1. *Movements of the puppets.* When a puppet speaks, they make it move appropriately, as a whole or often only its head. It is held out towards the audience or hidden away, while punctuating a comment of the puppeteer. When an action is carried out, the puppet is moved as a living character, sometimes with many refined gestures: it is made to walk, to dance, to appear and disappear, to slap another puppet, to show objects to the public, or hide things. The puppet applauds by clapping its hands and also blows out candles.

2. *Speaking and other sound productions.* In dialogue, the children show their awareness of the fact that they are not in an everyday situation of communication. They adopt a theatrical way of speaking and make the distinction between the play and real life quite clear. Certain puppets are characterized by a special way of speaking, for instance Grandmother often speaks in a high voice. Whenever children change puppets, they also change their voice. Their intonation changes as a function of the particular situation and of the person addressed: E. speaks in a deep voice when her puppet wants to frighten Little Girl and comments a little later, in a gentle appeasing voice, 'He's nice' with reference to Dog who at first frightened Little Girl. The puppets also produce noises and onomatopoeias, often in accompaniment to certain actions—chewing and swallowing noises when Dog eats or 'Sh—sh' when Clown sleeps. Sometimes, when an animal actually has to say something, the children use a special way of talking: they speak very softly or with their mouths half open.

In dialogue, the children use several procedures to *emphasize* important points or to contrast two utterances: repetitions, moralizing, or disapproving intonation or affectation both in the choice of words and tone. Vocal productions are frequently accompanied by other expressive means. The children show surprise, malice, amusement, disapproval, fear, and anger with their own face and body as far as it is visible, thus adding in a highly expressive way to the movements they can make the puppets perform: when a puppet says 'I'm cold', the child makes his own body shiver.

As we see, three-year-old children dispose of a wide range of expressive means which they use in constructing their scenarios and performing their plays. They show that they have already analyzed, to the point of being capable of producing them, many ways of expressing emotions, feelings, and judgments. They are perfectly aware of the particular demands inherent in the puppet situation, in particular they are constantly aware of the presence of the audience: this awareness of the overall aim of a show is obvious in the way the two co-producers use theatrical devices (emphasis, gags, appeals to the public) to keep the interest going, and in the way they give certain explanations in advance (announcements, introductions, commentaries) only rarely needed in free pretend play.

Awareness of social relationships and of social rules

Two different scenes are presented to illustrate the way we analyze the sequences from the point of view of the children's awareness of social relationships and of social rules. Among the puppets the children have at their disposal, there are animals (Big Dog, Little Dog, Cat, Bear) and human beings (Little Girl, Little Boy, Grandmother, Grandfather, Man, Clown). In some of the scenarios (often in those that are more emotional) the puppets are given a particular role, they play a specific character. In such scenarios the characters personified by the children have precise individual characteristics. Scene I falls into two parts:

Scene IA In the first part, Big Dog frightens Little Girl. The two puppets play complementary roles and the responses of each are directly dependent on those of the other (see pp. 138–9).

Elisa holds Big Dog (1) and Noeline, holding Little Girl, hides her (2); Elisa interprets this action as an indication that Little Girl is afraid of Big Dog, and that Little Girl believes Big Dog to be nasty or dangerous. Elisa at first refuses to give Big Dog an unpleasant role and says 'he's nice' (3), but when Little Girl continues to manifest fright saying 'Oh, I'm afraid, I'm afraid' which explains why she hides herself (4), Elisa accepts that Big Dog is a nasty dog and makes threatening noises (5). Noeline then gives a magnificent theatrical performance, acting Little Girl's fright (6) whereupon Elisa once again proposes that Big Dog is a nice Dog, 'he's nice' (7), but Noeline does not accept this proposal and continues to mimic Little Girl's fright: 'no, he's nasty, ooh!' and makes her hide again (8). Elisa then takes Big Dog away, which means the end of this part of the scene.

This sequence shows us the children's ability to play the part of an aggressor and particularly, the role of the person aggressed. As in other similar sequences the child who plays the aggressor's part is not very pleased; Elisa tries to modify his character ('he's nice'), without success in this example. Their two characters act a conflictual relationship in a tensely emotional scene. In the second part of the same scene we see how the children try to dedramatize the situation.

Scene IB In this sequence, Elisa works with another puppet, Bear, and Noeline continues to act with Little Girl. Bear will act the role of Daddy and the two children construct a scene representing the relationship between a nice Daddy and his capricious little girl. Once again the two characters have complementary roles (see pp. 139–40).

Elisa, with Bear, inaugurates the new sequence: Bear asks what has happened to Little Girl, who is still hiding behind the scene (1): Noeline, as Little Girl proposes that Bear should be Daddy, and she asks him to buy her some sweets (2). Elisa does not immediately understand this role given to Bear (3) but when Noeline repeats her request (4) she tries to comply and makes Bear offer sweets to Little Girl (5); Noeline immediately makes a clever use of the fact that Bear has touched Little Girl in his action of offering her sweets. Little Girl pretends to be hurt, Bear has hit her in her teeth, she hits Bear and resumes her role of the aggressed person (6). Elisa, put out by this response, takes the bear puppet off her hand, and offers a small pot to Little Girl (probably intended as another way of giving her sweets) (7). However, Little Girl prefers to take no notice (8). Elisa, still without a puppet (has she forgotten her task as a puppeteer?) tries to console Little Girl directly (9). Little Girl refuses to be consoled and Noeline takes up Little Dog, whom she makes eat from the pot Elisa has proferred: she seems to indicate that dogs, but not little girls, eat directly from pots (10). Meanwhile Elisa, not quite knowing what to do, looks for another puppet and breaks up the sequence (11).

In this part of the scene, the two girls play two typical, picturesque roles: that of a spoiled, capricious little girl who pretends to be victimized, and that of a doting daddy who is a bit at sea. They demonstrate their well established knowledge of certain relationships that may exist between children and parents.

Moreover, in this scene as in others, one part dealing with an emotional theme (fear) is followed by another part that tends to calm the tension; in a sense the second part constitutes a consolation.

Scene IA

Elisa (3;3) 'Big Dog' (was present at the previous presentations)	Noeline (3;2) 'Little Girl' (took part in several presentations before this one)
	Finishes the preceding scene holding Little Girl:
	'That's finished, till tomorrow.'
	Broad gesture with Little Girl as if drawing a curtain. Holds her arm in the air shaking Little Girl.

1 Puts on Big Dog.
Turns Big Dog towards Noeline:
'Bow, wow, wow!' (normal voice).
Looks at Noeline.

Hides Little Girl behind the coun- **2**
ter.

3 Looking at Big Dog, *'He's nice.'*

Keeping Little Girl hidden: *'Oh,* **4**
I'm afraid, I'm afraid' simulating
fear. Looks at audience.

5 Brings Big Dog nearer to Little
Girl. In a louder voice: 'Bow, wow,
wow!'

Lets Little Girl reappear while **6**
simulating fear; hides her again
behind the counter, brings her out
again, hides her again behind part
of the counter where the puppet is
still visible to the audience: *'Oh*
dear, is hidden there'.

(adjusts Big Dog)

7 In a gentle voice: *'He's nice'*. Moving him in an appealing way: *'He's*
nice'.

Shakes Little Girl behind one side **8**
of the counter: *'No, he's nasty,*
ooh!'
Then, hiding the puppet behind the
counter: *'I've hidden her.'*

9 *'Oh!'* retreating Big Dog and then
taking him away.
Looks for another puppet.

Scene IB

Elise (3;3) 'Bear'   Noeline (3;2) 'Little Girl'
Puts on Bear.

1 *'What happened to . . . children?'*
Turns Bear toward Little Girl:
'what happened?'

Makes Little Girl reappear and **2**
brings her closer to Bear. She chir-
rups: *'Oh Daddy, say, would you*
please buy some sweets?'

3 *'What happened?'* putting her hand
on Little Girl. Questioning
expression.

Implores: *'Sweets, Daddy, sweets.'* **4**

5   *'Oh!'* pulls Bear down behind the counter. Brings him out again with munching noises. Brings him towards Little Girl and touches her as if he was giving her something.

'oh' plaintively while moving back **6** Little Girl. *'You hurt my teeth'* Hits Bear with Little Girl. Turns Little Girl towards herself, looks at her sadly, lays her down on the counter and then lies down on top of her with pained expression.

7   *'Oh'* regretfully. Takes off Bear and without a puppet picks up a pot from behind the counter and holds it out to Little Girl: *'Look, there are!'*

Does not react.   **8**

9   Tries to raise up the head of Little Girl while still holding the pot.

Shakes Little Girl (it looks like a **10** refusal).
Then takes off Little Girl and puts on Little Dog.
'Little Dog'
'That's the Dog'· and lets him eat out of the pot for a long time.

11 Bends down to take another puppet.

Scene II. This scene is different from the preceding one in the sense that this time the two actors have similar roles, since both personify adult educators. The theatrical aspect of the sequence lies in the presence of a third puppet, the object of the adult's administrations (see pp. 141–2).

After the opening rituals (introduction of the puppet, opening dialogue with the spectators) and a short negotiation about the theme, Marie (holding Little Boy) declares that Big Dog (present on the scene but not held by either of the two children) has 'made a mess': 'I'll have to change him, he's done poo-poo' (1). Julie accepts this idea (2). A certain complicity is then established between the two educators-puppets, who both make their disapproval of the 'mess' quite clear (3, 4, 5) with an explanation of the 'mess' by Julie ('poo-poo in his pants' (6, 8)). Marie then administers the punishment using her puppet in a most adroit way and, almost in the same movement, consoles the victim, bringing both punisher and punished character close to her face and concluding in a rather surprising manner that 'now he's all clean again' (9). Julie is a little astonished (10), but all's well that ends well; victim and castigator dance and sing together in a newly composed moral chant: 'too-too, when you do poo-poo, when you do poo-poo . . . ' (11).

This scene almost looks like a parody of adult behaviour towards a child who has made a mess. Adults have to punish this kind of offence, but they may find such things almost amusing. The punishment should therefore not be too serious, and especially, it should immediately be followed by a consolation. The children show their knowledge of certain adult-child relations, and more precisely, their awareness of what adults think of certain acts done by children. In this scene as in others, the two puppet-adults are in total agreement about the event which forms the topic of their dialogue: both are made to express the same moral precepts.

Scene II

Marie (3;0) 'Little Boy'	Julie (3;0) 'Man'
	Puts on Man and addressing the audience: *'Good day, I'm called a Man who has glasses'*.
Squeezing Big Dog under her arm puts on Little Boy while looking attentively at Julie.	
	'What do you want?' while moving Man.
	'What do you want? Flopper?'
Pulls back her arm that holds Big dog: *'What is it you want?'*	*'Look, I'll go get him'* bending over as if to pick up Big Dog.
	'Flopper'
Moves Big Dog towards Man: *'Here I am.'*	
	Moving towards Big Dog: *'Go!'*
Pulls back Big Dog: *'Wait.'*	
	'Go!' louder, while moving still closer Big Dog.
1 *'Wait, will you wait.'* Starts animating Little Boy *'I have to change him. He's done poo-poo.'*	
	Interested, she turns towards the 2 audience and moves Man. *'He has done poo-poo, oh!'*
3 Indicates her approval with malice and complicity.	

'Yes, he's done poo-poo,' in a very   4
low voice.
'You've done poo-poo, Flopper,
oh, no!' and she beats him with
Man.

5   Looks at Julie and in turn dis-
approves: 'Oh, no!'

'No' with a severe expression.   6
Towards the audience: 'You can
. . . Flopper did poo-poo in his
pants'.

7   Mischievous expression looking at   8
Julie. 'Big Dog.'
Puts on Big Dog.

'Look . . .' towards the audience.
Turns towards Big Dog: 'In his
pants.'

9   'Wait, you'll be scolded' puts Big
Dog on counter and Little Boy pre-
tends to lift up and let down Big
Dog's pants while holding his head.   Looks attentively at Marie.
'Poo bang' and she hits him on the
snout. 'There, he's been scolded!'
Then brings the two puppets to her
face.
'There now, he's quite clean.'

'He is quite clean?' in a very small   10
voice. Man is turned towards the
other two puppets.

11   'Yes'. Brings forward the two puppets and sings: 'Too-too, when you
do poo-poo, when you do poo-poo.'
Looks at the camera as if waiting.

Summing up, the children's awareness of social relationships is evidenced in several ways in their puppet shows. In the first place, they are
able to characterize a personage, not only in a traditional social role (an
adult taking care of a baby, for example) but in a much more specific and
refined way, as in the scene where the doting daddy tries several ways to
please and pacify the little girl. In the second place, they are able to construct conflictual scenes on a general emotional theme such as fear. Each
protagonist complements the role of the other, and genuine little dramas
are created. Moreover, with great virtuosity, they then manage to resolve
the conflict and to provide a 'happy ending' for the sequence, making use
of a general constructive device we have often observed, i.e. the use of
contrasting structures. In the third place, the children are aware of degrees
in the transgression of moral or social rules. Some transgressions are
serious, others are commonplace, and their consequences are very differ-

ent: refusal to comply with a 'devouring' act (scene resumed p. 145), mild punishment and consolation for poo-poo.

Thus, the conceptualization of different interpersonal relationships which has been observed by other authors (Schwartzman 1978; Sutton-Smith 1979; Forbes 1982) is already found in children as young as three.

Our observations show that the children are capable of representing not only social interactions they may have experienced themselves but also interactions they have observed (e.g. the moralizing complicity between two adults). This double awareness (of one's own experiences and of one's observation of other people's behaviour) also seems to be referred to by Schwartzman (1976; 1978) when she suggests that, in fantasy play, children can not only interpret their own relationships to one another, as experienced in real life, but also experiment with new social roles.

Task sharing modes between the puppeteers

In the puppet situation, *two* children have to improvise the show. How do they share this task? Do we observe a common production where each child has an equally active part? Does one child act as the main executant, whereas the other plays only a secondary part, just approving or assisting the first? To sum up, can we speak of symmetrical relationships in the construction of the scenario?

1. In the setting-up phase, the two puppeteers try to reach agreement about the content, the theme of the scenario. They start a *negotiating process*, progressing in two possible ways:

—one child's idea is immediately accepted by the other, whose response completes the proposition and directs the setting-up of the theme. In this case, the agreement between the two partners is immediate;

—less frequently, the first proposal suggested by one of the partners is rejected by the other and the negotiating phase is a little longer. Thus, in Scene II (p. 141), Julie, once again, proposes the 'devouring' theme produced at the beginning of the show. Marie refuses by delaying the action: 'Wait!' and does not let herself be influenced by Julie's insistence. Repeating 'Wait, wait!', she proposes a new idea which is approved by Julie. In this scene, in spite of the initial disagreement both show their capacity to take into account the other's desires and ideas.

We intend to look at a larger number of examples, to see whether there is a link between the richness of the scene's content and the type of negotiation during the setting-up phase.

2. After the setting-up phase, the two puppeteers engage in the development of the theme and the building-up of the scenes. We have seen that all the scenes develop around a central 'plot', which leads to the question: do

both children contribute to the elaboration and progression of the plot equally, or is it constructed more by one than the other?

In all the scenes presented here, the story has been elaborated by the two puppeteers: in scenes IA and IB (pp. 138–40) the two puppets have been given *complementary* parts. The two children have then to adjust mutually in their replies, as each character's actions narrowly depend on the other's. In Scene II (pp. 141–2) the two puppets have *similar* parts and the two children elaborate the story simultaneously as well. The plot is structured as follows: Big Dog has done poo-poo (1); first disapproval ('oh no!') (4); he has done something naughty (poo-poo in his pants) (6); he has to be punished (but not too severely) (9).

The four replies marking the progression of the story are well distributed between the two partners who both play the role of an educator. Replies 1 and 9 are Marie's, replies 4 and 6 are Julie's. Thus, all along the development of the scenes, there is a sharing of the task, and the two children complete each other's replies: we obviously have symmetrical productions.

We also observed a few non-symmetrical scenes where one of the partners elaborated the content of the scenario, while the other just encouraged and approved the other's ideas, but such scenes are not frequent.

We should like, in conclusion, to insist upon the existence of these *shared productions*. These young children are able, through negotiation, to achieve a shared task which has been elaborated in common, and results in a coherent whole.

Some logical aspects of the scenarios

The scenes that form part of scenarios all show a coherent construction, as has already been noted. A theme is decided upon, is developed and, often, the final remark resembles a conclusion. Utterances and actions follow one another in a coherent way. What is the underlying logic of these sequences? What sort of logical links exist between the partners' utterances?

Different types of relations can be distinguished and the most significant appear to be the following:

a) *Clarifications and requests for clarification.* While they are building up the scenes, children try to make them more intelligible. In Scene IA (pp. 138–9) Noeline uses a series of theatrical procedures in a crescendo sequence to make Little Girl's fear clear to the public: first she hides Little Girl behind the curtain; in the next reply she verbalizes 'oh, I'm afraid, I'm afraid'; then she makes Little Girl appear and disappear several times symbolizing her fear; finally, just in case the spectators have not yet understood, she adds 'hidden here'.

b) We also note replies which are *explanations* or *justifications*: children explain what they have just said or done. Scene IB (pp. 139–40) illustrates a justification of a preceding action: Bear (Daddy) tries to console Little Girl, he offers her a pot; Little Girl refuses to take it and then justifies this refusal by taking Big Dog and making him eat in the pot . . . (indicating that it's dogs that eat directly out of a pot, not girls?).

c) Sometimes scenes have *conclusions* which figure as final remarks. In Scene II (pp. 141–2) the two partners act as educators and comment on the 'mess' made by the 'child'. Marie carries out the punishment and draws a moral conclusion (11): 'too-too, when you do poo-poo'; a slap on the bottom, that's what happens if you dirty your pants.

d) We also observed *extrapolations*, often in a challenge: but if—then what? In another scene Big Dog has devoured Clown, and the producers as well as the public are shocked. But how far would one dare to go? Who may eat whom? Marie holding Cat proposes to let Cat eat Man; Julie refuses and proposes that Big Dog should eat Man, but Marie rejects this and proposes instead that Big Dog could eat Little Dog; Marie refuses once again saying 'you know, you may not eat him, you know' and she proposes a socially acceptable solution: a sandwich! In this scene the relation 'x eats y' is explored, with different values given to x (Big Dog, Little Dog, again Big Dog) and to y (Clown, Man, Little Dog, and sandwich). Logical possibilities are confronted with socially acceptable ones and a common solution is found: a sandwich.

e) Analyzing our examples, several *contrasting statements* have been indicated. Another scene played by Julie and Marie illustrates a particularly clear contrasting structure in which several inferences are embedded (if x then y, if non-x then non-y). The two children start by stating that they are cold, and that when one is cold, one has to put on warm clothing, and they explain that a thunderstorm makes it go cold. The second part of the scene forms a contrapuntal contrast with the first: the children decide that the thunderstorm is finished and that therefore the warm clothes can be taken off.

The coherence of the children's constructions seems firmly rooted in their capacity to establish the types of logical links we have briefly illustrated. Inferential links between conditions and consequences, sometimes series of conditions, challenges and provocations, contrasting structures, and conclusions are elaborated by the two children together.

In our view the social interaction pattern is of fundamental importance for children's cognitive development in general. The construction of reasoning processes is consolidated by the necessity to take into account the intentions of the partner, to find ways of solving conflicts and to justify and explain their own thoughts. These constraints are inherent in all social interaction and lead one or other of the partners to introduce a certain distance between reality and ideas about reality, thereby forcing him or her to engage in a certain control of his or her own reasoning.

Concluding remarks

In our various studies of peer-relationships we try to analyze how young children organize their shared activities and the ways in which they regulate their cooperative social behaviour. Such analyses are interesting *per se*, as a contribution to our understanding of social development. However, our aim is more ambitious: since we believe that the mental growth of a human baby is that of an intrinsically social being, we now sustain the hypothesis that social interaction is an important factor in the acquisition and mastery of knowledge.

Within this context, we believe that the detailed study of interactions amongst peers can provide the developmental psychologist with data from which he can make inferences about both knowledge and about processes. However the study of peer interaction can, in our opinion, not only demonstrate already existing knowledge and processes, but can also show the on-line construction of new knowledge and new constructive processes. Specifically, we believe that a particular situation may place demands on children for which they have to elaborate together strategies that will also influence their own processes of thought.

The puppet show situation is highly specific. It has a formal structural framework, the puppets have more or less defined characters, a scenario has to be constructed, and the audience must be kept amused. The task is shared by two children. Our analyses of the different scenarios, of which we have given several examples, clearly show, in our opinion, that the children already at the age of three have interiorized some essential components of social relationships and especially of the rules existing in their particular micro-society, the day-care centre. They comment on what is permitted and what is not, and on the kind of sanctions to be applied to transgressions. We certainly agree with Forbes (1982) when he affirms that shared fantasy play 'may support the development of structured knowledge about the social world . . . Firstly, it supports such development because it provides the child with an opportunity for active construction of a social world, rather than merely an occasion for passive observation'. However, we would like to clarify the meaning of 'support'. Does play consolidate already existing knowledge? Or does it have a truly constructive role, i.e. does it generate new strategies and new knowledge? Though this question cannot yet be answered, certain of our observations seem to indicate that, indeed, shared play may generate new strategies. For example, the socially acceptable solution of 'having a sandwich', which concludes the devouring scene, appears to be an on-the-spot invention rather than the application of an already existing mode of resolving a difficulty. Similarly, the 'cleansing' punishment in the scene of the dirty pants also seems to result from the pressure of the *here and now* scene. Though the children's productions show many such on-the-spot inventions, these two seem to be examples of strategies leading to solutions that have larger social and intellectual implications than others. In other words, we consider it possible that shared play can actually structure the child's social world as well as consolidate and externalize what he/she already knows about it. Certainly play allows, as well as forces, children to take a certain distance from real situations and thereby fosters a reflection on actions which is less possible when they are carried out in everyday life.

The same kind of questions can be raised concerning our analysis of the logical elements in the construction of the puppet shows. How far do the contrasting structures, the inferences, and conclusions we have pointed out in our examples consolidate and/or structure the children's reasoning processes? Once again, the question cannot be answered directly, but shared play, especially in a highly structured framework such as puppet shows, certainly provides exceptionally favourable conditions for the observation of such phenomena.

However, it will probably be the analysis of the modes of interaction by which children construct shared meaning systems that will be most significant as regards the truly constructive influence of shared play. The distance between 'pretend' and 'real' (of which the children are perfectly aware) leaves the partners a certain freedom to act and provides the opportunity for thinking about the activities; on the other hand, the necessity for carrying out a shared project places particular demands on the partners. They have to adjust their actions, to regulate the turn-taking, and in the puppet shows, they also have to take the audience into account. The combination of a certain freedom from the pressures of 'real' situations, and of the particular constraints involved in the shared construction of a 'pretend' event, lead to the elaboration of surprisingly sophisticated social interactions: the children negotiate, they take the other's point of view into account, they clarify their intentions, they argue and justify their actions. A coherent structure, of which we have emphasized certain aspects, results from these interpersonal coordinations and, we suppose, plays a role in overall psychological development equally important to the intrapersonal coordinations highlighted by Piaget. We share the opinion of authors such as Inhelder, Sinclair, and Bovet (1974) and Perret-Clermont (1979), on the importance of socio-cognitive conflicts for the elaboration of knowledge: confrontation between different points of view and different strategies fosters the construction of new coordinations and confers a coherent aspect on the children's productions.

Acknowledgements

We are grateful to: Hermine Sinclair, Professor at Geneva University, and Christiane Royon, Cresas—INRP, for their helpful contribution.

Discussion

Marionettes and real-life

Kummer asked, 'How realistic are the skills shown with marionettes?' and, 'Do they generalize to natural, real-life situations?'

Stambak replied:

'In constructing scenarios, children use skills and knowledge previously acquired in daily experiences. In our opinion, shared play can structure and consolidate new skills. These new acquisitions will then be used in everyday life as well as in play situations.'

Tizard suggested that evidence of learning from marionette play could be sought also in other naturally-occurring contexts.

In reply *Stambak* emphasized:

'Our research has not been carried out in an experimental setting: children were observed in the everyday situations of a *jardin d'enfants*. But we should like to insist upon the fact that, for several years, researchers and educators shared opinions about the organization of the institution.'

11

The structure of peer relationships and possible effects on school achievement
LOTHAR KRAPPMANN

1 Socialization by maintenance of social relationships

a. Social relationships as a structured field of tasks

Socialization research is characterized by the attempt to trace social experiences that advance the developmental process in childhood. Yet when such social experiences are defined in terms of indicators of social inequality, as when social class is treated as an independent variable, the resulting explanations are not very satisfactory. We miss the links connecting social-structural factors with the mental potential of individual children.

One way to link the development of competence with factors operating on the level of social structure has been pointed out by Piaget. Piaget holds that, 'Human intelligence develops in the individual in terms of social interactions—too often disregarded' (1971, p. 224 f). Sociologists, in particular, could profit from this approach. For a high proportion of interactions take place within social relationships, and sociologists should therefore study not only the mere social interactions of children but also these interactions as embedded in social relationships. This approach is appealing because sociologists consider social relationships not only as a product of the common activity of the partners involved but also as a matrix provided by society to make social action possible (see also pp. xv–xvii, this volume).

The concept of social relationships has held varying positions in the history of sociological thought. Weber defined social relationships to mean 'the behaviour of a plurality of actors in so far as, in its meaningful content, the action of each takes account of that of the others and is oriented in these terms' (1964, p. 118). Later, the concept of social relationships was largely, if temporarily, superseded by the concept of 'roles', defined by Parsons (1964) as reflecting the reciprocal rights and duties of members of functionally differentiated and hierarchically ordered institutions. According to this view a role contains a well specified set of instruments for common action. By contrast, Weber repeatedly pointed out that relationships, rather than guaranteeing some common action, merely provide the possibility of agreements on mutually consented behaviour. Though Weber does not speak of competencies, let alone of their acquisition on the part of children, he does make it clear that the expectations of the partners to a relationship seldom fully match, and that the prerequisite condition for the maintenance of a relationship is the fact of being 'mutually oriented' or, in more modern terms, being capable of decentration (Piaget 1950) and coordination of perspectives (Selman 1980). Weber goes on to say that the partners to a relationship must be capable of negotiating plans of action. The continuity of a relationship depends on promises and obligations. Thus, relationships formulate requirements which have to be met by the participants and which thereby stimulate attempts to reach more satisfactory solutions when the strategy selected does not adequately correspond to need.

b. Socialization through active participation in meaningful social relationships

Early childhood relationships have frequently been investigated solely in terms of the extent to which they provide reliable partners who care for the children, raise and teach them, communicate values and norms, and especially, satisfy their emotional needs. Less attention has been given to the stimuli that participation in a relationship of a specified quality provides for the development of children's capacities. If this aspect—the development of child's capabilities through its relationships—has been largely overlooked, it may be because many observers tend to consider the child incapable of making an active, problem-solving contribution to the maintenance of the early mother-child relationship.

However, studies on the links between infant-mother attachment and the development of person and object permanence (Bell 1970; Levitt *et al.* 1984) and on the interplay between maturity demands and capacity shifts in mother-child dyads (Baumrind 1967; Cross 1977) have demonstrated clearly that children's capacities develop best when the active contribution of the child is accepted and the reactions between the partners are reciprocal. This indeed, is the prime characteristic of a functioning relationship. These reciprocal reactions would be impossible without the frame of reference provided by the relationship. In their study on the development of linguistic skills, Camaioni *et al.* (1984) have shown that it is not interactions *per se* that stimulate development most, but interaction episodes characterized by culturally defined and commonly accepted conventions and rules. Social relationships are important vehicles of meanings and rules which are

necessary in order for the participants to contribute adequately to interactions and to learn from them.

Stimulating impulses of this kind are contained not only in relationships
between children and adults, but in those among children themselves.
Hence the acquisition of cognitive coordinations is enhanced when children work out cognitive problems together (Murray 1972; Silverman and
Geiringer 1973; Doise *et al.* 1975). Following Smedslund (1966), Doise
(1978, p. 344) traces this effect of child-child interaction on cognitive
development to the fact that such interaction reveals individual-specific
centrations and that the resulting conflict encourages decentration. In the
view of these investigators children profit not only from correct objections
to their opinions but likewise from confrontation with false opinions
(Mugny *et al.* 1976). This may depend on the nature of the cognitive task.
If reasoning contributes to the solution, the initially incapable child may
profit from the verbal explanation of a more competent child (Russell
1979). If improvement of practical strategies is needed, 'two wrongs can
make a right', because the confrontation with the moves of the other dis-
rupts established inefficient strategies and leads to the testing of new,
superior strategies (Glachan and Light 1982, p. 258). But under what conditions can a child be expected to take such objections seriously? According to Glachan and Light (1982) an extended, structured interaction rather
than just a brief encounter is needed. We suggest that extended, solution-
oriented interaction is to be expected in every day life, when the child is
involved in a relationship that gives meaning to a partner's opinions or
actions.

c. The special character of child-child relationships

The data of Mugny's study indicate that children's reactions to adult's
objections are different from reactions to objections made by other children. While these may be wrong, they are on the same cognitive level as
their own opinions. Already in his early work Piaget (1971, p. 278) stressed
the asymmetry in parent-child relationships and the necessity for a
balanced coordination of views in child-child relationships. Youniss (1980;
1982) has adopted this view, elaborating the significance which the symmetrical reciprocity of child-child relationships has in terms of the construction of shared meanings. In Youniss' view it is characteristic of peer
interactions that 'self and other are equals in the right to understand and
the power to construct knowledge' (1980, p. 19).

No doubt, this description represents an ideal type of peer relationship
(in the Weberian sense). It should not be interpreted as meaning that peer
relationships are always reciprocal. The idealization emphasizes the special
promise of peer relationships. Their structure provides the possibility that

children can exchange thoughts and intentions and can try to fit together plans of action without being emotionally dependent on the other as a beloved caretaker and without institutionalized means of coercion or sanctions at their disposal. Actual reality, however, may be full of struggles for dominance, quarrels about who is right, punishment for breaking rules, as well as resistance, avoidance, and retreat. These phenomena have to be understood as the strenuous and often futile side of the difficult and sometimes excessively demanding processes of coordinating perspectives and actions.

Cognitive development can be expected to be enhanced by social relationships among children only when children have the opportunity to test conflicting ideas and explanations, discuss them, and decide to accept, reject, or modify them. Mugny *et al.* (1984) have shown that social regulations may either facilitate or impede productive solutions to socio-cognitive conflicts in relationships of same-aged children, as well as in adult-child relationships. Such social regulations are certainly very much present in the relational patterns that determine interactions among children, and they largely prescribe whether and how any given problem will be resolved.

d. Research findings on the quality of peer relationships

To our knowledge there is no study that observes the various patterns of children's relationships or the types of group in which same-aged children are involved. Most research is designed to investigate social acceptance and status, uses sociometric measurements, and generally specifies the categories of popular, isolated, and rejected children (Peery 1979 adds 'amiables' to this list, Gottman 1977 includes 'teacher negatives' and 'mixers', called 'controversials' by Coie and Dodge 1983). Much attention has been devoted to the development of the concept of friendship (e.g. Selman 1980; Kliegl 1983). Although there is some evidence from interviews about how children think friends do behave and ought to behave (Keller 1984), the question of how friends actually behave towards one another still remains open (Hartup 1983, p. 140). Furthermore, it is still unclear whether the conception of friendship characterizing any particular stage of development can be realized in different patterns of interaction.

Reviews like those of Hartup (1978, 1983) show that only rarely have children's groups been studied in natural social settings; most investigations have been conducted in situations of instruction, in summer camps, or under laboratory conditions. Furlong (1976) believes that classroom associations of children are too changeable to be called groups, and similarly Hallinan (1979) found no cliques in a third of the classes she studied, while those she did find turned out to be highly unstable. Davies (1982) and Meyenn (1980) observed children's groups and their social life, but neither they nor Hallinan were able to distinguish such groups qualitatively, though the different types of group may well have different effects on the development of their members' interactional and cognitive capacities.

2 An exploratory analysis of social experience in peer relationships and its effect on school achievement

a. The study 'Everyday life of school children'*

The research project 'Everyday life of school children' is designed to promote a better understanding of what same-aged children learn from one another. Its focus is on interaction strategies that children aged 6 to 12 adopt in problem situations, such as arguments, hurt feelings, negotiations, seeking help and rejection; and on the kind of relationships and social networks formed by children. New or differentiated interaction strategies are considered to represent responses elicited by the specific properties of the interaction arrangement and the relational involvement of the children. The project combines a cross-sectional study of children from three classrooms (grades 1, 4, and 6; age 6, 10, and 12) with a longitudinal study (the 4th graders were studied again in 5th and 6th grades). It is based largely on qualitative methods of data collection and interpretative data analysis. Interactions in natural settings (primarily classrooms and school playgrounds) were recorded in field notes and, partly, on videotape. To round off our data we conducted semi-structured interviews with the children and their parents and teachers. Data were also collected about life events, school achievement, and the family situation. Data collection took place between September 1980 and June 1983, each age group being studied for half a year. The school was located in a metropolitan housing area with predominantly upper-lower and lower-middle class residents. The following presentation is based mainly on the data from grade 4 and the follow-up study of the same children in grade 6.

Originally, our study was not designed to examine what relations emerge between participation in different kinds of relationships and cognitive development. However, we can use the marks on the children's report cards as an indicator of their cognitive development. We were not able to conduct school-independent cognitive tests due to the authorization conditions imposed by the school administration.

* This study is a joint project of the Max Planck Institute for Human Development and Education and the Free University of Berlin, conducted by Hans Oswald and the author. We are grateful to Lisa Wassmann for her assistance in preparing the data for this analysis and to Eberhard Schröder for advice regarding the statistical analyses.

b. Indicators of social experience in peer relationships

Crucial to any analysis of the links between children's social experience and behavioural patterns of whatever kind, are good indicators that differentiate intensity levels and qualities of social experience. Since we assumed that the type of relationships among interaction partners will structure their mutual tasks and acceptable solutions, we designed our categories to reflect the different ways in which children relate to others of the same age. Since it is still uncertain which dimensions of social relationships deserve emphasis (cf. Hinde 1976, 1979), in the course of our study we adopted various approaches to describing the children's relational ties. These approaches were not independent but merely represented different vantage points from which to view children's participation in peer relationships:

(i) *Closeness of relationships*. This categorization approaches sociometric scales measuring close relationships. However, we do not count choices but rather interpret information from different sources. We distinguish between two main groups of children:
— participants in close and frequent relationships, including personal friendships; and
— participants in primarily weak, sporadic relationships.

Although children who were able to form only weak or sporadic relationships were not without intensive experiences (e.g. when attempting unsuccessfully to obtain the close friendship of another child), we can assume that children who maintained several close relationships will profit most from a broad range of peer experiences. As partners to such relationships they learn to take others' opinions into account, to justify their own intentions and suggestions, to monitor means of mutual influence, to achieve a working consensus with others, and to accept obligations towards them.

(ii) *Constancy and change in children's close relationships*. A comparison of the relationships maintained by the 4th and 6th graders indicated that the overall amount of stability and change in close relationships was almost equal. However, individual children demonstrated extremely different ratios of stability and change (Oswald and Krappmann 1984). We distinguish among the following categories:

Type I.   'Marginals': these were children who had only very few or no close relationships both in grade 4 and grade 6, and who were relegated to the margins of the relational network.

Type II. 'Changers': these children had close relationships at both periods of data collection, but almost all relationships had broken off or recently been constituted by the time of our follow-up two years later.

Type III. 'Consistents': these were children who had several stable relationships, most of which they maintained over the two years between observation periods.

Type IV. 'Intensives': these children maintained close, stable relationships from one period of data collection to the next, and had given up some close relationships and entered new ones.

We proceeded on the assumption that the greatest range of social experiences was open to the children in the 'consistent' and 'intensive' categories (types III and IV). Besides the tasks mentioned above, these children were faced with the additional problem of ensuring the stability of their relationships. They had to develop initiative, an ability to convince others, and a willingness to negotiate concessions and reach agreements. Children of type IV must make the additional effort to hold their own, end' unsatisfactory relationships, deal with disappointments, and present themselves to new partners and win them over. Hence for the purpose of the present analysis, we have contrasted the children of types III and IV with those of types I and II.

(iii) *Participation in social formations.* The term 'peer group' is often employed quite loosely. Our research into the types of grouping revealed three distinct social formations called group, association, and interaction field. These three formations differ in terms of the stability of internal relationships, the distinctiveness of borders, common interests of the children, and the internal structure of the formation (cf. Krappmann and Oswald 1983a; Oswald and Krappmann 1983).

'Groups' provide a stable framework for personal relationships and demand reciprocal support for a specified common aim. The groups we observed proved to be quite tension-laden, since a contradiction was apparent between internal role and status assignments and the promise of equality and frankness that peer relationships outside the group held. 'Associations' also build a stable frame, but allow greater leeway for the children's changing interests and relationship preferences. Their greater flexibility appears to give associations greater stability than groups. We emphasize that association is not a weak copy of a group, but a social formation in its own right. In both types of social formation, the children in our sample were confronted with the necessity of dealing with others' expectations, and sometimes even of facing a majority consensus; they had to adapt their personal aims to those of the formation, and to establish themselves as partners capable of reaching agreements and defending their own actions.

The children assigned to the 'interaction field' category were restrained from a broad range of interactional experiences because they formed no stable relationships with others. They were rarely asked to participate in decision making, seldom had to balance give and take, and, unlike members of a group or an association, were not compelled to think about ways to maintain good relationships. But also among the members of groups and associations, some children were in marginal positions, which reduced the range of experience open to them. Thus the children in our sample may be classified in two categories:

— those who were solidly anchored in a group or association and played an active part in the undertakings of these two social formations; and

— those who had little or no influence on the activities of the social formation to which they belonged, or who were not members of a group or association.

The entire range of social experience, we can conclude, was open mainly to children of the first category.

(iv) *Modes of integration into peer society*. Though children usually refer to all of their relationships as friendships, the predominant way in which they relate to one another can vary widely between different relationship contexts. We distinguish the following patterns of affiliation (cf. Krappmann and Oswald 1983b). These patterns we term integration modes.

Mode A: Integration in the mode of friendship. The child prefers relationships in which children appreciate each other as individuals liked for their specific personality make-up. The child expects that children related to one another in the friendship mode share favourable and unpleasant experiences and can count on each other in emergencies.

Mode B: Integration in the mode of partnership. What guides the children's mutual orientation in this mode is the other as appropriate company for the common pursuit of certain appreciated activities—soccer, fishing, etc. They tend to value each other's personal qualities not so much because they are appealing individual traits, but because they promise greater success, or more fun, in pursuing the common interest.

Mode C: Integration in the mode of play comradeship. These children orient their interactions with other children in accordance with their willingness to observe accepted rules and agreements in play and games as equals in rights and duties. This means that playmates are chosen less for their personal qualities than for a certain general sociability that is expected of all children, although its form differs considerably.*

---

*These modes of integration may obviously be influenced by the development of mutual understanding and the concept of friendship (Selman 1980). As yet we have not been able to examine the independence of the mode-of-integration categorization from the development stages of the friendship conception, although our interview contains some questions that are relevant for the assessment of children's friendship concepts. We assume, however, that

Children who comply with a preferred integration mode must stick to a certain pattern in order to maintain the relationships they feel are significant. Thus, a friend's secret must not be given away, nor must a partner's interests be jeopardized. Hence, each of these integration modes involves problems that demand a high degree of awareness of the other and his intentions, as well as thoughtful consideration of the consequences of one's own actions.

There are two further modes of integration, however, modes D and E, in which children face less stringent demands and are required to give less consideration to others—and to themselves:

Mode D: Integration in the mode of rambling. The interactions of 'ramblers' are seldom planned and usually occur as the opportunity arises. They wander around, joining whatever activity seems most attractive at the moment—generally some risky game or troublemaking—more or less exploiting the other for selfish action or fun.

Mode E: Integration in the mode of isolation. The children of this category have no relational attachment at all over considerable periods of time, or at most occupy the position of an extreme outsider. However, as our observations indicated, they maintain interactions, which are likewise shaped by their respective social situations—sometimes reflecting anticipated frustrations, sometimes absorbed by the wish to increase the chance of being accepted.

We expect the main contrast in peer experience to exist between children involved with others in integration modes A, B, and C, and children in the integration categories D and E.

To sum up: different typologies were constructed in order to distinguish types of children according to the quality and intensity of their involvement in peer relationships. Each of the various approaches allows us to identify a group of children with a great range of challenging peer experiences as opposed to a group of children with incidental or poor experience with others of their age. Apparently some of the strategies adopted by the children in order to coordinate actions within the enduring framework of intense relationships include cognitive components, e.g. taking into account the perspectives of others, reflecting one's own expectations, analyzing mistakes and failures, voicing good reasons, and giving sound evidence. Therefore, it seemed worthwhile to examine the relations of peer experience to school achievement.

development takes place in all of these modes of integration; further, that their importance may vary in the course of the socialization process; and that practice of each mode prepares for relationship systems in the adult world, where friendship, partnership, and comradeship exist side by side without one preceding or following the other.

c. Quantitative analysis of differences in school achievement relative to experience in peer relationships

The statistical analysis of differences in school achievement manifested by groups of children with different experiences in peer relationships is based on data from one classroom. This class was attended by 34 children in grade 4 (first period of data collection), and 30 children in grade 6 (second period of data collection). Twenty-seven of the children belonged to the classroom both in grade 4 and grade 6. For various reasons (missing data, special situation of foreign workers' children) not all of the children could be included in all comparisons. School achievement was measured for fourth-graders by the average mark based on three subjects (native language, mathematics, science) on report cards and for sixth-graders by the average mark based on four subjects (native language, mathematics, English, biology) on report cards. The following hypotheses were statistically significant according to t-tests of the differences between means:

(i) That half of the children in grade 4 with closer, more frequent social relationships (n = 16) obtained better average marks than the other half (n = 16) whose peer relationships were looser and more sporadically realized (t = $-2 \cdot 429$; p < $\cdot 02$).

(ii) That half of the children with closer relationships in grades 4 and 6 (n = 13) obtained better average marks on their report cards at the end of grade 6 than the other half (n = 13) whose relationships in grades 4 and 6 were less close (t = $-1 \cdot 836$; p < $\cdot 05$).

(iii) Children who were predominantly involved in consistent or both consistent and changing relationships within the two year period between grade 4 and 6 (n = 14), tended to obtain better average marks on their report cards at the end of grade 6 than children who had almost no close relationships or predominantly changing relationships only (n = 12; t = $-1 \cdot 495$; p < $\cdot 10$).

(iv) That half of the children in grade 4 who were more closely affiliated with and held a more influential position in a group or an association (n = 16), obtained better average marks than the half of the children who belonged to no group or association, or who were in an outsider position only (n = 16; t = $4 \cdot 635$; p < $\cdot 01$).

(v) Those fourth-graders who were integrated in relational networks in the mode of friendship, partnership, or comradeship (n = 18), obtained better average marks than those children assigned to integration types 'rambler' or 'isolate' (n = 14; t = $-3 \cdot 785$; p < $\cdot 01$).

(vi) Those sixth-graders whose relational integration was in the mode of friendship, partnership, or comradeship (n = 20), obtained better average marks than those assigned to the integration types 'rambler' or 'isolate' (n = 10; t = $-2 \cdot 337$; p < $\cdot 05$).

Due to the small size of our sample these results can claim only exploratory significance. Nevertheless, all of these statistically supported statements reflect a single clear-cut tendency: social experiences in peer
relationships—whether defined by closeness, intensity, or type of integration in peer relationships, or by a solid and influential affiliation with a
group or an association—do correlate with school achievement.

The numbers of children in subgroups are too small to conduct complex
statistical analyses. However, individual tests were computed to test relations between experiences in social relationships and social class background as well as family structure. The results suggested that middle class
children maintained closer relationships, more frequently experienced consistent or both consistent and changing relationships, held firmer and more
influential positions in a group or association, and were more frequently
integrated in relational networks in the mode of friendship, partnership, or
comradeship. None of these differences, however, were statistically significant at an acceptable level.

Children who did not live with both their physical father and mother
showed a tendency to participate in looser relationships or to be in a marginal position in social formations. Again these relations are not statistically significant by conventional criteria.

Finally, the relation between school achievement and social class background as well as family structure was investigated. The grades of the few
middle-class children in our sample (n = 8) were better on the average than
those of the lower-class children (n = 24). However, these differences were
not statistically significant. The same holds true for the relation between
fourth-graders' better report cards and living with both parents (n = 17 vs.
n = 15).

In summary, we can state that only the different measures for participation in the social world of peers yielded consistent and statistically significant relations with school achievement. In spite of the caution warranted
by results obtained from only one classroom, these relations indicate that
serious consideration must be given to the varieties of experiences in peer
relationships as an important factor in the socialization process.

3 Discussion of causal directionality

a. Ambiguity of the data

Our sample did not permit us to check statistically whether it is experience
in social relationships that influences school achievement, or whether
achievement has an effect on children's position in peer relationships, or
whether both have a common cause. A single school class is simply too
small in size for complex statistical analyses. Moreover, the lack of school-

independent cognitive tests and the relative homogeneity of the children's social background would have limited the possibilities of analysis. However, our approach did permit, with the aid of the comprehensive information available about the children's social life, to estimate the plausibility of certain assumptions concerning the direction of causation.

The interview data enabled us to determine whether children were excluded from relationships because their poor achievement had given them a bad reputation. The sociometric choices of the 4th graders indeed showed that most of the boys who were most often rejected belonged to the lower third of the class in achievement. However, several other boys with poor marks received no negative votes, and none of the poorly achieving girls was rejected to any notable degree. A number of verbal statements about the rejected boys clearly indicated that underachievement was not the reason for their classmates' negative votes. All complaints involved behaviour in social relationships (e.g. rule-breaking, unprovoked physical attacks, broken promises, cheating at games, etc.).

Thus, social exclusion does not appear to be a consequence of poor achievement in school. Neither the children's statements nor our observations suggest that good achievement automatically warranted intensive social relationships. The good students in the class were indeed respected because of their marks. But children maintaining close relationships with them tended to speak critically about their success in school, accusing them of bragging or being too competitive, and emphasized other positive social qualities of their well-achieving friends rather than achievement.

Moreover, the data provide some indication that the good students' achievement might even have had an adverse effect on their social relationships. Both 'groups' among the social formations we observed in the class consisted primarily of high achievers, and they both broke up between 4th and 6th grade. In the case of the girls' group, which was particularly school-oriented, the subject of achievement, a source of continued competition, may even have contributed to the group's dissolution. As far as the observed class is concerned, one might go so far as to maintain that the high achievers managed to maintain good social relationships, not as a consequence, but in spite of their performance.

Another measure of the link between school achievement and social relationships was the children's statements concerning the personality characteristics they valued most in their best friends. These statements represent the obverse of the criteria by which they had justified their rejection of others. Easy-going, generous, understanding, helpful, and not arrogant were the positive attributes mentioned in various combinations. Behavioural characteristics related to school achievement were hardly mentioned at all.

Hence the data analyses reported above allow us to conclude that in the elementary school class studied, school achievement did not directly or consciously influence the social relationships the children maintained. It may very well be, however, that children are unconsciously influenced by the achievement of others when forming their close relationships. A comparison of the differences in marks obtained by friends and the average distance between the marks of all members of the class revealed that most of the children tended to be more similar in achievement to their friends than randomly selected dyads in the classroom. These figures seem to indicate that school achievement did, after all, exert a certain influence on the formation of close relationships and groups in the class observed.

An influence of this kind has been frequently postulated. Specht (1982) concluded from his data on 9th and 10th grade students that evaluation of school achievement was an important factor in the formation of friendships and in the marginalization of poor achievers. Apparently the correlation between school achievement and sympathy votes increases with the emphasis a particular school places on achievement. In an elementary school class however, the criterion of achievement does not appear to play such a significant role. Still, influences from that direction cannot be completely discounted, although the children did not explicitly refer to this criterion in relationship formation.

b. Approaches to further clarification

Does an unconscious orientation explain the observed relations between peer experience and school achievement after all? The gist of this explanation would be that, without being aware of it, highly achieving children are attracted to each other and offer each other opportunities of intense social life in close relationships and stable social formations, while low achievers do not like to deal with each other and, therefore, do not share in satisfactory peer experience.

It is significant that children aged ten to twelve deny that they are influenced by the marks of their classmates. Therefore, we suggest that the relation between peer experience and school achievement and the greater similarity of friends' marks than random pairs can be better understood by reference to a third determinant. The two results can be combined if we assume that in their dealings with others, children acquire behavioural skills which, on the one hand, enable them to maintain intensive relationships and, on the other, help them to cope better with cognitive problems. Since socially competent children can enter more satisfying mutual relationships, they can be expected to relate more to one another than to less competent children. We can expect, as a side-effect, that all these children would be high achievers as well. Yet the children would be quite correct in stating that marks are no criterion of selecting a friend.

We would go far towards an explanation if we could shed light on the underlying processes that may be responsible both for the formation of intense social relationships and for higher cognitive achievement. Since the analysis of the data is not complete, we must limit ourselves here to the discussion of two approaches to explain better the relations observed. The first approach focusses on children's negotiating strategies; the second on strategies of identity maintenance.

Negotiating strategies. This approach is based on the proposition that children will profit cognitively by being confronted with divergent views in the course of an interaction (Doise 1978). It is important, however, to distinguish whether children merely comply with a prig or indeed deal with proposals and objections; and to examine which strategies they adopt to come to terms with different ideas and intentions. Children who have acquired types of strategies that enable them to cope with diverging perspectives, will presumably be in a better position to make friends and maintain relationships, and also to work out decentered cognitive solutions. At present, we are involved in a detailed study of children's negotiation processes, and intend to test whether the presence or absence of successful strategies correlate with children's positions in social relationships and with their achievement in school.

Identity maintenance. This approach is predicated on the assumption that secure children will find it easier to relate to others, and will also tend to face the risks involved in seeking cognitive solutions to problems (Krappmann 1982). By contrast, children who are unable to maintain a stable identity will probably both jeopardize their social relationships and experience difficulties in cognitive development. To test these hypotheses, we are investigating whether children who have worked out successful strategies for protecting their identities even in critical situations tend to maintain intensive social relationships and reach good cognitive achievements.

Our material contains impressive examples of the consequences engendered by presence or absence of both negotiating strategies and strategies of identity maintenance. There are children whose behaviour better fits the first and others whose behaviour better fits the second approach. So far, it is not possible to foresee which of the two approaches will better explain the interlocking phenomena of social relationships and cognitive problem solving. However, there is no denying the fact that there are other children who combine intense involvement in social relationships with poor achievement or vice versa and, therefore, do not fit any of the patterns discussed so far.

Due to the lack of school-independent tests, we do not know whether the low marks of some socially well integrated children represent cognitive deficits or whether school performance was inhibited by other factors. The common assertion that broken homes lead to lowered school performance of children was not confirmed by our data, although this may explain single cases. Another effect appears to be more salient: an increasing differentiation of the children's life orientations which lead children to differing concerns and activities. These preferred activities, to varying degrees, elicit cognitive strategies for coping with problems.

The few children who criticized their classmates' poor achievements tended to have high achievement aspirations themselves. Possibly, these represent children who, as noted above, in later school years tend to select their friends largely on the basis of achievement criteria. Yet they conveyed the impression that they did not judge others so much by the criterion of achievement *per se* as by evaluating others according to subcultural affinities that take on increasing importance with age. They prefer children with which they can share topics of talk and reading as well as special hobbies. Other children more and more choose activities resembling the focal concerns of working-class youth emphasized by Miller (1958): 'trouble, toughness, smartness, excitement, fate, and autonomy'. Such focal concerns are definitely attractive; children may very well choose them, not out of disappointment at being rejected by their better achieving classmates, but for their own sake.

In the class observed there were children who chose not to continue relationships with high achieving children, but instead turned to more exciting social formations which increasingly disregarded school problems. The school achievement of these children suffered in the process. Had we been able to apply school-independent cognitive tests, we could have determined whether these children regressed cognitively; however, we think that this was not the case. More likely, their capacities were not used or stopped to develop because, as their relationships changed, their interests shifted as well.

Here also a correlation is apparent between relationship experiences and school achievement. But the explanatory model has changed. While cognitive competencies emerge from the operations of basic social processes, performance-related factors in the children's relationships decide whether cognitive competencies are used at all. Certainly a more differentiated analysis of cognitive achievements would be helpful. Possibly these children achieved poorly in terms of global achievement measures precisely because their interests had turned elsewhere, beyond school. These children may nevertheless do well in tests of specific problem-solving abilities that reflect the cognitive content of their focal concerns.

The opposite case, social marginality in conjunction with high achievement, appeared in our data as well. A tentative assessment of the interviews with these two children suggests that they had a highly developed conception of friendship (stage 3, according to Selman 1981) and were apparently looking for a partner whom the class could not provide. It is likely that the more highly developed a concept, the more advanced the requirements of a relationship, the more difficult it is to realize it. In the cases both of children no longer using their competencies and of children not establishing intense relationships we find sociocultural and ecological constraints whch produce conditions of performance. These conditions can conceal existing relations between intense peer experience and developing cognitive capacities.

c. Concluding remarks

In spite of the exploratory nature of the study, the quantitative analysis and the interpretation of children's social behaviour support the hypothesis that intense social relationships of same-aged children generate experiences that influence cognitive development. These effects appear to depend on the specific character of social relationships as a framework of meaning and concern. Depending on their quality, these relationships both define tasks for the individuals involved and ways to deal with them. Thus, they favour the emergence and employment of social and cognitive strategies that come to be elements of lasting competencies.

These fundamental processes can be modified by motivational and situational factors. This is particularly evident when we investigate school achievement. It may either represent performances that are adequate to the cognitive competencies of the child or influences of factors that conceal or even distort the cognitive-developmental base. Social relationships are influential not only on the level of emerging developmental competencies, but as conditions of performance as well.

On the basis of the present analysis we do not expect one single model of causation to be valid. We may end up with several models that combine and weight factors in different ways which are plausible each for a number of cases. In-depth studies of children who do not fit conventional expectations may help to elucidate constitutive processes and influences on the performance of social and cognitive capacities.

Acknowledgements

I would like to thank Wolfgang Edelstein, Hans Oswald, and Maria von Salisch for their helpful comments on an earlier draft of this paper. The paper was translated by John Gabriel.

Discussion

Other possible influences: similar academic achievement and asymmetry

Given that social relationships affect cognitive development, the question of whether some sorts of relationship are more effective than others becomes crucial (see for example Chapters 5, 6, 8, and 22). Rutter raised the further question of whether the nature of the partner, rather than the quality of the relationship, might not be important.

Rutter commented:

'As I understand it, your view is that it is the *intensity* of social relationships that is most important as an influence on cognitive development. But your data show that children tend to choose as friends those of similar academic achievement. Is it not possible that the crucial feature is not the intensity of relationships but rather the characteristics of the friends chosen? In other words, may it not be that friends of high achievement foster cognitive growth whereas friends with low intellectual abilities inhibit such growth? The peer group influence may operate through the quality of conversational interchanges, through attitudes to education, through the content of shared experiences, or through effects on self-esteem. However, with the possible exception of the last mechanism, it is likely that the characteristics of the peers will be influential. In short, do your data show a greater correlation with academic achievement for the intensity of relationships or for the level of achievements shown by a child's friends?'

Krappmann replied:

'On the basis of the data now available I am not able to compare the correlation of academic achievement with intensity of participation in peer relationships with the other correlation of academic achievement with the level of achievement demonstrated by a child's friends. The relative power of these correlations is interesting. However, it will not put an end to the discussion about the direction of causality. Data from our follow-up study indicate that the influence through attitudes and preferences of friends may grow with age (and is more clearly observable for girls than for boys). Through which mechanisms are these attitudes and preferences transmitted? One answer, in line with our search for the contribution of relationships, could be the following: attitudes are transmitted in the course of interactions, which are shaped by the kind of relationships between the interacting partners. In my paper I outlined another explanation for the existence of both correlations. Our observations suggest that basic interaction strategies help to make friends and also promote cognitive development. We are well aware that established relationships and advanced cognitive capacities react upon each other and upon basic interaction strategies. We assume that the main factor in the network of interlocking influences changes in the course of socialization. The data from this exploratory study encourage us to look for more evidence supporting the hypothesis that, especially in middle childhood, the structure of peer relationships has far-reaching implications for the development of interactional and cognitive competencies.'

Children's reasons for choosing friends

Rutter asked:

'In your paper you discount the saliency of academic achievement in children's choice of friends because children do not report that characteristic as a reason for choosing or not choosing friends. But, is that valid? Children also did not mention age and sex as reasons for their choices, but presumably there was a strong tendency for them to choose those of similar age and the same sex.'

Krappmann replied:

'When explicitly asked, children did mention age and sex as reasons for friendship choices, but denied selecting friends by the criterion of school achievement. When asked what characteristics of their friends they like or reject, appreciation and criticisms of school achievement were articulated, but very seldom, and rather independently of friendship choices.'

Hartup asked:

'Why did you combine Groups A, B, and C for contrast with D and E? It seems to me that this contrast is essentially between "socially involved" children and "less involved" children, thus suggesting that the correlation with achievement may reflect group differences in terms of general competence.'

Krappmann replied:

'Due to the small size of our sample and for statistical procedures, we had to combine groups of children which had earlier been differentiated. Nevertheless, I computed some statistical analyses using the differentiated categories and these also produced results in the same direction. I shall repeat these analyses when more data are available.

Our categories do not only contrast children with high social involvement with those less involved. Children assigned to Mode D ('rambling'), Type II ('changers') or to the social formation 'interaction field' are not without intense social experiences. We even suspect that sociometric investigations would erroneously label these whirling children as popular. We believe, however, that their 'intense' kind of involvement does not produce incentives which advance their interactional and cognitive competencies. Our qualitative analyses are underway to identify which specific behavioural strategies children from different categories of peer experience utilize.'

Section D

12

Emotion, cognition, and social interactions
EDITORIAL

This section emphasizes that the reciprocal influences of cognition on social interactions and of interactions on cognitive functioning cannot be considered in isolation from other aspects of the individual. Especially important here are the emotions. As implied in nearly every chapter in this volume, the application of social skills in the management of interpersonal relationships involves an awareness of the other person's (as well as one's own) emotions, and the avoidance or resolution of conflicts. Yet the emotions have been neglected by most workers on cognitive development. Even Piaget (see p. 175) gave little consideration to specific emotions, though he regarded affect (and especially 'interest') as an energizer for cognitive activity.

This is a context in which we must recognize that the psychologist's concepts, though necessary for his analysis, do not necessarily fit nature precisely. That there is an heuristic value in distinguishing cognition and emotion cannot be doubted, yet at the same time emotion cannot be considered independently of cognition, or cognition of emotion (e.g. Candland *et al.* 1977; Lewis *et al.* 1984). The Darwinian view that emotional behaviour involved the mere 'expression' of internal states neglects the role of cognitive processes in evaluating the situation and monitoring the response in most emotional behaviour, and the role of the emotions in instigating and directing cognitive processes. Indeed it has been suggested that emotional behaviour involves a continuum from that which is primarily expressive to that which involves negotiation between individuals, the latter having a strong cognitive component. For instance the cry of the newborn is presumably primarily expressive, while the ingratiating smile involves negotiation (Hinde 1985).

The interplay between cognitive and emotional components is in turn related to the behaviour shown in interactions and relationships:

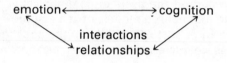

A classic example of such an interplay is the affective aspect of the attachment relationship of a child to a mother-figure. 'No form of behaviour is accompanied by stronger feeling than is attachment behaviour. The figures towards whom it is directed are loved and their advent is greeted with joy. So long as a child is in the unchallenged presence of a principal attachment-figure, or within easy reach, he feels secure. A threat of loss creates anxiety, and actual loss, sorrow: both, moreover, are likely to arouse anger' (Bowlby 1982, p. 209).

Whilst the demonstration of capacities such as that for object permanence are now seen as being more closely dependent on the social situation than was formerly the case (pp. 246–7, 315–19), there is an heuristic value in the attachment theorists' attempts to draw parallels between the stages in the development of the attachment relationship and Piagetian levels of cognitive development (see Ainsworth *et al.* 1978; Bowlby 1982).

1. The initial *pre-attachment phase* lasts from birth to a few weeks. Here Piaget's 'reflex schemata' may be related to the infant's behaviour in signalling without directing the behaviour to a particular person.

2. The phase of *attachment-in-the-making* relates to Piaget's second and third stages of sensori-motor development, when signals become directed to particular persons.

3. The phase of *clear-cut attachment*, from six months, coincides with Piaget's fourth stage of sensori-motor development and is characterized by locomotion and goal-corrected behaviour. There is evidence for a relation between attachment and object permanence (Bell 1970) and between attachment and means-end development (Bates *et al.* 1977). Intentionality also begins in Piaget's fourth stage and Frye (1980) suggests that at this age attachment behaviour must be regarded as intentional. Like Bowlby, he stresses the infant's *goal* of seeking proximity under stress and the immediate *function* of proximity-seeking behaviour. This emphasis on a goal rather than the means to that goal accommodates the lack of correlations involving specific behaviour patterns across time (Sroufe and Waters 1977) and the way in which 'old' patterns (e.g. crying, smiling) come to be used in 'new' ways (Bell and Ainsworth 1972). Defining intention as 'doing something in order to brings about something else' (1980, p. 316), Frye views 'mutual intentionality' as the 'central feature of social relations' (p. 328), including attachment relationships.

4. The *phase of a goal-corrected partnership* is evident by four years of age. In Piagetian terms, egocentrism has declined. Four-year-olds are 'less egocentric, more capable of perspective taking, and more able to sustain a relationship on the basis of communicative skills, sharing of mutual plans, and internalized models of self and mother and their relationship' (Ainsworth *et al.* 1978, p. 205). In the attachment relationship, their behaviour is characterized by communication across a distance

rather than by maintenance of physical proximity, and during brief sep-
arations from mother there is seldom separation distress. Direct support
of links between socio-cognitive development and attachment behaviour
comes from the study of two, three, and four-year olds by Marvin (1972,
1977). He found that behaviour in specific tasks (e.g. perspective-taking
and delay of gratification) was indeed related to behaviour indicating
security of attachment in the strange situation.

Furthermore, one may postulate that with the formation of object permanence comes the beginnings of an 'internal working model' of a relationship, which in turn influences future interactions within that
relationship. This working model is not just an objective picture of the
parent, but rather of the entire parent/child relationship, including interactions, intentions, and outcomes. Main *et al.* (1985) outlines the characteristics of such a working model and then uses this cognitive concept to
tie together a variety of results within the attachment literature. For
example, Main found a number of adults who, despite unfavourable
childhood experiences, were judged currently to have *positive* models of
those early relationships. She suggests that whereas early working
models may require concrete experiences for any alteration, with the
onset of the stage of formal operations, this may no longer be necessary.
By then, the individual has the ability to step outside a system and see it
operating, to imagine alternative models, and to modify models by drawing upon other experiences.

In the first chapter in this section, Radke-Yarrow and Sherman present
a variety of theoretical views on cognition, emotion, and their interplay.
Results are then presented which demonstrate that emotional
expressions of others are salient stimuli, even for very young children.
What develops is the child's competence, both cognitive (understanding
and responding appropriately) and emotional (expressing, experiencing,
and regulating emotions). Such competence may be improved by adult
modelling and explanation, coupled with affective involvement.

Whilst Yarrow and Sherman are concerned with situations in which
another person is distressed but the child (initially) is not, in the second
chapter Shure focusses on conflicts of interest between the child and
another, as when both want the same toy. In the former case there is no
real problem to be solved, and modelling may be sufficient to increase
prosocial behaviour. Where there is a problem to be solved, Shure
argues that an adult should not provide a model or a solution for the
child, but rather should encourage the child to think for him or herself
about the nature of the problem, possible solutions of it, and their consequences. Shure and Spivach have developed a training programme for
adults (mothers and teachers) which is successful in teaching Interpersonal Cognitive Problem Solving (ICPS) skills to children.

Although the focus of Shure's chapter is on cognitive problem-solving,
the complexity of the emotion-cognition interaction here should perhaps

be stressed. Thus the adult's recognition of the child's cognitive performance involves interpretation of the child's emotions, and has emotional consequences for the child which may affect future cognitive performance.

The third chapter also deals with conflict, both within the self and with another. Selman presents a developmental model of negotiation strategies and goes on to relate this to a technique which he calls 'Pair Therapy'. Its aim is not to induce purely regulative behaviour but a more reflective control from the patient himself. Reflective control requires some kind of symbolic process that develops with age according to circumstances: humans have the capacity to express power, control, threats, and gratifications symbolically. These communicative competencies, connected with adequate interpersonal perceptions, permit interindividual negotiations. The 'Pair Therapy' is designed to improve the child's ability to develop effective communicative skills and strategies and to form a repertoire of possible alternatives. Such an aim has similarities to Shure's ICPS. As with ICPS, the children have to work out their own solutions. However the therapist's role is a less intrusive one, involving the setting up of stimulating but safe interactional contexts in an 'atmosphere of warmth', and the recognition of the progress of the negotiation that then starts to take place between the peers. This seems to offer to these children an opportunity to become conscious of some of their previously unrecognized social postures. Thus, emotional involvement is part of the treatment, as indeed it was in Yarrow and Sherman's method for increasing prosocial behaviour. Perhaps emotional involvement is also an unspoken feature even of the ICPS training advocated by Shure. We shall see in the following chapter that cognition cannot be fully understood independently from emotions. As Radke-Yarrow and Sherman point out, 'Systems whose function is determined by their interaction can only be understood by studying them in interaction' (p. 174).

13

Interaction of cognition and emotions in development

MARIAN RADKE-YARROW AND TRACY SHERMAN

Full understanding of the nature and origins of human cognitive abilities requires examination also of cognition in interaction with emotions and social behaviour. Although understanding of cognitive processes has benefited from extensive systematic research, the interface of cognition with emotions and social behaviour is territory that has only just begun to be mapped in theory and in empirical data. In this paper we examine how the currently available theories and data address this issue.

The history of research in developmental psychology provides a perspective on current views and knowledge about the relation between cognition and emotion. Research interests in American developmental psychology can be described in epochs. Behaviourism reigned in the 1940s to 1960s. At that time it was heresy to speak of affective processes or, for that matter, of social perception and cognition. The succession by cognitive psychology (beginning in the late 1960s) brought changes—the influence of Piagetian theory, with its compelling developmental point of view, and information-processing theory, less tuned to development. The child whose behaviour had been characterized as passively determined by reinforcement contingencies became the child whose behaviour was indicative of an active developmental process whereby the child was able gradually to make more and more sense of both the object and social worlds. Again, however, there was little interest in affective processes.

In developmental psychology, however, one outgrowth of learning and cognitive research was the burgeoning study of the human infant. Out of this field of research has come an amazing richness of information and ideas, and evidence of the availability, even in the human neonate, of a rich repertoire of skills. Infant research also moved the discipline in some new directions, particularly as regards the study of affective processes. For example, the *quality* of the *relationship* between mother and infant, which could be measured with some specificity, became a strong focus. Here was

Support for this work was provided by the John D. and Catherine T. MacArthur Foundation and the National Institute of Mental Health.

an *affective bond* (hardly a traditional concept of learning or cognition) with undebatable consequences for the infant's and child's psychological development.

Other influences on developmental psychology also brought affective issues into consideration. Adult social psychology introduced questions about children's interpersonal behaviour, social cognition, and moral development. Animal studies dramatized how rearing conditions could affect development. Investigation of maladaptive behaviours, disturbed thought and affect in children reflected psychiatric influences. These broadened questions of behaviour have required reconsideration of development and reexamination of questions and paradigms in cognitive research.

Although these historical trends could have wedded the various influences, this has not been the case. Cognition, social behaviour, and emotion have been investigated, usually, in isolation.

There is reason, then, for some excitement in focusing attention on the interface of the processes underlying cognitive, affective, and social behaviour, considered developmentally. This excitement stems not merely from the novelty of the enterprise but from a deep-seated belief that full understanding of cognition, social behaviour, and emotion can come only from considering them as functioning interactively. In 1970, Reitman warned that the study of 'decoupled' systems would not yield the result psychologists had in mind. Systems whose function is determined by their interaction can only be understood by studying them in interaction. While the problem can be subdivided at times, and main effects studied, the student of statistics knows that no main effect is interpretable until the interactions have been analyzed as well.

We have composed our paper (a) to present, in brief review, theories of cognition and affect that have played a role in developmental theory, and to highlight their positions regarding the interrelation of cognition and affect, and (b) to examine empirical data on young children and ask how the theories aid in interpreting the data and how the data enrich the theories.

Cognitive theories

A number of distinct theories have evolved to explain cognitive development. In American research, the two guiding frameworks stem from the Piagetian and the information-processing approaches.

Piaget's theory of cognitive development emphasizes the child's active involvement in the construction of reality. Motivated by the affect 'interest,' the child engages the world of objects and of people and constructs meaningful perceptual experience. Present at birth are a small set of behavioural reflexes (e.g. sucking, prehension, looking) and the biologically-based tendency to make sense of the environment and adapt to it via the processes of assimilation and accommodation. From these few givens,

according to Piaget, the infant proceeds to build the schemata of the sensori-motor period, which in turn will form the bases for the development of
the schemata of the next stage. The transition from stage to stage is marked
by qualitative reorganization of the child's way of making sense of his world.

The information-processing view of cognitive development is not a single
unified theoretical position, as is the Piagetian. Rather it is a means of describing cognitive development that derives from a metaphor that compares
human cognition to computer processing. It has as its goal the modeling of
human cognitive processes. This emphasis on understanding *cognitive processes* stands in contrast to a more traditional goal of psychology which was
to understand the functional relations between experimental conditions
and outcomes. In fact, the scientist working within an information processing framework is seeking to build theories that 'extend over a whole range
of cognitive tasks and experimental settings and that handle the interaction
of several cognitive mechanisms (e.g. short term memory, attentional
mechanisms) rather than single mechanisms in isolation,' (Simon 1979,
p. 364). What changes with development is:

(i) the set of cognitive processes (e.g. rehearsal strategies, retrieval
strategies) that the child has available to him in order to make inferences or
solve problems;

(ii) the rate at which various processes can be performed; and

(iii) the amount, quality, and organization, and therefore, the usability
of his knowledge.

These changes in turn affect the quality, speed, and likelihood of future
cognitive activity. In contrast to the Piagetian position, in which cognitive
processing undergoes qualitative change, for information-processing theorists knowledge acquisition is gradual and incremental. What appear to be
qualitative shifts are the result of a variety of quantitative changes that in
combination produce the observed behaviour.

For cognitive theorists, the issue of affect and its interaction with cognitive functioning has not been of central concern. What little interest there
is has stemmed from some need to motivate cognitive activity. For Piaget
(1981) affectivity is viewed as an energizing process. Interest, an affect, is
the fuel that motivates all cognitive activity. To the extent that he discussed
affect, he viewed cognition and affect as not dissociable; neither caused the
other, neither preceded the other. 'There is no behaviour pattern, however
intellectual, which does not involve affective factors as motives; but reciprocally, there can be no affective states without the intervention of perceptions or comprehensions which constitute their cognitive structure.
Behaviour is, therefore of a piece even if the structures do not explain its
energetics and if, vice versa, its energetics do not account for its structures'
(Piaget and Inhelder 1969, p. 158). Despite the equal status implied,
Piaget's work on cognitive development, which spanned approximately

half a century, made little reference to affective development or to the reciprocal influences of cognitive and affective processes.

Similarly, information processing theorists have not been particularly concerned with studying the interface between cognition and affect. In fact, as summarized by Simon (1979), the mainstream of information processing theory has not even been concerned with what motivates the human to perform cognitive activity. One important recent contribution to psychological theorizing in this regard comes from the philosopher, Dennett. He has made explicit the line of reasoning that expands the behaviourist's law of effect to explain the evolution of adaptive *cognitive* behaviour. Quite simply the law of effect states that: 'Actions followed by reward are repeated' (Dennett 1978). He argues that it is quite meaningful to postulate the evolution of an *inner environment* which provides feedback for events in the brain, in a manner analogous to the feedback the external environment provides the organism for action in the world.

In summary, other than acknowledging that there must be some motivation for cognitive activity, cognitive psychology has paid scant attention to the influences of affect on cognitive behaviour or development. Whatever has been said about the interface of cognition and affect has come from theorists concerned with understanding and exploring affective behaviour, not cognitive behaviour.

Theories of emotion

Despite a rich history of investigation of emotions and a recent revival of interest, there is no single dominant theory of emotion and emotional development. There are a number of current theories, variously labelled cognitive theories, organizational or systems theories, and somatic theories. The explicitly cognitive theories of emotion view emotion as a function of cognition (Schachter and Singer 1962). Cognition comes first in the flow of information, and only after evaluations are made is there a subjective experience of emotion. One would anticipate that any expansion of this point of view to incorporate development would emphasize the dependence of emotional behaviour and emotional development on development in the cognitive sphere.

The idea that the emotional experience is purely the outcome of cognitive evaluation is rejected by other emotion theorists (e.g. Campos and Barrett 1984; Ekman *et al.* 1972; Hinde 1972; Izard 1984; Tomkins 1962, 1963). Emotion is viewed in terms of biological origins, multiple components, and multiple functions. Izard (1984) stresses genetic and biological factors, defining the existence of emotion by neurochemical and sensory processes, and viewing emotion as a primary motivator of cognition and action. He assumes a one-to-one correspondence between the facial expression of a specific emotion, and the specific feeling or experience of emotion, with the feeling having cue-producing and motivational functions. Cognitive representations, he assumes, can change emotional experience, and can and do interact reciprocally with the systems of emotion and action. Campos and Barrett (1984) summarize and elaborate on the organizational approach that includes in it a description of the affective system. Emotion is a construct organizing neurophysiological processes, facial expressions, feeling or experiencing, action tendencies, appraisal, and coping behaviour. Emotions are elicited by, and are regulators and determinants of, multiple intrapsychic and interpersonal processes, they affect or organize perception, cognition, and behaviour. For organizational theorists then, the affective system plays a very broad integrative role in all aspects of psychological functioning, and is in no way merely subservient to, or dependent on, the output of the cognitive system.

Yet another view has been presented by Zajonc (1980; Zajonc and Markus 1984). He argues that affect should not be treated as unalterably post-cognitive and that the arousal of affect may only minimally—or not at all—depend on cognition. 'Feelings may be aroused at any point in the cognitive process: registration, encoding, retrieval, inference, etc.' (1980 p. 154). We can like or be afraid of something without *knowing* what it is. Zajonc is concerned basically with feelings of the approach-avoidance kind, not the entire range of emotions (guilt, pride, anger, etc.). He buttresses his position that affect is a separate system with phylogenetic arguments as well as ingenious empirical demonstrations. Affect is controlled by the limbic system of the brain, an adaptive system phylogenetically *preceding* the development of the neocortex and the evolution of language and cognitive processes dependent on language. It is a system not highly subject to attentional control, very rapid in responding and relying primarily on nonverbal channels.

Zajonc and Markus (1984) point out that the construct, emotion, has been associated with the arousal of the autonomic and visceral systems, the activity of the motor system, and cognitive experiences of emotion. Nonetheless, they claim, the theories have, in fact, dealt with only one component, the *experience* of emotion (the cognition of having one). Even in the somatic emotion theories, for instance, it is assumed that the kinesthetic feedback of muscular acts to the cognitive system is the basis for the representation of affect and that this processing and encoding is much like other processing. They propose 'a simpler route for affective processes': that the motor system may be also a point of contact between affect and cognition. Instead of regarding all motor processes as secondary, they propose that a representational role of motor processes be considered. 'For if affect is not always transformed into semantic content but is instead often encoded in, for example, visceral, or muscular symbols, we would expect information contained in feelings to be acquired, organized, categorized, represented, and retrieved somewhat differently than information having direct verbal referents' (1980 p. 158). Zajonc is the first to say that additional empirical evidence is needed. However, at the very least, his ideas are a stimulant to the exploration of the nature of cognitive-affective interaction.

Developmental theories of cognition-affect-behaviour relationships

Only recently have the issues of emotional development and the developmentally changing interface of cognition-emotion-social behaviour become the focus of investigation (by Campos and Barrett 1984; Cicchetti and Hesse 1983; Emde 1980; Hoffman 1984; Kagan 1984; Lewis *et al.* 1984; Sroufe 1979, among others). Some of the main dimensions of consideration are summarized here. One issue is the question of the innateness of emotional equipment in the human infant. There is little doubt that specific prewired configurations of facial expressions of 'emotion' appear in the first months. However, what experience is represented in these very early patterns is a matter of debate. Izard (1984), not a developmentalist, takes a clear position: 'the ontogeny of emotions is completed in infancy.' The infant is equipped at birth with a set of basic emotions (sadness, pleasure, fear and rage). This theory attributes to young infants feeling states congruent with discrete facial expressions which are assumed to be 'involuntary' expressions. The theory does not explicate developmentally changing relations of facial emotional expressions and experienced feelings. Others (e.g. Emde 1980) interpret these initial expressions as early subcortical reactions that evolve into psychological reactions.

For theorists who postulate developmental changes in the affective system, a second question is how best to characterize these changes. The current Zeitgeist is to conceive of the affective system as a complexly interacting system made up of multiple components (e.g. neurophysiological, facial expression, feeling state, action tendencies). The theories differ in how many and which components are specified. The theories appear to share the view that the set of components that contribute to the functioning of the affective system does not change with age. Rather it is the functioning of the separate components and therefore concomitant changes in the nature of the interaction between these developing subsystems that determine the course of affective development.

Sroufe (1979) for example, has particularly stressed that changes in the affective functioning of the infant and young child are derivative of changes in the cognitive system. At birth and for some weeks the infant is basically a discharging system able to communicate general distress. Following

Spitz's formulation, this phase merges into rudimentary consciousness, an awareness of 'in here' and 'out there', and the infant 'relates' to the surroundings with coordination of attention, motor activity, and smiling. It is in this period (the first three or four months) that 'true' emotions, pleasure, and disappointment emerge. The succeeding developmental transformation is marked by the infant's manifestation of recall memory and anticipation behaviours and of affectivity (fear and anxiety) with regard to objects not visually present. By the last quarter of the first year, a range of specific emotions is experienced by the infant, affective components are represented in the infant's memory, active affective engagement, and participation in the environment are achieved, and affective bonds to significant social others are formed. In the second year, affect is further organized by increased capacities for internal representation, the development of language, and an awareness of self.

This rapid march through infant development does not do justice, of course, to the research and theory in this recently revitalized field. In summary though, developmentalists concerned with emotions have, like Sroufe, focused primarily on the age at which the average infant reaches certain cognitive competencies, and they have explained, by and large, the development of emotional competencies in stage-like terms, as deriving from earlier-appearing cognitive competencies, the guiding assumption being that 'cognitive factors underlie the unfolding of emotions' (Sroufe 1979, p. 491).

Although certain cognitive abilities may be necessary precursors for the development of particular emotions, so that charting the orderly appearance of cognitive and emotional milestones is an important first step for understanding the nature of the interaction of cognitive and emotional development, this does not complete the theoreticians' task. Recall memory may be a necessary cognitive capacity underlying fear, yet, of two infants who have recall ability, only one may show stranger fear or separation anxiety. Why? An exclusive reliance on a normative approach will not give answers to such questions.

Changing the emphasis

In order to begin to answer such questions, we suggest a shift of focus to the cognitive and affective *performance* of young children, i.e. their *modal* behaviours, as well as their developmentally expected competencies, i.e. their highest levels of performance. To this end, we shall consider empirical data on young children—cognitive and affective data—which raise specific issues with which theories must deal. By looking at typical performance as well as optimal performance, we will attempt to derive some further understanding of affect-cognition interrelations in early development. We will present a series of studies that permit a close look at performance and competence in young children.

Our research focuses on (a) how young children respond to social-emotional stimuli, i.e. emotional expressions of others, and (b) how young children express and regulate their own emotions. We shall draw on data from a number of studies which allow us to look at the interface of cognition, affect, and social behaviour. The children studied ranged in age from 10 months to 8 years. In these studies we have taken advantage of both experimental and naturalistic methods.

Developmental transformations in responding to social-emotional stimuli

A naturalistic approach was used in the first study to observe developmental transformations in children's cognitive, emotional, and action responses to emotions in others. Our objectives were (a) to obtain an inventory of young children's response repertoires upon encountering social-emotional stimuli in naturally occurring contexts, and (b) to look for a possible developmental time-table for these responses (Radke-Yarrow and Zahn-Waxler 1984; Zahn-Waxler and Radke-Yarrow 1982). To achieve these objectives, a high density of data on each child was needed, over time. The impossibility of repeatedly injecting emotion-elicitors into a child's environment immediately ruled out laboratory experiments as primary data sources. Methods were developed which permitted access to the child's reactions to the range of emotions (anger, joy, pain, sadness, love) inherent in the daily experiences of children. Mothers were trained as research assistants, observers, and reporters. Their task was to record their child's responses (or absence of response) to naturally occurring emotion events. Following a prescribed format, mothers dictated as soon after the event as possible a description of the event, the emotional expression, by whom, in what context, the nature and sequence of the child's attention, and overt reactions (motor expression, language, social behaviour), and the responses of others to the child. To ensure certain comparable stimulus situations for all of the children, mothers were also coached in the performance of simulations of specified events (e.g. accidentally hurting themselves, being sad) to be performed at designated times. Some children were studied in cross-sectional designs, others in a longitudinal design. To satisfy ourselves of the veridicality and reliability of the data, a number of verifying measures were taken (See Zahn-Waxler and Radke-Yarrow 1982).

For the purposes of this paper we will emphasize the normative data, the order of appearance of responses to emotional distress expressed by others. A summary is presented in Table 13.1. The ages represent the times at which all or almost all of the children had demonstrated that the given kind of response was in their repertoires. Thus, for example, at the time the study began (when the children were 10 months of age), all of the children showed awareness of the social-emotional event by attending to it. Generally they were slightly older before disorganizing distress was manifested by their facial expressions, cries, body freezing or stiffening: this form of behaviour waned in subsequent months. At the time they themselves were showing what appeared to be distress, the children began to look to their mother for information, and at about the same time moved *toward* the distressed person (touching, stroking, etc.). Somewhat later the children imitated the expressions of the distressed person (although, some investigators, e.g. Meltzoff and Moore 1977; Field *et al.* 1982, have reported imitation of facial expression even in neonates). By $1\frac{1}{2}$ years children could follow through on their processing of distress with a meaningful (though sometimes incorrect) positive intervention (giving, helping, comforting, etc.). A little later they were able to test hypotheses about their interventions; if one attempt on their part did not seem to help the other person, they tried another, and another.

Table 13.1 *Developmental characteristics of children's responses to emotional distress in others**

Responses to other's emotions	Approximate Month–age of appearance
Child as bystander	
Fixed attention	10–11
Disorganized distress (seen in face, cries, freezing, stiffening)	12–14
Seeking mother—visual reference or contact	12–14
Tentative positive approach to person in distress (touching, patting)	12–16
Imitation of other's expression	14–16
Inspection of other's distress	18–22
Affectively positive functional approach	18–22
Labelling of affect	18–24
Affectively positive mastery attempts (trying alternatives to repair/console other)	18–24
Child as cause of other's distress	
Reparation attempts	18–24

*Adapted from data published in Radke-Yarrow and Zahn-Waxler 1984

The developmental transformations in one child's record, shown in Table 13.2, provide an example of the data upon which Table 13.1 is based. The data show that affective stimulation in the environment is very arousing and very salient even for young children, and in addition, indicate a predictable progression in children's competencies in responding to other's affect. We will use these findings as a developmental standard of competencies as we try to interpret data from our other studies.

Table 13.2 *Developmental transformations in a child's response to distress emotions in others**

Response	Age in months
Child watches crying baby intently; begins to cry; looks to mother.	12
Child sees her mother who is reacting in pain having scalded her hand. Child comes to mother; hugs her mother; then nestles with her mother.	15
Mother accidently bites her cheek and winces. Child's face is 'an exact mirror of the pain'; child then touches mother.	16
Child observes a crying baby. She approaches the baby, strokes his head. His crying continues. Child concentratedly hugs and pats him; then tries giving him toys; then brings her own mother to the baby.	23

*From individual data files, from research reported in Radke-Yarrow and Zahn-Waxler 1984

Studies of interaction of affect and cognition

The common conclusion illustrated in the following studies is that the child's behaviour cannot be predicted simply from knowing either his/her competence to understand a particular situation or competence in expressing, experiencing, or regulating his/her emotions. It is only when we consider the interaction of cognition and social-emotional factors that we can predict and understand behaviour.

Developmental competence and variable performance

In this analysis, we are interested in why young children ($1\frac{1}{2}$–$2\frac{1}{2}$ year olds), all of whom have demonstrated their competence to recognize and respond positively to someone in emotional distress, show vast differences in their inclination to respond when social-emotional events present themselves. Some respond in a positive prosocial way in 5 per cent of the events, some in 70 per cent. It was hypothesized that mothers' behaviours in certain critical situations of emotional distress were significant factors contributing to this variability (Zahn-Waxler *et al.* 1979). Of particular importance might be the mothers' reactions to situations in which their children caused distress in others. Mothers' descriptions of their own behaviour in such situations were examined. Mothers could be categorized as follows:

(i) Mothers who treated children's transgressions by issuing an affectively expressed prohibition ('Stop that!'), without further explanation;

(ii) Mothers who prohibited and also provided explanations of the consequences of their child's behaviour for the victim, in a neutral manner
('Tommy is crying because you bit him and made his cheek hurt'); and

(iii) Mothers who prohibited and offered similar explanations but
embellished their statements with intensity of feeling ('it must *never, never*
happen again').

Children's variable rates of responding both in situations in which they had
caused the other child's distress and in which they merely observed
another's sadness were related to their mother's behaviour.

Specific knowledge provided by the mother's explanation is crucial for
children of this age to learn the basic principles of prosocial behaviour. The
mother's statement to the child, in situations in which the child was the cause
of the 'hurting', influence the child's behaviour not only in that type of situation but also in situations involving merely the observation of 'hurt', i.e. the
bystander situation. Children who were told only, 'No' but received no
explanation, as compared to children who had received explanation, showed
fewer acts of reparation toward the person they had hurt and less frequently
behaved altruistically when they were bystanders to another's distress. However, the mother's offering of information is not the total explanation. The
mothers who neutrally communicated explanations had children who did not
differ in their behaviour from the average of the group of children: rather it
was the children who received explanations embellished with the mother's
emotion who showed high levels of reparation and altruism. The child's prosocial behaviour is best predicted by the *interaction* of the information content of the cognitive message and the affective involvement of the mother.

Cognitive competence and social-emotional behaviour

In another study (Barrett and Radke-Yarrow 1977) we were interested in the
relation between a specific cognitive ability (social inferencing) and children's altruistic acts. Five- to eight-year-old children interacting with peers
on the playground were observed over a 6 week period. Inferencing ability
was measured by the child's ability to interpret a series of videotaped social
episodes involving interactions between child and adult, adult and adult, or
child and child, and to explain the motivations which led to a given person's change in behaviour or mood. Interestingly, children's scores on the
social inferencing measure were not correlated with chronological age, $[r
(76) = 0.16]$. Moreover, for each age group and for the total group of chil-
dren, social inferencing ability was not correlated with the child's prosocial
responses to peers in distress; $[r (33) = 0.02$ for boys, $r (36) = 0.06$ for
girls]. This total lack of relation between altruism and what has been

hypothesized to be its cognitive base was explicated when a variable of temperament or sociability was considered. Specifically, when the children with high inferential ability were also high in assertive sociability (i.e. qualities of leading and directing), they were more altruistic than average, but when the children had high inferential ability and low assertiveness they were less altruistic than average. Level of assertiveness made no difference to the behaviour of children with low inferential ability. Here again cognitive competence plays a role in determining the likelihood of the child's behaving altruistically, but in and of itself, cognitive competence is insufficient to explain or predict interpersonal behaviour. Rather it is the interaction of a specific cognitive capacity (social inferential skill) and a general interpersonal behavioural style (assertiveness) that in useful in predicting a specific type of behaviour (prosocial behaviour).

The affective environment, learning, and child performance

The interface that we have been exploring was investigated in another way, by experimentally varying the affective surroundings of the child (Yarrow et al. 1973). Adult caregiving characterized as warm, nurturant affect was compared with watchful but aloof and detached caregiving. Children's performance under the contrasting conditions was investigated. Two nursery school teachers were trained to carry out both roles with different groups of children. In the course of their regular nursery school morning, small groups of children spent half-hour sessions, over a two-week period, with one or the other type of experimental teacher. In the third week the teachers embarked on identical training programs, again over a two-week period. Using modelling, reinforcement, and labelling, their objectives were to teach the children awareness of and concern for the welfare of others. Attractive toy-size dioramas depicted a diversity of ordinary situations in which need or distress was quite obvious. For example, a child has fallen off a bike; a monkey strains to reach a banana outside its cage. The teacher discussed the dioramas with each child. She verbalized the cues to which she was reacting, the inferences she was making about the victim ('I think he's hungry', 'He's trying so hard to get the banana') and her own motivation and affect in aiding the victim ('I want him to feel better', 'I'll help him so he feels better', 'There, I feel good about that'). For some children pictures were also used in a similar training format. Some groups, in addition, were exposed to scripted events in which distress incidents were enacted by an adult confederate in the course of the session. The caregiver again modelled altruism in both precepts and behaviour.

The children's learning was tested in each training medium two days and two weeks later. Regardless of the affective qualities of the environment, training with dioramas and pictures resulted in enhancement of children's altruistic behaviour expressed in dioramas or pictures. This learning of altruism at a symbolic level generalized to new and different distresses within a medium but did not generalize to a new medium. In other words, if children were trained only with dioramas, they performed well only with dioramas. In contrast to the results of training in the two other media, the results of training in the live situation were more complex. Only those children who had *both* specific training with real persons and had been trained by a nurturant caregiver showed an increased likelihood of performing altruistic acts. In this study, then, the general finding of the two earlier studies was replicated: the child's knowledge of appropriate behaviour is necessary but not sufficient for determining whether the child will perform. Success in training children to act prosocially in novel, live situations was achieved only when a nurturant environment was established by the adult.

Now looking back over the data sets that we have presented, two seemingly disparate conclusions have been drawn. First, we were able with reasonable accuracy and inclusiveness to chart a normal developmental progression in children's reactions to emotional stimuli. The majority of children reached these behavioural milestones at approximately the same ages. Second, we have illustrated that these competencies are not predictive; that behaviour is complexly determined by the interaction of cognitive, social, and emotional factors.

We suggest that for the majority of children from normal samples, early development is an interaction of maturational factors and environmental factors, both of which differ within a relatively narrow band from family to family. The infant and young child are well cared for, a secure attachment relationship is formed between mother and child, and the child is exposed to a full range of human emotions. These environmental conditions allow the unfolding of what appear to be 'universal' patterns, and in some regards obscure the multiple determinants of these developmental achievements.

In order to highlight the complexity of the interactions among cognitive, emotional, and social development, we will now turn to research from our laboratory in which we have examined development in children whose hereditary endowment may distinguish them from the normal sample and whose childrearing environment most certainly is outside the narrow band of 'the average expected environment' (Hartmann 1964). These children are from families in which the mother or both parents suffer from a major affective disorder and at least one of the parents was diagnosed as manic-depressive. Translated into research variables, these are families in which the environment of the child is chronically affectively disturbed. The child is exposed to episodes of sad affect, blunted affect, irritability, apathy, hopelessness, and self-preoccupation by the unipolar depressed mother and, in the case of manic depression and mania, there are additional episodes characterized by exuberance, grandiosity, and uncontrolled energy. These data will contribute to our understanding of how intricately dependent on each other are the cognitive, emotional, and social behavioural systems, for determining both behaviour at a given moment in time and the course of development.

Affect, cognition and social behaviour in children of depressed parents

Children of depressed parents and normal controls were studied. One sample was observed at 12, 18, and 24 months; a second sample, at $2\frac{1}{2}$ years (Gaensbauer *et al.* 1984; Zahn-Waxler *et al.* 1984a; Zahn-Waxler *et al.* 1986; Zahn-Waxler *et al.* 1984b).

The longitudinal sample was seen in the laboratory at 12 and 18 months, and in the home at 12 and 24 months. The laboratory sessions consisted of mother and child play, a modified strange situation (Ainsworth 1982), a portion of the Bayley Scales of Infant Development, and some items intended to be mildly frustrating.

At 12 months, children from disturbed backgrounds and children from normal families were quite similar to one another in their behaviour, and in the quality of the mother-child attachment. Their performance on the Bayley test was appropriate to their age. Children of the ill parents showed divergence from the children of normal parents in some affective reactions, namely, they expressed more fear and sadness in play with their mother and during testing. Also, the heightened fear and sadness that was elicited in all children by maternal absence remained high in the children of depressed parents even after reunion with their mothers. In their homes the children were given tasks assessing their understanding of object permanence, their self-awareness, and the maturity of their play. Children of the two backgrounds had similar scores. Thus, in an overall normative sense, the children of the depressed parents appeared to be functioning well.

From 18 months on divergent patterns of social and emotional development began to appear in the two groups. The quality of the attachment relationship as measured in the laboratory between the affectively-ill mothers and their children had deteriorated. Also, this group of children expressed less pleasure and more anger during the testing procedures, and showed slow recovery from frustration and anger. The children of the affectively-ill mothers continued to differ from the control children when they were again seen in the laboratory at two years of age. When confronted with the negative emotions of other persons, these children 'froze', remained riveted to the event, and were slow to resume normal activity. The norms presented in Table 13.1 place this pattern of responding at a very early developmental level. In addition, unlike the children of normal parents, these children rarely looked to their mothers for information in these emotion-laden situations. Whether this failure to reference the mother and to benefit from any information she might have to offer in these arousing situations was due directly to prior experience (e.g. the blunted affect of a depressed parent may severely restrict the information content conveyed by her facial expression), or whether it was mediated by a deteriorating attachment, could not be determined. What was developing however, was a pattern of responding to affective stimulation that was very immature and nonfunctional for the child at that moment in time and had negative implications for the child's continued emotional development.

Data from the home observation at two years also reflected group differences. From a cognitive developmental point of view, there was no reason to expect that these children, who were developing symbolic and representational skills (as indexed by the tests of object permanence and self-awareness, in an age appropriate manner) would show a deficiency in the development of their play skills. Yet this was the case. At 12 months, both groups of children had showed more self-oriented (pretend feeding of oneself) than other-oriented (feeding the doll) play. Between 12 and 24 months, both groups of children showed increased ability to perform other-directed play. However the offspring of depressed parents did not focus their play time on practising this new skill as did the normal children. Rather, they persisted in spending the majority of their time engaged in the more immature form of play. Thus, although they had the competence to play in an other-directed manner, they persisted in devoting most of their play to self-directed play.

The importance of this finding becomes magnified if one considers the developmental implications. The children of normal parents were spending at least six months more than the offspring of the manic depressed parents predominantly practising the more mature form of play. Would this tend to put the children of the depressed parents at a disadvantage for the continued development of symbolic and cognitive competencies? It is reasonable to conclude that if affective factors affect a child's practicing of developmentally appropriate behaviours then this may, in turn, affect future cognitive development.

Effects of a disturbed affective environment on children's social-emotional competence

We believe that the significant disturbance in the emotional and cognitive development that we have seen in the children of affectively-ill parents can, in large measure, be attributed to what these children have learned from, and the adaptations they have made to, living in an affectively disordered rearing environment. We have begun to test this hypothesis by experimentally creating emotionally charged environments (a partial analogue of rearing environments) in order to observe how background climate influences the behaviour of normal $2\frac{1}{2}$-year-old children as they play with a peer (Zahn-Waxler *et al.* 1984b). The environment in the laboratory was experimentally perturbed in three ways:

(i) By having an affective exchange (angry or friendly interaction) enacted by two unfamiliar adults. (Their interaction was with each other and they did not engage the child);

(ii) By exposing the child to a distressed peer. (This was achieved by having the peer's mother leave the room); and

(iii) By distressing the child. (Having his/her mother leave the room).

Each child who was the subject of our research came to the laboratory bringing along a familiar peer. The children were observed over an extended period of play. The affect-laden events we have described were sequenced in the following way: *period 1*, a neutral condition with both mothers present (5 minutes); *period 2*, a background of positive emotion—two adults interact in a friendly way (5 minutes); *period 3*, a neutral condition (5 minutes); *period 4*, a background of anger—two adults quarrel angrily (5 minutes); *period 5*, a neutral condition (5 minutes); *period 6*, a background of positive emotion—adults reconcile (2 minutes); *period 7*, peer's separation from mother (1 minute); *period 8*, subject-child's separation from mother (1 minute); *period 9*, both mothers return (4 minutes). Figure 13.1 summarizes the children's aggressive and prosocial behaviour in each experimental period.

In the initial pleasant peer situation, children showed many positive social behaviours. They also showed some aggressive behaviour. Friendly background affect did not disturb this positive balance. By contrast, the effect of background anger was strong, indicated by a large increase in the amount of time spent interacting aggressively at the conclusion of the adults' fight. There was not much diminution of positive interaction at this time. The effect of the background anger continued to be apparent even in the period of the adults' reconciliation when there were only aggressive interactions and no positive behaviour between the two children. When the friend's mother left, the subject-child's response was an increase of prosocial acts and a return to the baseline level of aggressive acts. When the subject-child's own mother left as well, there were almost no interactive episodes. Interestingly, within the reunion period itself, both the level of aggressive acts and prosocial acts returned to that of baseline.

What is demonstrated with this very brief background disturbance is that both the amount and quality of social behaviour were radically altered. Although the upset and inhibition of behaviour that were induced by one minute of maternal separation were resolved almost instantly by reunion, the effects of five minutes of background anger, created by strangers and in no way directly involving the child, were still not resolved for the children

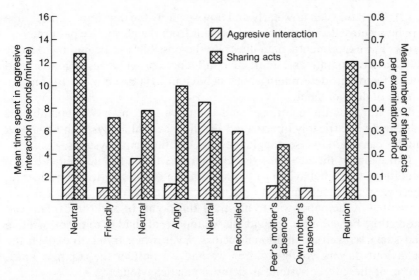

Fig. 13.1 Behaviour in changing affective contexts. Aggression and sharing in 20 children with normal parents. (Adapted from figures presented in Zahn Waxler *et al.* (1984b) *American Journal of Psychiatry* **141 (2)**, 239.)

even ten minutes later. One can only begin to speculate on how disorganizing anger expressed chronically by the mother (and sometimes to the child) is for a child of this age. Such experiences are part of the daily life of a child living with an affectively disturbed parent.

In order to provide some empirical basis for postulating cumulative and enduring effects of disturbed affective environments, a new sample of normal children was exposed to experimental periods 1 through 6 (i.e. the background affects of nurturance, anger, and reconciliation) on two occasions separated by one month. A finding of particular importance from this study was that a second 5 minute exposure to anger a full month later resulted in still higher levels of distress and aggression.

Summary and conclusions

The findings we have reported can be summarized as follows:

(1) It has been possible to take a normative approach to young children's emotional development: to present an unfolding sequence of reactions to social-emotional stimuli along a developmental time-table.

(2) As the studies have illustrated, the norms are produced by complexly interactive factors. Only when *both* cognitive and affective mediators are considered can the norms be understood, and can individual differences and deviations from the norms be explained or predicted.

(3) We have seen how early and how severely the developmental course can be disturbed and can become deviant from the normative pattern.

(4) The experimental data illustrate the possibility of beginning to study the specific separate and interactive influences of affective, cognitive, and social factors in determining both behaviour at a given moment and the course of development.

It is not that the emotional and cognitive systems are inseparable or hopelessly inextricably intertwined. Rather, careful analyses and complex studies can begin to disentangle these highly interactive systems.

Our claim at the outset of this chapter was that a full understanding of cognition, social behaviour, and emotion can come only from considering these systems functioning interactively. The intent of the empirical data presented in the paper was to illustrate this point. In order to explain and predict the behaviour of individuals we repeatedly had to refer to cognitive and social as well as emotional factors. Even when trying to explain the behaviour of very young children we had to consider the complex inter-action of these three systems in determining behaviour.

The theories we reviewed in the text provided broad outlines, and con-structs that were useful in explaining the data. The data did not prove or disprove any theoretical claim. Rather we found that to describe develop-ment in any one system adequately we had to draw upon our understand-ing of development in the other systems. The play behaviour of the child of an affectively ill mother could not be explained in terms of cognitive theory alone, nor could the quantity of prosocial behaviour of the school-age child be predicted by knowledge of the child's skill at social inferencing.

These studies and our selective review of the theoretical literature have demonstrated that the building blocks for an adequate theory of social-emotional-cognitive development are beginning to be available. The study of special samples and individual differences, and the use of more complex task situations will promote cross-fertilization of social psychology, cogni-tive psychology, and the psychology of human emotions. This we believe will speed the development of richer, fuller theories of human develop-ment.

Discussion

The discussion ranged from developmental issues, such as changes in security of attachment with age, to the independence and interplay of cognitive, emotional, and social variables. The points raised were either incorporated into other dis-cussions or into the paper itself.

14

Interpersonal problem-solving: a cognitive approach to behaviour
MYRNA B. SHURE

Children's observable behaviour, including social adjustment, inter-personal competence, and peer relationships can be predicted from ecological, familial, and predispositional factors and also, whatever their source, from the child's own skills—behavioural and cognitive. Interest in the latter has recently skyrocketed—particularly the overt demonstration of social skills, and of the ability to think about people and events in the interpersonal arena. The relevance of social cognitive skills to children's behaviour has been studied primarily by looking at how they think about others' feelings, thoughts, and motives, even when those feelings, thoughts, and motives may differ from their own. These skills, called role-taking or perspective-taking, clearly contribute as critical mediators in healthy human functioning (Shantz 1975).

This chapter will focus on another set of cognitive skills that have been found to contribute to important domains of children's behaviour—ICPS, or Interpersonal Cognitive Problem Solving skills—how a person thinks through and solves typical, everyday problems between peers and figures of authority. It has, so far, been possible to distinguish good and poor problem solvers, and to identify how such skills predict behaviour in children as young as four years of age (Shure *et al.* 1973; Spivack and Shure 1974).

What's the rationale behind all this? It is possible that an individual who has only one or two ways to solve a problem, or cannot plan steps toward reaching a goal, may be less likely to succeed than one who can turn to another (perhaps more effective) way. If one's initial need remains unsatisfied, and such failures recur, the resulting frustration could lead to forms of impulsivity such as persistent nagging and demanding, overemotionality, and/or forms of aggression. Or, such repeated failures could lead to a tendency to evade the problem entirely by withdrawing.

What are these skills, where do they come from, and how can they be nurtured, at home or at school? Following a discussion of specific ICPS skills identified as important to behaviour at various age levels, attention will turn to what we have learned about how the child develops these skills through a key source of acquisition—the child's own mother—and how helping children build these skills (through training) can enhance social adjustment, interpersonal competence, and peer relationships.

ICPS and social behaviour

Three skills which best predict social behaviour have so far been identified: 1) alternative solution thinking, 2) consequential thinking, and 3) means ends thinking. We have also learned that role- or perspective-taking plays an important part in the problem solving world of the child.

1 Alternative solution thinking

An individual's ability to think of different options (solutions) that could potentially be used to solve a problem defines his or her capacity for alternative solution thinking. In solving interpersonal problems, a four-year-old may for example, want her sister to let her play with her doll, or a ten-year-old may want her friend to accompany her to the playground after school. These children can ask, but should their request be refused, they may or may not be able to think of alternative ways to get their wish. Any further attempt to solve the problem is, in our view, largely dependent upon the child's ability to generate other ways to go about it.

If a person has only one or two options available, his or her chances of success are less than they might be for someone who can turn to alternative solutions if the first attempt should fail. If a person's requests were refused, and no other options were available, that person could become quite frustrated. Such frustration could lead to aggression or impatience (e.g. a child might lash out and grab the denied toy), or it could lead to withdrawal, and/or a tendency to give up too soon. Either impulsive or withdrawn behaviour could come to predominate if problem after problem remained unresolved.

Ability to generate alternative solutions is judged important for successful resolution of problems other than those of a want. For example, a four-year-old might have made his mother angry when he broke the cookie jar, however accidentally. An ICPS-deficient child may simply deny it, making his mother still angrier, which in turn may cause the child to retreat and cry. A more competent problem-solver may think of ways to avert mother's anger, such as offering to clean up the mess, or explaining how it happened.

Alternative solution thinking has been found consistently to be negatively correlated with behavioural impulsivity and inhibition, and positively correlated with prosocial qualities such as concern for or at least visible awareness of, others in distress, being liked by peers, and positive peer sociability. These relationships are not explained merely by IQ—in pre-schoolers (Granville *et al.* 1976; Olson *et al.* 1983; Schiller 1978; Shure *et al.* 1973; Spivack and Shure 1974), in kindergarteners (Arend *et al.* 1979; Rubin and Daniels-Beirness 1983; Shure and Spivack 1980, 1982b), in first- and third-graders (Elias 1978; Johnson *et al.* 1980; McKim *et al.* 1982), in second- to fifth-graders (Richard and Dodge 1982), in fifth-graders (Shure 1980; Shure and Spivack 1970), in eighth-graders (Marsh 1982), in adolescents (Platt *et al.* 1974), in adults (Platt and Spivack 1972a,b, 1973, 1974), and in the elderly (Spivack *et al.* 1978).

While some research (e.g. Rubin and Daniels-Beirness 1983) has shown that the content of solutions given to hypothetical problems correlates with the quality of social behaviours and peer relationships (e.g. rejected children think more of agonistic solutions than do popular children), most of the research cited supports our own theoretical position—that it is *how* one thinks, not what one thinks that is a basic issue for the quality of one's social adjustment. As early as age four, both adjusted and nonadjusted youngsters express forceful ways to obtain a toy from another child (e.g. hit the child, grab the toy). While most children can also think of some form of 'asking', the adjusted youngsters can consider a greater variety and range of nonforceful ways (e.g. 'trade a toy', 'be his friend') than the non-adjusted ones, a finding confirmed in both middle- and lower-socioeconomic level groups (e.g. Olson *et al.* 1983; Shure and Spivack 1980, respectively).

2 Consequential thinking

This refers to an individual's capacity to appreciate the impact of an inter-personal act upon him or herself and upon others. The four-year-old who was refused her sister's doll (described above) may hit her sister as a reaction to the frustration of having been denied her wish, or she may think through different options and decide that, at that particular moment, hitting is one of a number of alternative ways to get her sister to give in and let her have that doll. If she thought about it and decided to hit her, the question arises as to whether she also thought through the potential interpersonal consequences of hitting and whether having done so might have influenced her decision to hit. When four- and five-year-olds were asked, 'what might happen next?' if one child takes a toy from another, better adjusted youngsters could think of more different consequences than could those more behaviourally aberrant (Shure *et al.* 1973). But what we learned about consequential thinking at this age level is fascinating. Impulsive children are not aware of what might happen next. This is not surprising if one remembers young children warned, 'You can't hit, he'll hit you back' respond with, 'I know but I don't care, he won't give me that truck!' Awareness of such consequences does not stop them, because other than ask, which is often refused, these children cannot, or do not think of what else to do. Instead of pursuing a new course of action, they may well create a new problem with their quick and sure way to get what they want 'now'. Inhibited children, on the other hand, do not seem able to think of either solutions *or* consequences. Perhaps these children have experienced failure so often that they just withdraw from other children and from problems they cannot solve. Although not studied as extensively as alternative solution thinking, consequential thinking is also associated with behavioural functioning—in preschoolers (Shure *et al.* 1973), in ten-year-olds (Shure 1980), and in adolescents (Ford 1982; Spivack and Levine 1963).

Social perspective-taking skills and ICPS. Given the associations of alternative solution and consequential thinking skills to behaviours, we piloted a study in four-year-olds to see how role- or perspective-taking may play a part in all this. It turned out that impulsive youngsters were deficient in both role-taking *and* solution skills, and there are youngsters who, as we had observed, may not let awareness of potential consequences stop them from performing an impulsive act. If awareness or concern for another's distress is unknown, or known but disregarded, this deficiency in addition to lack of solution skills adds further possible explanations for why these children behave the way they do. Inhibited youngsters turned out to be better role-takers than impulsive children, but they were not better than adjusted ones. It is possible that their awareness or sensitivity to others' feelings without the wherewithal to deal with them (that is, other options) causes them to freeze and socially withdraw. It seems reasonable to assume that knowing how someone feels, while very important, cannot itself resolve a problem and may even heighten anxiety or frustration unless one also knows what to do about it. The question now is whether role-taking can enrich problem-solving abilities. Does appreciating the viewpoint of others open up a broader repertoire of solutions from which to choose? Does ability to draw upon this repertoire prevent, or at least diminish, continued frustration and subsequent need for impulsive behaviours or withdrawal?

Robin, seen exercising her skills one day, may support these speculations:

'She wanted Melissa to give her the water cup (containing plant seeds). When Melissa said, "No, I need them" (the seeds), Robin did not create a new problem by reacting impulsively. Her ability to think of other options led her to another tactic. "When I get the big bike, I'll let you ride it."

Defiantly, Melissa shouted, "I said NO!" Robin then asked, "What are you going to do with those seeds?" and Melissa answered, "Grow them." A few minutes later, Robin returned with a sand shovel and offered, "I'll bury some and you bury some. Two of the flowers can be yours and two can be mine. How's that?" Melissa and Robin began to count the seeds, each burying "their own" in the dirt.'

(From Shure 1981, pp. 160–161.)

Like other good problem-solvers, Robin may have *thought* about hitting, or grabbing the cup from Melissa. She may also have been able to anticipate the consequences of such acts. But her ability to find out about the other child's motives (a form of perspective-taking), and incorporate them into a solution that ended successfully, prevented Robin from experiencing frustration and failure.

Whether appreciation of the viewpoint of others is a skill associated with good problem-solving can be examined in ways other than whether a youngster has a broad repertoire of solutions. For example, Selman (this volume) has ingeniously created a two-factor classification system that places the *content* of some problem-solving skills (specifically, interpersonal negotiation strategies), at levels corresponding to their developmental stage of social perspective-taking (see also Brion-Meisels and Selman 1984). The question now becomes whether youngsters with the majority of responses at the highest level of negotiation strategies are also functioning at the highest level of perspective-taking, and whether these two cognitive skills, in combination, can predict a child's social adjustment and interpersonal competence.

Michael Chandler once challenged us by suggesting that one 'good' solution would really suffice (see Shure 1982). Perhaps it would. But we believe it is the process of turning to another that encourages one not to give up too soon. It may be that such cognitive flexibility is a style of thought that helps one consider multiple options from one problem situation in another. While in the short run it may be one 'good' solution that solves a given problem, in the long run the issue for social adjustment is the ability to generate the kind of thinking that results in resiliency instead of frustration.

The relation of cognition to behavioural resiliency has now been supported by Schiller (1978) and Arend *et al.* (1979) who found in the middle class (lower class was not studied), that competent solution thinkers were more likely than poor ones to behave flexibly, persistently, and resourcefully, especially in problem situations—a measure of ego-resiliency developed by Block and Block (1971). Robin's resilience (ability to bounce back) and flexibility (ability to think of and try a new solution) is particularly interesting in the light of findings by Wowkanech (personal communication). She found that four-year-olds trained to generate their own ICPS skills were, in real life, more likely to try more different ways to resolve conflicts than a behavioural modelling group, who in the face of conflict had been told what to do, and then shown how to do it.

Regarding content, it is true that Rubin and Daniels-Beirness (1983) found that rejected peers suggest more agonistic strategies (e.g. hit, grab), while non-rejected peers suggest more positive ones (e.g. ask, trade a toy). Still, they also found that Grade 1 peer-rated popularity or rejection was also predicted by the total *number* of relevant solution categories produced when the children were tested a year earlier in kindergarten. And Richard and Dodge (1982) found the quantity of solutions to be more deficient in aggressive and isolated boys (girls were not studied) than in popular ones. But importantly, they also discovered that, while the initial solution given by all boys was an effective one, in subsequent responses aggressive and isolated boys tended to increase the proportion of aggressive and ineffective solutions, solutions also characteristic of their behaviours. Returning to the Chandler-challenge, Richard and Dodge put it best when they say: 'Deviant boys are not deficient in offering a single effective solution to an interpersonal problem, nor are they deficient in evaluating the relative effectiveness of solutions which are presented to them. When more than one effective solution must be generated, however, deviant boys are relatively less skilled in doing so. It may be that the behavioural problems of these children occur in situations when the initial behavioural solution is not sufficient, and alternate behaviours are necessary. At that point, the cognitive problem-solving patterns of popular, aggressive, and isolated boys diverge.' (p. 232).

While our research is guided by a process rather than a content theory, with the view that habitual patterns of behaviour are guided (in part) by a general capacity to think, nevertheless, a closer look at content, and at the developmental stages of that content (à la Selman) can no doubt add more refined understanding to the role that social cognition plays in the quality of a child's interpersonal relationships.

3 Means-ends thinking

While alternative solutions are discrete, and unconnected, the process of means-ends thinking involves careful planning, step-by-step, in order to reach a stated goal. Such planning includes insight and forethought to forestall or circumvent potential obstacles and, in addition, having at one's command alternative means if an obstacle is realistically or psychologically insurmountable. The process implies an awareness that goals are not always reached immediately, and that certain times are more advantageous than others for action. While the content of stories may differ, the most striking relationship to adjustment revolves around the presence or absence of planning as part of the cognitive skills of the respondent. For example, in the story about how a child new to a neighbourhood can make friends, a normal ten-year-old tells it this way:

'First Al got talking to the leader. He found out the kids liked basketball but Al didn't know how to play. When Al got to know the leader better he asked him to get the kids down to the skating rink. The kids went and saw him practising shooting goals. So the kids asked him, "Would you teach us how to do that?" So he did and they organized two teams and the kids liked that and Al had lots of friends.'

(From Spivack *et al.* 1976, p. 66.)

This child conceptualized a story wherein the first means, talking to the leader, uncovered an obstacle—kids liked a game that Al didn't know how to play. The obstacle was overcome by stimulating interest in another game (hockey) that Al did know how to play. The child also recognized that it takes time to make friends, as evidenced by his statement, 'When Al got to know the leader better . . .'

A disturbed child more typically would think of the end goal rather than means to obtain it. For example, one eleven-year-old ego-disturbed girl related her story of how Joyce made new friends:

'She'll go out and meet some kids and then she'll have lots of friends. Then she won't be lonely any more and her mother will be very happy because she went out and made lots of friends. She was happy too because she wasn't lonely any more. She and her friends had lots of fun together because they played a lot during recess and after school.'

(From Spivack *et al.* 1976, p. 67.)

Typical of disturbed youngsters, this child revealed less ability to conceptualize means for achieving the satisfaction or goal. When motivated in fantasy, her thinking moved immediately to consummation with insufficient consideration of how to get there or awareness of obstacles that might have to be overcome. Most of her story described feelings and events surrounding the resolution of the problem, all of which occurred after the goal (making friends) was reached.

Though lack of means-ends planning is often clear in diagnostically disturbed youngsters (Shure and Spivack 1972), the less well-adjusted within more homogeneous groups are relatively poor means-ends thinkers as well: in third- to sixth-graders (Pellegrini 1985), in ninth- to twelfth-graders (Ford 1982), in young adults (Gotlib and Asarnow 1979), in older adults (Platt and Spivack 1972) and in the aged (Spivack *et al.* 1978). Based on these and our own (unpublished) pilot data with six- and seven-year-olds, this relatively sophisticated cognitive skill does not begin to guide behaviour until about age eight or nine, possibly because it requires the capacity to consider a complex series of possible events, including how more than one person may interact in a given situation. While the more poorly adjusted tend to reach right toward the goal (such as to make a friend merely by being introduced), the more socially competent think of plans to attract friends, recognize and overcome a potential obstacle (perhaps someone might not like him), and recognize that making friends takes time.

Other potential ICPS skills

To the extent that one understands that how one feels or acts, or that problems that arise have been determined by prior events, one is capable of social causal thinking. And interpersonal sensitivity, another skill, involves the cognitive ability to perceive a problem when it exists, and the tendency to focus on those aspects of interpersonal confrontation that create a problem for one or more of the individuals involved. If the girl who wanted the doll from her sister should hit her to obtain it, then recognize that her sister hit back 'because I hit her first', it also seems reasonable to assume that if she were aware that a problem or potential problem could develop once she decided to ask for the doll, her behaviour and/or problem-solving strategies might differ from what might ensue in the absence of such sensitivity. Further, if she were also sensitive to the fact that a new problem might emerge once she 'hits her' in retaliation for having been told no, her behaviour and/or problem-solving strategies might differ from the course pursued in the absence of such sensitivity. By themselves, causal thinking and interpersonal sensitivity do not relate strongly to overt behavioural adjustment. Yet, they do relate to both alternative solution and consequential thinking in young children, and to these and means-ends thinking in older children (Spivack *et al.* 1976; Shure 1980), and they probably play a significant role in the complex chain of thought necessary for interpersonal cognitive problem solving.

Considering all ICPS skills measured to date, perhaps the consummate problem-solver first identifies the problem (not always an easy task), has a tendency to look back and consider what led up to it (its causes), considers alternative solutions, or plans, taking the perspectives of others into account, looks forward in anticipation of potential consequences of an act, then changes his or her solution or plan, if need be.

ICPS and childrearing

We do not know the subtleties of how ICPS skills are first learned, but we have gained some information about why some children come into nursery school with relatively competent skills and others do not. It has become clear to us that parents can significantly affect the social adjustment, interpersonal competence, and peer relationships of their children by how (and if) they encourage the development of ICPS skills.

Two studies of low SES mothers and their four-year-olds (in Shure and Spivack 1978) have shown that girls' (but not boys') ICPS skills relate to mothers' ICPS skills and their problem-solving childrearing style. Childrearing style refers to the extent to which the mother guides and encourages her child to think of his or her own solutions to problems and of consequences to acts, how often she suggests solutions and explains consequences (called induction), the frequency with which she demands the child to do what she thinks best, and how often she belittles the child, avoids communication, or solves the problem for the child. The focus is on the style of communication, not the content of it. For example, advice given by the mother *to* hit back, and advice *not* to hit back and instead to tell the teacher is scored the same. Both forms of specific advice indicates that the mother is thinking for her child rather than stimulating the child's own thinking. In the natural course of childrearing, very few of these mothers ask their children for their ideas about what to do or what might happen next. But mothers who are more competent problem-solvers do tend to use inductive techniques, while poor problem-solvers are more likely to demand, belittle, or ignore the child completely. Though mothers' ICPS skills and child-rearing style are, in our research, linked only in the case of girls, it is still the child's own ICPS skills, and especially alternative solution skills, that best predict the child's behaviour in both boys and girls.

Why do mothers' child-rearing styles and thinking skills appear to affect their daughters' ICPS skills more than their sons, even though overall, boys are no more or less ICPS-deficient than girls? Minton *et al.* (1971) found that preschoolers who were punished more at home remained at a greater physical distance from their mothers in a waiting room. Though this behaviour was not assessed in our research, it is possible that since boys are generally spanked more than girls (Sears *et al.* 1957; Newson and Newson 1968), especially when they do not comply (Minton *et al.* 1971), they may become psychologically as well as physically more distant, and thus harder to influence and perhaps less likely to learn ICPS skills from their mothers. Or possibly, boys are normally more aggressive and resistant to influence than girls, especially when they do not have fathers (Hoffman 1971), the case in nearly 70 per cent of our youngsters. It is also possible that if children adopt parental characteristics to the extent that the parent is an important, relatively consistent source of nurturance and reward (Mowrer 1950; Mussen and Rutherford 1963), young fatherless boys not only may resist forms of influence and discipline but also may model their mothers' ICPS thinking style less than do girls.

If preschool boys do not model their ICPS thinking after their mothers, and many of them do not have fathers, where then do they acquire their skills? From peers? Young boys do appear to be generally more vulnerable to pressures and challenges from their peers than girls (Emmerich 1971). The importance of peer relationships notwithstanding (e.g. Hartup 1979), it is difficult to believe that peers would have an influence on boys' ICPS skills equal to that of mothers on their daughters. (For a more thorough discussion of this issue, see Shure and Spivack 1978).

Whatever the course of acquisition of ICPS skills during the natural course of child-rearing, the further question arises whether significant adults in the child's life could, through specific training techniques, help the child enhance his or her ICPS skills and subsequent behavioural adjustment.

Enhancing ICPS skills—at home

For four-year-old preschool children, a three month sequenced programme script was developed by Shure and Spivack (1978) for use by parents with their child at home, with lessons designed to help the parent learn to solve problems as well. As the mother helps her child think about his own and other's feelings and how to consider the effects of his actions upon others, she also thinks about feelings and how what *she* does affects others (including her child). As a mother guides her child to think of alternative solutions to problems relevant to him, she also thinks of solutions to problems relevant to her (particularly when a child creates a problem involving her, such as 'Mike won't do what I ask him to, lately'). Just as the child is never told solutions to problems or consequences to acts (the *process* is the issue), neither are the mothers. The value is not on what they think, but *that* they think. In addition to formal lesson-games, mothers are taught how to guide their child to solve real problems as they arise through a process we call 'problem-solving dialoguing', or 'problem-solving communication'. Problem solving communication increases the child's opportunity to exercise his/her ICPS thinking. The trainer, through leading questions, guides the child to see the problem, generate different possible solutions, explore consequences, and so forth. Perhaps it introduces a new *in vivo* quality to the training, and leads to a more effective utilization of ICPS thought when the child subsequently faces real problems on his or her own.

Before training, here is how one mother talked to her child, seen grabbing a toy:

Mother: Why did you snatch that truck from John?
Child: 'Cause it's my turn!
Mother: Give it back, James.

Child:	I don't want to. It's mine.
Mother:	Why don't you play with your cars?
Child:	I want my firetruck!
Mother:	You should either play together or take turns. Grabbing is not nice.
Child:	But I want my truck now!
Mother:	Children must learn to share, John will get mad and he won't be your friend.
Child:	But, Mom, he won't give it to me!
Mother:	You can't go around grabbing things. Would you like it if he did that to you?
Child:	No.
Mother:	Tell him you're sorry.

(From Shure and Spivack 1978, p. 32.)

This mother did talk to her child beyond demands and belittling. She also used inductive techniques (e.g. 'John . . . won't be your friend'). Yet nothing was communicated that would teach her child how to think. She was thinking about what was important to her, not what was important to him. When her demands and suggestions were met with resistance, this mother suggested a new solution to the problem while explaining the adversities of grabbing. Instead of encouraging James to think about the problem, the mother continued to impose her point of view.

Here is how a trained mother handled a similar problem. Four-year-old Ralph had just grabbed a racing car from his friend:

Mother:	What happened? What's the matter?
Ralph:	He's got my racing car. He won't give it back.
Mother:	Why do you have to have it back now?
Ralph:	'Cause he's had a long turn.

This mother just learned something that she would not have discovered had she simply demanded he 'share'. She learned that in fact her son had shared his toy. The nature of the problem now appeared different.

Mother:	How do you think your friend feels when you grab toys?
Ralph:	Mad, but I don't care, it's mine!
Mother:	What did your friend do when you grabbed the toy?
Ralph:	He hit me but I want my toy!
Mother:	How did that make you feel?
Ralph:	Mad.
Mother:	You're mad and your friend is mad, and he hit you. Can you think of a different way to get your toy back so you both won't be mad and so John won't hit you?
Ralph:	I could ask him.
Mother:	And what might happen then?
Ralph:	He'll say no.

Mother:	He might say no. What else can you think of doing so your friend will give you back your racing car?
Ralph:	I could let him have my match-box cars.
Mother:	You thought of two different ways.

(From Shure and Spivack 1978, pp. 36–37.)

Ralph's mother did not try to solve the problem from her point of view. She did not tell him to share his racing car or not to grab toys. In fact, the problem shifted from Ralph's grabbing to how Ralph could get his toy back. Although Tizard (this volume) has serious concerns about questioning-techniques that require a particular 'right' answer, this mother, by guiding Ralph to think about the nature of the problem, to think of his own solutions to the problem, and to evaluate the consequences of his act, was teaching her child how but not what to think. She was using a problem-solving style of communication. For mothers, the training goals were:

(i) to increase sensitivity that the child's point of view may differ from her own;

(ii) to help her recognize that there is more than one way to solve a problem;

(iii) to teach her that thinking about what is happening may, in the long run, be more beneficial than immediate action to stop it; and

(iv) to provide a model of problem-solving thinking—a thinking parent might inspire a child to think.

For children, the cognitive goal was to teach a set of skills that would enhance their ability to conceptualize alternative solutions and consequences relevant to interpersonal problems.

We learned that mothers' improved ability to solve hypothetical adult problems (such as how to keep a friend from being angry after showing up too late to go to a movie) did *not* relate to her child's improved ICPS skills, but her ability to solve hypothetical problems about children (or about children and their parents) did. Also having a significant effect on the child's ICPS skills was the mother's increased ability to guide her child to solve his own problems, and mothers who could best do this also improved most in ability to solve hypothetical problems between two children or between mother and her child. These relations suggest that increasing mother's ability to think about these kinds of problems is intimately related to how she guides her child to solve real problems that arise, and together, both have a significant impact on the child's ICPS skills.

Mothers who received systematic ICPS training of their own had children whose ICPS skills improved more than an earlier group of trained mothers who did not. While both of these mother-trained groups of youngsters improved more than controls, these findings suggest that there is a greater impact on the child when the mother as well as the child are taught how to think. Given the significant impact of mother's ICPS skills and childrearing on the child's ICPS skills, it was, in the end, the child's own acquired ICPS skills that had the most direct impact on his or her behaviour, behaviour which had generalized from home to school (Shure and Spivack 1978). The mediating effects of a child's ICPS skills on his or her behaviour have clear implications for optimal mental health programming for mothers and their young children. Importantly, we learned that when mothers are specifically trained to go beyond suggestions and explanations, and guide children to generate their own solutions to problems and to consider the consequences of acts, both boys and girls could benefit, a crucial finding in light of the pretraining correlations of mothers' ICPS skills and child-rearing style with the ICPS skills only of daughters. If boys, or at least fatherless boys, are more resistant to mothers' communication before training, it is possible they are less resistant to it when first guided, and then freed to think for themselves.

Enhancing ICPS skills—at school

Three to four month programmes similar in contents and style to that developed for use by parents have also been developed for use by teachers in school—for preschool (Shure and Spivack 1971; also in Spivack and Shure 1974), for kindergarten (Shure and Spivack 1974), and for the intermediate elementary grades, ages 9 to 12 (Shure and Spivack 1982a).

In our early research with four- and five-year-olds, we learned that teachers can effectively train ICPS skills, and that such gains could significantly reduce observable negative impulsive and inhibited behaviours, as well as increase positive qualities such as concern, or at least visible awareness of peers in distress and of how much the child is liked by his or her peers (Shure and Spivack 1982; Spivack and Shure 1974). Within a wide Binet IQ-range, the gains in ICPS skills that best related to improved social adjustment were alternative solution thinking, and secondarily, consequential thinking. We also learned that:

(i) ICPS impact on behaviour lasted when measured one and two years later;

(ii) if a child was not trained in nursery, kindergarten was not too late; and

(iii) one exposure to the four month programme had the same behavioural impact as two.

Further, well-adjusted children trained in nursery were less likely to begin showing behavioural difficulties in kindergarten and first grade than were comparable controls, highlighting implications of the ICPS approach for primary prevention.

If kindergarten and first grade are critical transition points for younger children, we know that junior high is for older ones (e.g. Blyth *et al.* 1983).

We thought if we could intervene before that time, we might be able to prevent mental health dysfunction as youngsters make that move. We have now completed our first attempt at ICPS-intervention for an older age group—inner-city public elementary school fifth-graders, ages 10 and 11.

Our major finding was that, overall, gains in ICPS scores correlated with gains in behaviours, most consistently with positive, prosocial behaviours, in both boys and girls (Shure 1980). In both sexes, it was improvement in the number and range of solutions that best related to these behavioural gains, most consistently to observed concern or awareness of others in distress, positive peer sociability, and the degree to which the child was liked by his or her peers (as judged by both teachers and classmates). In boys, though not girls, gains in consequential and means-ends skills were also related to prosocial changes. The relationships between ICPS gains with decreased negative, aberrant behaviours were less clear. Although all changes were in the predicted direction, only gains in solution skills related to decreased negative behaviours, and in girls only. These negative behaviours were:

(i) overemotionality in the face of frustration; and
(ii) aggression.

ICPS gains and decreased inhibition did not relate to each other. Nevertheless, the indication that solution thinking is a behavioural mediator in girls is important. Increased prosocial behaviour was accompanied by decreased impulsive factor scores of overemotionality plus aggression, with partial correlations showing it was the cognitive *solution* skills which directly mediated both.

The possibility exists that in contrast to preschool and kindergarten youngsters, ICPS training of fifth-graders affects prosocial behaviours sooner and more consistently than negative impulsive and inhibited ones. Perhaps longer and/or more intense training is required to reduce significantly aberrant behaviours, and to help children associate their newly acquired ICPS skills with what they do, and how they behave. Given immediate ICPS gains, that ICPS and behaviour are correlated phenomena, and the possibility that negative aberrant behaviours are simply more habitual in older than in younger children (and therefore more resistant to change), the necessity for more intense or extensive ICPS intervention seems reasonable. It is also possible that training within a single school year does affect negative aberrant behaviours, but the effect only surfaces later. Gesten and his colleagues (Gesten *et al.* 1982) found that while trained second- and third-graders showed significant immediate ICPS gains (called SPS [Social Problem Solving] by this research team), both positive and negative (acting out) behaviour gains showed up in ratings of teachers one year later. In the controls, there was immediate behavioural improvement, but those gains were lost at the one year follow up. Importantly, this group never showed cognitive (ICPS/SPS) gains. While it is clear that any (immediate) behaviour gains were not due to cognitive gains, it is still impossible to ascertain for sure whether the delayed behaviour gains shown in the trained group were or were not directly caused by the increased ICPS/SPS skills obtained immediately following the intervention. Whether these gains can be linked to immediate or latent ICPS/SPS gains is not yet clear. What is clear is that among the controls, there were immediate behaviour gains, but they were lost at the one year follow up. Importantly, the ICPS/SPS skills of the controls never significantly improved. Therefore, the possibility exists that the immediate ICPS/SPS gains of the trained youngsters did mediate the (delayed) behaviour gain.

It appears that one cannot assume that within a three to four month time frame, ICPS intervention will affect all interpersonal behaviours equally at different ages, or that specific ICPS skills will have the same impact on both boys and girls. In addition to looking also at the quality of solutions and plans, we know that other, unstudied ICPS skills must exist as well. Before concluding that it is too late for ICPS intervention to ameliorate more substantially the behaviour of school-aged children, the consistencies and the questions raised above must be probed. We are encouraged to learn that Elias (1984) has found that ICPS intervention has helped to reduce the intensity of stressors experienced at transition to middle school (grade 6), and that the amount of stress experienced was directly related to the length of training (the more training, the less intense the stress). This is remarkable considering the range of stressors encountered, 'from serious public and mental health problems as becoming involved with smoking or drinking to issues of coping with peer pressure, academic requirements, and the logistics of being in a large, unfamiliar school' (ms. p. 11).

While we have not, perhaps, demonstrated with certainty that enhanced ICPS skills *caused* gains in behavioural adjustment, we believe we have addressed the issue of the 'tertium quid' (Bryant, this volume) in at least two ways:

(i) any measured cognitive impact on behaviour was a function of ICPS and not IQ, as there were no significant behaviour gains as a function of initial IQ or IQ change; and

(ii) we have had two placebo-attention groups which received, as much as possible, the same experiences as the ICPS-trained groups except for the cognitive skills being taught (Shure *et al.* 1972; Shure, in progress).

We do believe that enhanced ICPS skills led to greater peer acceptance, as the youngsters learned to consider the needs of others, and to try another way before behaving impulsively if the first solution to a problem should fail. It is also possible, as Hartup (this volume) notes, that friendship may support social problem-solving in ways that interactions between non-friends do not. It would be a theoretically important experiment to test whether intervention to increase the number and strength of children's friendships would, in turn, increase ICPS skills, being careful not to inadvertently train ICPS skills in the process of helping the children make new friends. It would also be important to ascertain more fully how (and if) impersonal cognitive skills such as academic achievement (see Krappmann, this volume) affect interpersonal cognitive abilities (or the reverse), how interpersonal cognitive abilities affect academic achievement (or the reverse), and/or how either affect the quality of peer relationships (or the reverse).

It is extremely encouraging to learn that children can benefit from ICPS training at school, and how cost-effective it is when one stops to think that training just one teacher can reach 30 youngsters in one year. But we have only just begun our efforts to understand the complexities of what it would take to reach maximum behavioural impact. Some insights provided by Radke-Yarrow and Sherman (this volume) may add to our understanding of those complexities. They argue convincingly that cognitive competence alone may be insufficient to predict a particular behaviour. For example, they cite Barrett and Radke-Yarrow (1977) who showed that in 5- to 8-year-olds, it was the level of inferential ability (a cognitive domain) *and* assertiveness (a behavioural domain) which best predicted altruism (another behavioural domain). Whatever the content of the variables, analyses of their interactions can potentially give new importance to specific variables previously undervalued in the interpretation of a particular result.

In our ICPS research, we may be able to account for more of the ICPS/behavioural variance by examining how various patterns of ICPS skills and behaviours combine to predict optimally the criterion behaviour in question. Perhaps then we will also gain significantly more knowledge about what to emphasize in our interventions—and thus increase our chances for maximal impact on that criterion behaviour. Further such analyses as these, and of data such as those of Elias, will help us more fully to appreciate which indices of mental health and effective peer relations for various age and socio-economic level groups are (and are not) mediated at least in part by ICPS abilities, and therefore are (and are not) alterable by the ICPS approach. They will also begin to shed light on the efficacy of the ICPS approach through questions as: Who does it help? How does it help? How long does it take? How long does it last?

Discussion

Where do thoughts arise?

Selman felt that Shure was 'suggesting a necessary but not sufficient model of the relation between thought and action, namely that thought was necessary but not sufficient to produce a change in behaviour. If so, where do thoughts arise, if in fact they precede behaviour?'

Shure replied that 'Selman is really asking three questions: 1) do (social cognitive) thoughts precede behaviour? 2) If (1) is true, where do those thoughts come from? 3) Are ICPS skills (thoughts) sufficient to produce a change in behaviour? We certainly do not believe that thought precedes all behaviour, but rather, we are in agreement with Selman, who proposed that we look "not at a matching of social cognition and specific behaviours, but rather, the reciprocal relationship between children's interactions and their (general) understanding of (them)" (quoted in Shure 1982, p. 133). That understanding may develop from experience in a culture, beginning in the family, and enhanced from experience in interaction with peers. To what extent interactions that involve interpersonal problem situations create a disequilibrium that promotes further development of ICPS skills, and to what extent childrearers communicate in ways that encourage the exercise of such thinking (see my chapter), is still unknown. But we do believe that ICPS skills are a significant (albeit, certainly not the only) mediator of behaviour, and that enhancing those cognitive skills (through training) has a significant impact on social adjustment and interpersonal competence.'

Doise added, 'The question where do these social cognitions come from is a very important one. It is similar to the problem of generality of cognition in the physical area, for instance the conservation problem. In Piagetian terms, do "general structures" exist in the social area? If so then the problem is how do children acquire them and how do they apply them to different interactions in different situations?'

Shure replied, 'We believe that in regard to ICPS skills, generalization occurs when these skills are guided by a *process* more than by a context. A style of "I'll think of another way" or "I won't give up too soon" is situation free. That is what the use of problem solving dialogues in our training is designed to help the child do. And when children are able and allowed to think for themselves, they are more likely to act on that thought than when they are told what to do, and how to do it.'

15

The use of interpersonal negotiation strategies and communicative competencies: a clinical-developmental exploration in a pair of troubled early adolescents
ROBERT L. SELMAN

Introduction

One important aspect of the child's initiation into long-term and continuing dyadic peer relationships is the way the child, growing up, develops an understanding of and uses strategies for dealing or 'negotiating' with other persons around naturally occurring interpersonal conflicts. Such strategies are used at times and places I call 'contexts for interpersonal negotiation'. These are contexts where individuals simultaneously experience conflicts both within the self and between the self and a significant other around some interpersonal issue.

One commonplace but nevertheless challenging example of such a context is where two close friends both want to interact with the other but in different activities. Here the conflict between the self and other is clear. However, in an ongoing relationship, there is also a within-self conflict, one that may even go undetected by the participants themselves. This is an inner conflict between such concerns as wanting to do what one wants (to serve the selfs' interests) and at the same time wanting to do what the other wants (e.g. so as to keep the other's positive interest, affection, or commitment to the self). Another example is a context where one of two friends wants to change the nature or intensity of the relationship, i.e. make it either more or less close or exclusive, but the other is not sure this is what he or she wants.

Formally speaking, the types of negotiation strategies that here I am specifically calling 'interpersonal' emerge in dyadic interactions at the confluence of deep and intense streams of self and social relational interests. They occur in common everyday interactions where conflicting tides rush against each other, not only between persons, but within persons as well.

For younger children, the complexities of these conflicts may go unrecognized or feel fleeting; for adults they are often continuing underlying currents, difficult to disentangle. During early adolescence the coalescing awareness of the dual nature of these kinds of interpersonal conflicts makes for great intensity.

The goal of this chapter is to explore several topics related to this aspect of interpersonal behaviour and development. On a *theoretical* plane I attempt to show how these strategies have both structural and functional components and how they may relate to communicative aspects of interpersonal development. Here my interest is in developing a general model; yet it is expressed with specific reference to the period of early adolescence, the threshold of an age when strategies that can sustain intimacy ordinarily develop.* At a *methodological* level, this chapter describes a clinical-comparative 'hands on' approach to the study of developmental processes of interpersonal interaction and social regulation. It does so through the observation, over a period of one and one half years in a form of treatment we call Pair Therapy, of two early adolescents who have obvious, and very troubling, disturbances in their ability to relate to and deal with another person for an extended period of time, to be a friend or a chum with a peer. This case provides us with data on, and suggestions about, the association between inter-individual interactions and relationships on the one hand and the facilitation of individual development, both cognitive and social, on the other.

This chapter begins with an attempt to locate on a behavioural science 'map' the coordinates of our approach to the study of interpersonal relationships as it relates to several currently lively approaches to the study of children's interpersonal development and interaction: ethological and functional on the one hand, and structural-developmental on the other. In so doing, it attempts to locate where borders exist, and it tries to show how both these traditions have influenced our work.

To what do I refer by the terms structural and functional aspects of interpersonal behaviour and development? In the context of recent research on social skill development in childhood (e.g. Renshaw and Asher 1982) contemporary functional models have been developed to analyze the various social performance skills that need to be utilized and the sequence of steps an individual might take to deal with (adapt to) interpersonal challenges,

---

* The term, intimacy, is fraught with meaning. In this paper, one focus is on how early adolescents prepare for the intimate negotiations involved in heterosexual relations. However, there are other contexts for intimacy as well. In fact, in the pair's interaction described in this study, it can be seen that in addition to the concerns each girl has for the nature of heterosexual relations, there are also forms of intimacy that can be observed between the two girls, ranging from a primitive sharing of descriptions of bodily functions to shared self-reflection, a mutual sharing of the self's feelings with another.

e.g. first identifying an interpersonal conflict, next thinking of alternative social actions for its resolution, then evaluating the consequences of actions to be taken, etc. This type of analysis is somewhat analogous to thinking about what makes an automobile 'go' by examining at a given point in time the 'horizontal' sequence of functional relations among the ignition system, the system for the delivery of fuel to the engine, and the transmission of energy to the wheels. It can be contrasted with the structural-developmental approach which attempts to understand what makes an auto go by looking across time at the deeper or 'vertical' transformations in the 'design' features among each of the related systems of the engine, e.g. the 'developmental' shift in complexity from the structure of the carburettor to that of the fuel injection system for performing the same basic function, the delivery of fuel to the engine.

Any superficial survey of research approaches is a risky business. It is almost sure to please few of those researchers and/or theorists whose work is being surveyed. Nevertheless, it is an essential step if researchers with differing approaches to a similar problem are to come together mutually to understand the coordinates of any particular body of work being proposed. Figure 15.1 provides a rough 'map' of current approaches to the study of interpersonal development which may be helpful in understanding similarities, differences, influences, and disagreements among various approaches.

The map is in the form of a full 'equatorial' circle. This is meant to emphasize several points. First, it signifies that ultimately the relations among approaches to this area of inquiry come full circle, so that there is a connectedness among approaches to the left and far left, as well as to the right and far right. Second, not all recent integrations of approaches fall on the perimeter of the circle: there may be new connections made 'over the poles', so to speak. Finally, the designation of the map's centre is an arbitrary one, similar to the designation of the Greenwich Meridian as longitude 0. Consciously, and 'ego' centrically, I have located my own recent work along the meridian, not as an absolute or self-centred statement, but as a reflection of my belief that one's own work is typically experienced as being a central point against which to compare and contrast. This is so even if it is experienced as having moved or departed from the tenets of another, perhaps more settled land somewhere to the east or west. Indeed, the map suggests that as one thinks of work as moving in one or another direction, one needs to be cognizant of the research 'digs' already being worked in the 'new' territory. Ultimately, placing one's work in the centre of a research map of a particular field of study provides one with a better understanding of the border disputes and conflict on all sides. This improves understanding on both conceptual and methodological ground.

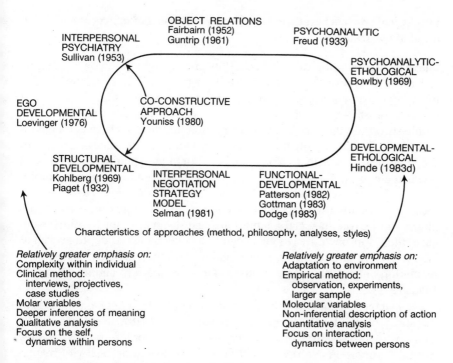

Fig. 15.1 A selective map of various behavioural science approaches to the study of interpersonal development and behaviour.

With reference to Fig. 15.1, because of their almost exclusive focus on intrapsychic or internalized aspects of interpersonal relationships, I will not provide discussion of those intriguing, but foreign and exotic, approaches to the study of interpersonal behaviour and development that fall on the far side of the 'globe', e.g. the Object Relations theories of Fairbairn (1952) or Guntrip (1961), or even the many and varied psychoanalytic approaches. Instead I will provide a brief report of approaches to our near left and right, with particular emphasis on how they have influenced our current theoretical model and methodological approach. I will begin with some considerations of the implications of 'developmental' ethology, and move leftward as far as the structural developmental approach, the area where our own current approach was spawned. As also can be seen in Fig. 15.1, I have listed some of the 'culturally relative' scientific characteristics of approaches to the left and right of our arbitrary 'centre', along philosophical, methodological, and even stylistic lines. These characterizations are not to be taken as gospel, but rather should be interpreted as ideal typologies that provide a picture of some of the similarities and differences among approaches.

Ethological approaches: their influence on our model of interpersonal negotiation strategies

I have located ethological approaches toward the right based on my characterization of approaches to my left as structural and those to my right as functional. The ethological tradition and emphasis is of importance to our research on strategies for interpersonal negotiation in several ways: in its emphasis on the descriptive study of natural interactions between individuals, in its insistence on taking into consideration the characteristics of social contexts, and in its deep concern for understanding processes of social regulation. Clearly, their concern with adaptation suggests ethologists are heavily weighted toward the functional orientation. However, the recent interest of ethologists such as Hinde (1983d) in the ontogeny of cognition and its role in social adaptation suggests an attempt to 'dimensionalize' their approach, i.e. to give cognitive developmental depth as well as interactive breadth to their analyses and to move beyond analyses of social regulation only at the plane of physical actions and interactions (Blurton-Jones 1972).

For instance, Charlesworth, in a discussion of ethological approaches to the study of children's emotional development (1978), using descriptive analyses of facial expressions of threats as an example, emphasizes how careful observation can reveal the early emergence in ontogeny of regulative aspects of human social behaviour. In our research on interpersonal negotiation strategies, we find the verbal expression of threats to be a common and normal step in the ontogeny of a repertoire of explicit negotiation strategies. Verbalized orders and threats emerge in the samples we have observed around age five to seven. In about this age range, these 'threat' strategies often replace less differentiated impulsive, physicalistic, negotiation strategies (impulsively grabbing or running away) as the predominant form for resolving interpersonal disequilibrium. At a later age, perhaps eight to twelve, they too will be displaced, at least to some degree, in the repertoire of most children, first by reciprocal trades, influencing, and persuasion, and still later by more truly collaborative approaches. But as Charlesworth points out, long before the age when children begin to use language to deliver articulated and reflected-upon verbal threats as a way to deal with interpersonal conflict, toddlers are already jutting their jaws and using other emerging threat postures to deal with (regulate) conflicts with peers in contexts involving competitions over toys, space, or persons (see also, Hay 1984).

This suggests developmental approaches to the study of interpersonal behaviour need to clarify the similarities and distinctions between unreflective regulative behaviour and superficially similar but more complex behaviour over which the individual has increasing reflective control. Conversely, ethological approaches need to consider the possibility that with age, humans restructure their understanding of the social world so as to become conscious of some of their previously unrecognized yet functional social postures, i.e. they make use of language and thought to express power and control symbolically and hence become able strategically to control or display behaviours such as threats on a new plane.

Ethological approaches provide another important influence for our developmental interest in interpersonal regulation; this is their consideration of the adaptive function dominance and submission patterns play in dyadic interactions. Personality theorists and researchers traditionally have been interested in *within-person* bipolar traits, such as assertiveness/submissiveness, introversion/extroversion, acting out/withdrawal, or passivity/aggressiveness. An interactive focus on dominance and submission patterns suggests the importance of parallel factors in the relationships *between persons*. Because strategies for interpersonal negotiation are often utilized in relatively ambiguous contexts (even acknowledging the powerful effects of background variables such as culture, class, gender, etc.), it is important to see what happens when individuals are asked to relate to each other in contexts where there may be few or ambiguous social norms and unclear power relations between them (Abramovitch and Strayer 1978).

Functional approaches to social interaction and development in children

Although some 'new world' developmentalists have been attracted to the power of the ethological approach (Savin-Williams 1976; Strayer 1983; Zivin 1977), other North American researchers, firmly wedded to stochastic approaches, still seem wary of purely descriptive accounts of social interaction, perhaps finding it not totally within the purview of an empirical developmental psychology. However, in the last ten years or so a number of child psychology researchers have attempted to integrate careful descriptive methods of behavioural observation with quantitative methods appropriate to the sequential analysis of repeated interactions of individuals in dyads or groups (Dodge 1983; Gottman 1983; Rubin and Krasnor 1983). Correspondingly, these researchers have appeared to feel most comfortable with models that break down the social actions of individuals into the functional sequencing of component social and cognitive skills.

An important influence on this brand of research has been the work of Gerald Patterson and his colleagues in Oregon on the interaction processes of aggressive children in family contexts (1982). Patterson's research suggests that one significant difference between troubled children and their better adjusted peers is that the aggressive acts of troubled children are more likely to be continued or escalated (unregulated) beyond the initial forays. Patterson suggests that these children are raised in families where the natural adaptive processes by which aggressive actions are socially regulated have broken down. In our developmental terminology, we would say that the naturally occurring impulsive and physical 'negotiation' strategies of early childhood have not, for whatever reason, been adequately transformed into higher level strategies that require the control of affect and the implementation of self or shared reflection.

Although Patterson's work comes out of a quite different theoretical tradition from our own—social learning versus structural developmental—it has at least one other important implication for our own research activity. Patterson notes that although problem children exhibit aggressive behaviour to a much greater degree than do their better functioning counterparts, even among such children, aggressive acts are still a low absolute percentage of all acts. As mentioned earlier, it is the likelihood of a sequence of escalation that is much greater. When later we look at the absolute number of interpersonal negotiation strategies observed in the course of Pair Therapy, we will see that they also occur seldom. However, this does not mean that they do not have a powerful influence on the tone of the interaction or the quality of the relationship.

Structural-developmental approaches

Whereas in Fig. 15.1 the language spoken in the lands to our right is of behavioural observation, social interaction, adaptation, and moving leftward, social information processing; to our left the focus is on in-depth interviews, complexity of social-cognitive structures, the ontogenesis of socio-moral understanding, and levels of personality and ego development. In this area the search is primarily for evidence for the inner structural unity within the person that is hypothesized to determine to a large extent how that person relates to others (Loevinger 1976; Kohlberg *et al.* 1983). The hypothesis here is that there exists in each individual a central tendency, or in structural developmental terms, an invariant ontogenetic sequence of central tendencies (levels, stages) that makes meaningful, and gives organization to, the form, if not the function, of these interpersonal behaviours.

A major feature of the structural developmental approach is the concern with whether certain forms (levels) of social development are qualitatively distinct from one another. This issue is manifest in attempts to reconcile:

(i) whether there is continuity or discontinuity in the development of structures of social thought or behaviour (Damon 1983); and

(ii) whether there are universal sequences in the ontogenesis of social or moral thought that hold across major demographic distinctions, such as gender or culture (Kohlberg *et al.* 1983).

A number of differentiating positions have arisen within the structural developmental camp, e.g. around issues such as whether the generality or universality of stages is more apparent than real when cuts are made along lines of gender (Gilligan 1982) or along lines differentiating among domains of social knowledge (Turiel 1983). Others, such as Youniss (1980), have argued that the structural developmental approach to the socialization of social knowledge is too individualistically oriented; that social understanding is not primarily constructed by individuals through processes of self-reflection, but rather is more often co-constructed by children in collaborative interactions with their peers (refer to Fig. 15.1). Nevertheless, structural-developmental approaches have guided our work by demonstrating how the form (level) of social understanding can be distinguished from its content. It has led us to attempt to analyse social actions (and interactions) along structural lines as well.

A two factor model of interpersonal negotiation strategies: structural developmental levels by functional interpersonal orientations

Although my own work in social-cognitive development (the structural analysis of the relation between the ontogenesis of levels in the coordination of social perspectives and the ontogenesis of interpersonal understanding, e.g. developing conceptions of friendship, etc.) was initially rooted in cognitive structural-developmental soil (Selman 1980), it has felt the limitations within the parameters of this approach and has sought a more fertile land in which to grow. One of the constraints of the structural developmental approach is that its focus is on the ontogenesis of social knowledge development (*competence*), but not on the structural-developmental analysis of the organization of social or interpersonal *behaviour* (Selman 1981). Another limitation, which is not to discount its contribution, is that while the constructivist approach deals with the interaction of the structural level of the subjects' current conceptualization with structures of 'external' knowledge (environmental input), it is not readily applicable to the study of the interaction between individuals, to subjects who 'rub' against one another. According to most structural-developmental approaches, once a level (form or structure) of social understanding is constructed, it is not easily lost, even if it is not usefully applied. But a developmental model of the organization of social behaviour can not rest easily on a model of social-cognitive competence alone (see also Radke-Yarrow and Sherman, this volume). It must expand to allow for regressive as well as progressive movement, and must account for the influence of external as well as internal factors of the moment on the level of performance expressed.

Our developmental analysis of interpersonal negotiation strategies focuses on the ways that individuals, growing up, deal with the inner and interpersonal aspects of disequilibrium, i.e. as experienced both within the self and between the self and other, as it is generated in contexts for interpersonal negotiation. As such, the analysis draws upon both the structural developmentalist's interest in levels of social understanding (increasing cognitive complexity) and the ethologist's and functionalist's concern for adaptation and social regulation. As a first factor, it is concerned with how individuals develop the *capacity to coordinate in conduct* the understanding of another's thoughts, feelings, and intentions with their own in the course of attempting to balance inner and interpersonal disequilibrium. This is the developmental factor: as a tool for its analysis, we used our earlier description of levels in the coordination (differentiation and hierarchic integration) of social perspectives (e.g. Werner, 1948).

As a second factor, our model is concerned with how, at each level in the coordination of social perspectives, social actions are taken to resolve both intra- and interpersonal disequilibrium and restore a balance, both within the self and between the self and other. Here we are interested in both how and upon whom—either the self, the other, or both—each actor attempts to act, to deal with or resolve the interpersonal conflict. The definition and integration of this factor has been influenced by interactive or ethological constructs around the relation between dominant and submissive behaviour. We call this second factor the *interpersonal orientation*. We consider its manifestations to be classifiable as either primarily self-transforming, i.e. taking actions that change the self in some way to meet the perceived needs of the other; as primarily other-transforming (taking actions that attempt to change the position or perspective of the other to meet the needs of the self) or as collaborative (actions that attempt to deal with others through a balancing of orientations).

Figure 15.2 provides a graphic depiction of this two-factor model. It basically proposes that, theoretically speaking, the development of certain types of strategies for interpersonal negotiation rely upon the construction of certain levels in the coordination of social perspectives. Thus, for example, truly 'collaborative' strategies (level 3 strategies on our heuristic model) require the level 3 cognitive capacity for the mutual coordination of social perspectives, i.e. the ability to step outside a dyadic interaction and view it from a 'third person' perspective. However, even if an individual has the level 3 social-cognitive competence, this does not mean that he or she will *use* this level of competence in all contexts for interpersonal negotiation. This is depicted in the model by the bidirectional arrows suggesting that strategy use may be progressive or regressive.

Classification of an interpersonal negotiation strategy at one of the four developmental levels (0–3) depends on an analysis of three components operating in the conduct of the moment: the construal of self and other's perspective, the primary purpose, and the perception and control of affec-

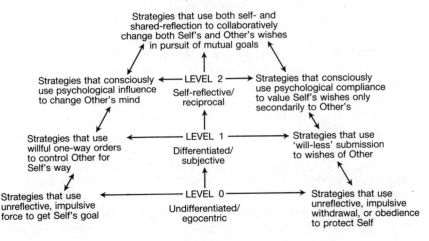

LEVEL 3

Third person/mutual

Strategies that use both self- and
shared-reflection to collaboratively
change both Self's and Other's wishes
in pursuit of mutual goals

Strategies that consciously ←— LEVEL 2 —→ Strategies that consciously
use psychological influence   Self-reflective/   use psychological compliance
to change Other's mind   reciprocal   to value Self's wishes only
                                                          secondarily to Other's

Strategies that use   ←——— LEVEL 1 ————→ Strategies that use
willful one-way orders   Differentiated/   'will-less' submission
to control Other for   subjective   to wishes of Other
Self's way

Strategies that use   ←——— LEVEL 0 ————→ Strategies that use
unreflective, impulsive   Undifferentiated/   unreflective, impulsive
force to get Self's goal   egocentric   withdrawal, or obedience
                                                        to protect Self

Classification of interpersonal negotiation strategies in the *other-transforming orientation*	Ontogenetic levels in social perspective coordination competence	Classification of interpersonal negotiation strategies in the *self-transforming orientation*

Fig. 15.2

tive disequilibrium. The *self–other construal component* involves the conception of self and other which is operative at the moment, i.e. in the particular interactive context. Development in this component moves from the least differentiated level, where self and other are construed as non-psychological objects, with inadequate distinction between self's and other's needs and interests, to an increasingly differentiated and integrative appreciation and valuation of the thoughts, feelings, and wishes of both self and other (i.e. from a 'third person perspective') at the higher level.

The *affective perception and control component* of a negotiation strategy refers to the way the individual perceives and controls inner emotional disequilibrium arising in a specific interpersonal context. At the most undifferentiated level inner emotional disequilibrium is experienced as diffuse, all-encompassing, and not controllable by the self acting directly on his own feelings. At higher levels of development, specific feelings, both within the self and between the self and the other, are differentiated in perception, and the self can control inner disequilibrium by putting various feelings into the perspective of a broader social context. For example, whereas at a lower level strong feelings might direct an action pattern of

impulsive and immediate flight, at a higher level, recognition of disequili-
brated feelings may dictate a self-imposed 'time out' to gain time to
become more in control of feelings when the context is re-entered.

The *primary interpersonal purpose component* of the strategy is the
dominant, reflected-upon purpose for which the action is taken. At the
least differentiated level, the strategy's purpose is seen as the sole and simple pursuit of the self's own immediate wants. Moving to higher (more
integrated) levels, the actor better discriminates and coordinates the purposes of both self and other, giving increasing valuations to each, such that
at the higher levels the purpose is understood to be gaining a differentiated
focus on both the attainment of the self's needs and on the pursuit of
mutually satisfying needs.

Working in coordination with the developmental factor and its three
components is the (earlier mentioned) *interpersonal orientation* (functional) factor. This factor refers to upon whom and how the actor predominantly *acts* in his attempt to deal with disequilibrium between self and
other. In the *other-transforming orientation* the individual primarily
attempts to change the thoughts, feelings, and actions of the other;
whereas in the *self-transforming orientation* he predominantly works on
changing his own thoughts, feelings, or actions. At higher developmental
levels of interpersonal negotiation the individual's actions are increasingly
integrated between the two orientations, both distinguishing self and
other's needs and giving more emphasis to the collaborative process by
which the thoughts, feelings, and actions of both self and other can come
into equilibrium. Therefore, as Fig. 15.2 depicts, development in interpersonal orientation incorporates movement from rigid, isolated, use of one
or the other of the two orientations at lower levels to an increasingly differentiated and integrated interplay between orientations at higher levels.

Using this model, our two major working hypotheses are:

(i) that both developmental level and interpersonal orientation factors
need to be considered simultaneously in the diagnosis of a strategy; and

(ii) that, within developmental constraints, the strategies utilized in a
particular context for negotiation by one member of the dyad are intricately tied to the strategies used by the other (e.g. it may be difficult to function at a more differentiated level in a dyadic interaction when one's
partner is not; on the other hand, it is possible that by using a higher level
strategy, one may 'pull up' a partner whose tendency is to use a lower level
one).

Figure 15.2 depicts the ontogeny of strategies for interpersonal negotiation as manifest in the individual; from the use of unreflective, impulsive
strategies to pursue immediate, material, and one-person focused goals at
one or the other of two unrelated and distinct orientations to the use of
self- and shared-reflection to pursue mutual or relational goals (ideally in

collaboration with other) with a balance between action oriented toward changing the self and those oriented toward changing the other.

Table 15.1 *Some prototypical interpersonal strategies coded at developmental levels 0–3 in each orientation*

Other-transforming orientation	Self-transforming orientation
Level 0	
A Forcefully blots out other's expressed wish	A Takes impulsive flight
B Unprovoked impulsive grabbing	B Uses automatic affective withdrawal
C Absolute repulsion of Other	C Responds with robot-like obedience
Level 1	
A Uses one-way threats to achieve Self's goals	A Makes weak initiatives with ready withdrawal
B Makes threats of force	B Acts victimized
C Criticizes Other's skills as a rationale for Self's activity	C Appeals to source of perceived power from a position of helplessness
Level 2	
A Uses 'friendly' persuasion	A Asserts Self's feelings and thoughts as valuable but secondary
B Seeks allies for support of Self's ideas	B Follows but offers input into Other's lead
C Goal-seeking through impressing Others with Self's talents, knowledge, etc.	C Uses Self's feelings of inadequacies as a tool for interpersonal negotiation
Level 3	
A Anticipates and integrates possible feelings of Others about Self's negotiation	
B Balances focus on relations with focus on Self's concrete goal	

Table 15.1 provides a sampling of prototypical strategies—as categorized by level and orientation. The examples are meant to illustrate the underlying structure of each category but are by no means exhaustive. The model emphasizes that various strategies within an orientation may appear at first glance similar, but developmentally speaking (e.g. using the component analysis) be at different levels. For example, leaving the actual physical area (context) of a negotiation as a way to resolve a conflict, may reflect different developmental levels, depending on how self and other are construed in the situation, what the primary purpose of leaving is, and how leaving helps the actor to deal with feelings. Conversely, the two orientation components identify how strategies that on the surface may appear quite different, i.e. fall into different orientations (e.g. fight versus flight), may be at the same developmental level of organization.

It should be stressed that assigning kinds of strategies to levels and orientations is a theoretical heuristic. The categories can stand reliably whether organized developmentally or not; they describe forms of negotiation and can be related to such validating criteria as age, degree of adjustment or pathology, variations in stress, or general social maturity. Assessing an observed strategy at any one level or orientation is an inferential process based on a particular developmental perspective and theory.

A clinical-developmental research context for relating strategies for interpersonal negotiation to developing communicative competences

There are many roads down which one can take the model just described. The most clearly marked empirical path, and the one most often travelled, is normative-descriptive research. This includes, but is not limited to, the construction of reliable measures to assess level and orientation (Selman and Demorest 1984), the cross-sectional and longitudinal study of the relation between strategies utilized and factors such as social-cognitive competence, age, gender, social class, culture, and ethnic background (Selman *et al.* 1983), the study of the adaptive nature of strategy choice, including comparisons between better and poorer adjusted individuals, and an often neglected, but most important type of developmental research, studies designed to look at the interdependence of the strategies used by two individuals at a given point in time and over time. This latter approach may yield important information on how processes of social regulation relate to developmental variation.

The research activity reported here has taken a slightly different turn. Because of our dual interest in the relation of child psychopathology to normality, and in the role negotiation strategies play in the broader development of interpersonal relationships, we have begun a series of studies of social interaction patterns as played out within the context of a clinical intervention for children and young adolescents called Pair Therapy (Selman and Demorest 1984; Lyman and Selman, in press). The clinical aim of pair therapy is to provide a therapeutic context in which two children, whose social relationships have proven problematic and ineffective, can work (together) to gain the skills, rationale, and *inner capacity* to relate to peers. The treatment is not limited to one circumscribed aspect of social development. Rather an attempt is made to improve the child's ability to develop, and to use flexibly and effectively, strategies for interpersonal negotiation from a repertoire of possible alternatives. To this extent the goal of therapy is directly related to the theoretical model of interpersonal negotiation strategies. Yet attention is paid also to improving skills in self-reflection and communication, in anticipating, planning, and problem-solving, in sharing and playing interactively for an extended period of time, in trusting self and other enough to develop a sense of effectiveness and a willingness to be vulnerable.

The therapist's role in pair therapy is important, yet the goal is to minimize this importance so that the children eventually learn to relate to each other with relative autonomy. The therapist attempts to facilitate the children's interaction in ways that are developmentally functional, setting a safe context for negotiation in an atmosphere of warmth and possibilities, but also of control and limits. For research purposes, it is stressed that the context of pair therapy provides several conditions that facilitate study of the relation between peer interaction and cognitive-communicative development. For one thing, it allows for an in-depth analysis of repeated interactions between the same dyad over a time span long enough to allow for the possibility that broad scale developmental shifts will occur and be observable. In addition, it allows the observation not only of peers in interaction with each other, but also with an adult who frames the nature of the social interactions in a powerful and particular way. Finally, in looking at the social interactions of children and young adolescents who have great difficulty in their social relations within a context designed to allow for the correction of *pathology* and the *development of social skills*, pair therapy provides a canvas on which is painted in bold strokes the nature of social interaction and regulation processes that otherwise might be more subtle and difficult to detect in the observation of better constituted children.

a. Heather and Jennifer: some speculations on how changes in communicative competence relate to changes in negotiation strategies

The Heather and Jennifer pair was observed from its onset, in January of 1983 until June of 1984. The focus of this case-study is on how the communicative competencies that are utilized during the course of pair therapy relate to changes in patterns of negotiation strategies, also observed during the treatment process.

To describe this case, I will use the following format. First, using the interpersonal negotiation strategy model as a clinical heuristic, I will provide a selective overview of the relationship between Heather and Jennifer as it evolved over the course of treatment. This course is divided into two phases, the demarcation is based upon an abrupt shift in the pair's focus that occurred following the 33rd session. I will then describe a preliminary functional classification system used to characterize certain striking shifts in the communicative interaction during this course of treatment. I close this section and the paper with a summary of the implications of the findings for hypotheses about the nature of normal social developmental and adaptational processes during early adolescence.

Phase 1: the game of 'Life'

Prior to their inaugural pair therapy session, Heather and Jennifer were introduced to some of the basic tenets of the approach: the pair must contract to meet for at least the school year, the pair must stay together during sessions, no physical violence or continual verbal abuse is allowed, etc. Then, as is usual during the first session of a pair, Jennifer and Heather were asked to decide together upon a game or activity from a finite list of options (e.g. crafts, board games, an activity in the gym). As in all pair therapy, they were informed that it was not required that they do things together at all times, but that they stay together. However options for shared play were made available and encouraged. In this initial session, and in all sessions to follow, it was Heather who moved to dominate the interaction. She was more active in scanning the room, in exploring the available materials in greater depth, and in unilaterally choosing that the pair play a relatively challenging board game called 'Life'. Heather claimed some familiarity with this game. Jennifer did not, but she accommodated to Heather's wishes without posing an alternative activity for the pair's consideration. During this initial session a dynamic interaction pattern was quickly established that lasted throughout each of the thirty-three sessions taking place between January 1983, and February 1984. This period was identified as phase one.

Each session during this phase began with the two girls running ahead of the therapist to the pair therapy room and, at Heather's insistence, playing the board game, Life, for the entire session. In this game each player alternatively moves her piece a designated number of spaces, based on the random spin of a wheel. Depending in part on chance and in part on skill, each player acquires a profession, with a regular salary, a family, material possessions, and capital. As in real life, losses are also possible due to a combination of poor judgment and bad luck. As each player reaches the end of the board, their accumulated wealth is counted, and according to the rules, the winner is the person who has earned the most money in her travels across the game board.

Looking in on this pair from the vantage point of the observation room, one gained the impression that the game had little shared meaning for either girl in either a competitive or a collaborative sense. Each girl appeared to be playing a separate game. Heather appeared obsessed with attaining for herself as much wealth as possible, but this attainment appeared independent of how well or poorly Jennifer did. For instance, Heather expressed neither dismay nor excitement in reaction to any of Jennifer's fortunes. Correspondingly, Jennifer also seemed to be unconcerned with her own fortunes, and totally disconnected from the competitive intent of the game as well. When it was her turn, she spun the wheel and moved her token in a rote, almost affectless fashion; little expression of feelings was attached to either success or setback. Nor did she seem concerned with Heather's fortune either. However, Jennifer did seem quite concerned with avoiding any possible confrontation with Heather or in taking any action that would incur Heather's displeasure.

During these thirty-three sessions of phase one, Heather appeared to be very impulsive, aggressive, and bossy; all examples of what we call other-transforming on our functional-adaptational dimension. Jennifer, on the other hand appeared to be fearful, overcontrolled, timid, and submissive; or in our theoretical terminology, self-transforming. Seldom during this phase of treatment did either girl spontaneously engage in any direct conversation with the other. For instance, seldom if ever did either reach out to connect with the other by inquiring as to what 'life' might be like for the other outside of the sessions. Nor did either reveal easily to the other what psychological reactions were taking place within. Neither shared or inquired into what relationships were like, for the self or other, with significant other persons in their lives. In other words, there was little inter-penetration between the two girls.

With respect to specific strategies observed in contexts for interpersonal negotiation, e.g. choosing an activity, or sharing snacks, from the opening seconds of the initial session Heather continuously and ruthlessly asserted herself using 'other-transforming' strategies both quickly and totally. And Jennifer complied, equally quickly and totally. She did nothing to challenge Heather's near total dominance. It was Heather who would define the day's major activity, and even within the chosen activity it was Heather who took steps to wrest total control. For example, Heather would spin the wheel to determine how far she could move during her turn, and then, impatient to get to her next turn, she would impulsively spin the wheel for Jennifer without asking, even moving Jennifer's piece without observable objection on Jennifer's part. Or Heather would 'allow' Jennifer to spin the wheel for herself, but then would not wait for Jennifer to take her move before she was spinning the wheel for her next turn.

As part of the therapeutic structure of this pair, each week the therapist would bring in a snack, two chocolate bars, for the girls to share. Typically Heather would wolf down her own bar, and then go through a sequence of (codable) negotiations with Jennifer in a blatant attempt to acquire whatever snack Jennifer had not yet finished. A typical sequence was as follows: first Heather would ask Jennifer for part of her snack, and in the early sessions Jennifer would usually comply, saying she did not like chocolate. Later, when Jennifer began to show some resistance, Heather would plead, then demand, and sometimes even attempt to grab. At such junctures the therapist would come to the girl's support, setting a limit on Heather's actions, and encouraging Jennifer to take limit-setting action on her own.

Thus although Jennifer was less overtly disruptive or aggressive, her own immaturity and insecurity colluded with Heather's to produce a very rigid, low level, form of interaction between the two girls during this phase. To couch the summary description of this initial phase in the terms of the developmental negotiation strategy model, at those not frequent but critical times in each session when Heather and Jennifer faced each other in explicit contexts for interpersonal negotiation, Heather most often utilized threats or forceful physicalistic strategies in an other-transforming orientation (levels 0 and 1), and Jennifer, conversely, exhibited automatic obedience or will-less submission in a self-transforming orientation (also levels 0 and 1), generating an apparently 'stable' (but rigid) pair. This clinical impression is quantified in Table 15.2, which reports the distribution of all detected interpersonal negotiation strategies used by each girl by level and orientation for the 33 sessions of phase 1 and the 14 sessions of phase 2.

Table 15.2   *Level and orientation of interpersonal negotiation strategies (INS) used by Heather and Jennifer during phase 1 and 2 of pair therapy (absolute number)*

	Level	Phase 1 (33 sessions)		Phase 2 (14 sessions)	
		Orientations		Orientations	
		S.T.[a]	O.T.[a]	S.T.	O.T.
HEATHER	0	0	12	0	4
	1	2	54	5	41
	2	0	0	0	2
	3	0		0	
JENNIFER	0	14	1	7	2
	1	39	2	16	3
	2	0	0	1	2
	3	0		0	

[a] S.T. = Self-transforming orientation; O.T. = Other-transforming orientation

It is striking, both from a qualitative and a quantitative perspective, that the ground for the development of more age appropriate (in this case level 2) strategies (reciprocities, trades, barters, manipulations, uses of reasons, etc.) hardly seemed to be broken during this first phase: most eight to twelve or thirteen-year-olds are able to use strategies that fall at level two and are beginning to discover those representative of level three. However it would be incorrect to assume that the seeds of social development have not been sown, nor that the ground has not been at least to some extent softened. In fact, I would speculate that two important conditions have been established for later growth; one is a sense of the continuity of the relationship, and the other is a sense of its safety. This is the important work of the therapist who provides assurances that the girls will meet each week in safety, setting clear limits and not judging, blaming or taking personally the girls' actions.

Phase 2: Life; no longer a game, but not quite real

At the beginning of the 34th session, Heather announced, 'Jennifer, I want to ask you a question about a boy in your class, Donnie A'. With this question the girls switched to a new track. For the next fourteen sessions, till the end of the school year, the girls put aside the boardgame 'Life', established a small fort in the corner of the room, surrounded it by chairs, and held a weekly dialogue, the primary topic of which was the activities and interests of Donnie. Initially, Heather's inquiries were almost exclusively about how Donnie felt toward her, a somewhat strange line of inquiry, for according to our own observations and discussion with the staff of the school which both girls attended, it appeared that Donnie was completely unaware of Heather's infatuation with him and showed no signs of having reciprocal feelings. Adding to the bizarre nature of the interaction between Heather and Jennifer around the topic was the way Jennifer provided for Heather, in her role as classroom reporter, fabricated romantic fantasies about how much Donnie cared for Heather. It was difficult at times to tell to what degree one or both of the girls entered completely within the boundaries of these fantasies and became so embedded in them as not to be able, at the time of their communication, clearly to differentiate them from the 'reality' of Donnie's interests.

Nevertheless, the switch in the quality of the communicative interaction between the two girls from phase one to two was striking. From little direct communication between partners, the session turned exclusively to a setting for communication, even though the communication seldom focused directly on the relationship between the two girls. To try to capture this dramatic shift in a systematic and empirically meaningful way, I have defined three category types of communicative competencies, and have scanned the corpus of videotapes from each of the 47 sessions to get a sense of their distribution. The communicative processes identified as salient to this pair are summarized in Table 15.3. They are as follows:

1 *Communication for exploration of self and other.* This category includes verbal attempts to reach out to inquire about the psychological life and activities of other and/or attempts to reach in and reveal the actions or feelings of the self to other. (The absence of evidence of this function suggests a kind of self absorption with neither an expressed interest in other nor the impetus to open the self to other.)

2 *Communication to test the reality of other's expressive behaviour.* This category attempts to capture a phenomenon that is fairly important, but also idiosyncratic to the interaction between these two girls; the use of

verbalization to test the 'truthfulness' of the other girl's expressions. (The converse of this category is either collusion with one's partner in fantasy or loss of self's control in the face of other's losing control.)

3 Expressions of shared meaning between partners. This is a dyadic category, requiring interaction between the partners in order to be scored. In addition, as can be seen in Table 15.3, the subclassifications of this category may form a hierarchy from a primitive shared focus on bodily functions to a more advanced focus on shared reflection.

What, if anything, does an examination of the rough-cut and preliminary quantitative summary in Table 15.3 suggest for our understanding of Heather and Jennifer's relationship as well as for interpersonal interaction and cognitive development, generally, in early adolescence? Not surprisingly, during phase 2 we see a considerable increase in the usage of each of these three communicative processes. With reference to the category, 'exploration; reaching out, reaching in', it is interesting to note that seldom did either girl spontaneously use this type of communication during phase 1. For example, whereas Jennifer was able, at times, to respond to the therapist's inquiries about her home life, Heather was unable to reveal even the simplest forms of information, e.g. the name or age of her brother or of her pet kitten. During the second phase, however, she was able spontaneously to reveal personal information such as, 'there is a boy in my class who likes me, but I don't like him'. In addition, she was increasingly able to reach beyond her own preoccupations to make spontaneous inquiries into Jennifer's world as well, e.g. asking Jennifer if there was any boy that she liked.

Consideration of the category, 'communication for reality testing' provides us with a glimpse of how the (hetero)sexual drives of early adolescence may not only channel same sex peers into closer interaction, but also indirectly provide the fuel to speed growth in their communicative and cognitive skills. Driven by her infatuation with Donnie, it was observed that as phase 2 progressed, Heather made increasing demands on Jennifer for more valid information and reports regarding his thoughts, feelings, and actions. Jennifer, on the other hand, driven perhaps by her own need to be liked by Heather which had developed in the course of pair therapy, attempted as best she could, to provide the material she sensed Heather desired. At first, Jennifer simply made up elaborate fantasies ('He kissed the teacher; He wrote you a love note') that Heather 'buys' ('Oh, Donnie, you're so cute. What did he do next?'). There was no clear sign, however, that either girl was clearly aware, in the sense that Gregory Bateson (1955) has made so well known, that 'this is play'. Eventually, as Heather reached beyond this initial (narcissistic) need to be fed evidence of Donnie's feelings for her, she made increasing demands on Jennifer for more realistic material ('No, really, what did he say?'). In fact, it was this aspect of their

Table 15.3 *Distribution (absolute number) of three types of communicative competencies used by Heather and Jennifer during phase 1 (P1) and phase 2 (P2) of pair therapy*

		Heather		Jennifer	
		P1	P2	P1	P2

I. INTERPERSONAL EXPLORATION: REACHING OUT TO OTHER; REACHING IN TO REVEAL SELF TO OTHER

	Heather P1	Heather P2	Jennifer P1	Jennifer P2
a) Inquiries into the real-life social interests of the other member of the pair (e.g. 'What kind of movies do you like, Heather?)	4	15	2	21
b) Spontaneously expresses one's own interest or feelings to the other (e.g. 'I don't like David Potter, even though they say I do')	6	11	2	5

II. REALITY TESTING: VERBALLY TESTS THE VALIDITY OF OTHER'S EXPRESSIVE COMMUNICATIONS

	Heather P1	Heather P2	Jennifer P1	Jennifer P2
a) Directly attacks the veracity of other's statements (e.g. 'You're lying'), or threats to check if other's statements are untrue (e.g. 'You better be right. I'm going to check after school.')	8	11	0	0
b) Makes direct inquires as to the factuality of other's communications (e.g. 'Come on! Did he really say that?')	2	10	0	1

III. FORMS OF EXPRESSING SHARED EXPERIENCES

	Phase 1	Phase 2
a) Dyad shares interest in expression of bodily functions (e.g. makes burps together)	12	6
b) Dyad embeds together in a fantasy experience without either member having a clear sense of the fantasy/reality border (e.g. Heather: What did Donnie do today? Jennifer: He wrote a poem about how much he loves you in his Algebra book (untrue). Heather: Oh, Donnie, you're sooo cute!)	1	29
c) Dyad shares in a discussion of social facts, situations, gossips, information, etc. e.g. J: Do you know what David did on the overnight? H: No, what did he do? J: He was so scared, he couldn't sleep. H: What did Donnie do? J: He was telling people to be quiet.	14	39
d) Dyad shares in humour, sarcasm, teasing as a way to relate to each other. e.g. J: If Donnie likes girls so much, the teacher could bring him a Barbie Doll (in a kidding tone of voice) H: Yeah, a life-size Barbie doll (adds on) J. and H. both laugh together	1	6
e) Dyad shares in self-reflection e.g. J: I'm going to miss Mr. H. (a counsellor who is leaving) H: Yeah, me too!	0	1

interaction that this category was meant to identify. In turn, this appeared to have a positive effect on Jennifer's social-communicative competence. Mostly abandoning her more bizarre early fabrications, she actually became a better observer, a more sophisticated reporter of social inter-action, and a more proficient discussant regarding psychological inferences about the meaning of others' (Donnie's) behaviour. In turn, this appeared to have a reciprocal effect on Heather who, shedding her unrealistic focus on what Donnie thinks of her (powerfully little), began to emerge from her narcissistic orientation to probe with Jennifer in a more realistic way what kind of person Donnie really is ('Does he like algebra?').

Perhaps the most striking shift in communicative competence was cap-tured by the dyadic interactive category, 'expressions of shared meaning'. During phase 1, instances of this category were almost entirely in the form of shared expressions of bodily functions. Often the two girls would become involved in burping contests or see who could make the loudest fart-like sounds (by pressing their lips to their arms and blowing force-fully). During most of phase 2, this form of linking up with one another was replaced by what we have called earlier 'shared embedded fantasy'. How-ever, toward the end of this phase, a displacement of this second mode was observed to make room for shared gossip and even some mutual joking. For instance, at one point Jennifer and Heather began to explore together whether Donnie was interested in girls. Escaping from the embedded fan-tasy mode, Jennifer says to Heather, 'If he does like girls, and the teacher catches him thinking about them, she can get him a Barbie doll to hold'. (Note the conditional form of speech, rare in the conversation of these two girls.) Heather builds on this idea, 'Yeah, a life-size Barbie doll for his very own.' This interaction is said with some real humour; indeed, the signal 'this is play' is now clear, as confirmed by the girls' laughter together in an all too rare moment of collaboration.

b. Change in negotiation strategies: is there any sign?

Table 15.2 suggests that very little change has occurred over time in the way the two girls negotiate (deal with) each other, either in developmental level or interpersonal orientation, even though there have been some strik-ing shifts in their mode of communication. This should not be too surpris-ing when one considers that their communications are about a third party, whereas their negotiations require them to deal with their own immediate feelings and motives in a context of face to face interaction. And yet there is a small clinical sign that a shift upward in level of negotiation strategy may eventually occur. During session 45, in the middle of a conversation between the two girls regarding Donnie's algebra work, Heather, as is typi-cal, makes the switch in topic. She shifts the focus to the distribution of candy between them, an important part of the tradition of this pair. She says to Jennifer, 'Where's your hands, Jennifer?' When Jennifer shows them to her, Heather proceeds, 'Do everything that I do.' This communication is delivered as a request rather than the more typical and familiar threat, demand, or order. Then Heather raises her hands, and in a reciprocal fashion, rather than the more usual obedience orientation (self-transforming level 1), Jennifer follows. Now laughing, Heather makes what appears to be a playful grab for Jennifer's chocolate pieces, different from her more typical approach. Jennifer also reacts differently from before. Rather than reacting with either wild screeches or totally will-less abandonment of her claims to her own candy, Jennifer responds verbally in an untypically controlled way. 'Stop it, Heather', she says in a controlled and forceful voice, as she pulls back her candy. Heather yields but proceeds, 'I'll give you this magic piece of candy (a very small piece) for what you have left'. 'Nope', responds Jennifer, 'I've already given you some before, and I want to finish what is left myself'. Heather accepts this reason and switches the topic back to its previous focus on Donnie. In this brief interaction we see the type of bargaining and reasoning that is typical of strategies we classify as level 2, a level usually well consolidated by children this age. For Jennifer to give a reason for not yielding implies that she understands that the reason may be listened to (by Heather) and a belief that it will have some force. For Heather to abide by Jennifer's reasoning, rather than to respond only to the control of the therapist, or to capitulate to her own felt needs, suggests that she too feels more in control and therefore is in a better position to engage in the barter and trade that is part of a higher (more differentiated) level of interpersonal negotiation than she is used to utilizing.

Conclusion

I conclude this paper on two notes, the first a speculation focused on the substance of the research data and the second, a statement on the integrative interest of the heuristic model. The speculative note is this: in the transition from middle childhood, through puberty, to early adolescence, new (some biologically induced) interests appear that require the implementation of communication skills like those identified in the last section of this paper. Peers now have greater motivation to learn how to reach out and explore other's feelings, as well as to reach in and articulate to others their own. They need to find new ways to share feelings about meaningful aspects of their lives, such as sexual attraction for others. Through intense dialogue with a trusted friend, using shared gossip, humour, or reflection, they rehearse the face to face negotiations and communications they will need to undertake in dealing with the intimate tasks of later adolescence and adulthood. For most early adolescents, these communicative competencies serve as the basis for the implementation of negotiation strategies for mutual collaboration, those strategies characterized as level 3 in our model. For Heather and Jennifer who are troubled and immature children, communicative interaction may be the way to set the stage for the self-reflection and reciprocity skills characterized as level 2 in our model, and more typical of the strategies of middle childhood. Even so, by observing troubled children like these, under the intense, continuous, and rich context of a process like pair therapy, we can begin to get a sense of what peer interaction may be like for more typical early adolescents, and how the needs of this period drive individuals together to develop mutually their communicative skills and negotiation strategies.

My second concluding note refers to the integrational intent of the proposed model. When behavioural scientists use the term 'better', they are not necessarily making statements of value. Developmentalists using structural approaches usually take 'better' to mean greater cognitive complexity, higher level mental processes, or greater behavioural differentiation and hierarchical integration. Those using primarily functional models often use the term 'better' to mean more adaptive, a better fit into some ecological niche, or as we have used to depict social interaction with others, a 'better' balance between self and other. The heuristic model that I have proposed here to capture strategies for interpersonal negotiation suggests that these notions of 'better' are at once distinguishable yet ultimately integrated. Structural level and functional orientation are two sides of the coin of interpersonal development.

Acknowledgements

I would like to express my appreciation to the members of the educational and clinical staff of the Manville School of the Judge Baker Guidance Center for their support of the Pair Therapy Project. Dr Pauline Hahn served as the 'pair therapist' in the study reported here; her insight into the meaning of the behaviour of the two subjects in this study has been relied on heavily herein. In addition, I would like to thank several members of the project who have read and provided me with their comments on this draft: Catalina Arboleda, Keith Yeates, and Mira Schorin.

Discussion

Normal vs. abnormal trajectories

Sherman commented:
'I would suggest that we consider that there may be common principles of devel-

opment in normal and abnormal samples within the separate lines of development (i.e. physical, social, cognitive, emotional, etc.), but when the rate of development within one line of development differs from that in the others—so that as a system the development across the separable spheres is out of synchrony—the resultant behaviour, or patterns of behaviour, is not one seen in normally developing children. Synchrony is meant to imply "at equivalent rates of development", so that all domains would be slowed or accelerated to the same degree. This is never true with abnormal development. There is always some asynchrony across the different domains.

The implication of this claim is that what one may learn from data on abnormal populations that is informative about normal development, may be quite limited. So, for example, Shure's observations on disturbed 4-year-olds and what intervention helps to reduce their extreme aggressiveness, may reveal little if anything about how normal toddlers learn to control their aggressive impulses.'

Selman replied that in his view 'synchrony' is a relative matter. 'Although even in "normal development" there is not likely to be lock-step synchrony, widespread asynchrony is one way to characterize pathology, another being fixation or retardation. Nevertheless I believe much can be learned about both normal and abnormal development by examining each in relation to the other.'

Hartup asked:

'Is it the case that the "developmental trajectory" upon which troubled children are embarked is qualitatively different from the negotiation trajectories of normal children, or is the difference between these two groups one of delay on a common trajectory? Or perhaps there is a third model that best describes the differences between such groups?'

Selman held the view that children whose difficulties emerge from a troubled interpersonal environment are neither simply delayed nor simply different, but act in ways that exemplify the interaction of delay and difference. And since certain biological functions continue to develop, whilst psychosocial ones may not, such children display the marked asymmetry discussed by Sherman.

Level of response

Shure raised a question about Selman's model:

'Could you address how you probe to place a response at a particular level. For example, is threat of force always level 1, or is it a higher level if say, done in defence of a friend (maybe level 3, A anticipates and integrates possible feelings of others).'

Selman replied that whether a strategy is high or low depends in part on the level of reflection, purpose, and affective control that underlies the strategy. Strategies may 'look the same' but be at different levels in terms of the underlying components (see Selman and Demorest 1985).

Strategies

Hinde referred to Selman's view that 'strategies at a particular level of conduct may indeed precede reflective understanding at the parallel cognitive level'. He then asked what controls *conduct*?

In reply, *Selman* said that this question often arises in the research literature in the form, 'Are measures of (social) cognitive level tapping reflective thought, i.e., conceptions about something, or are they tapping cognition-in-conduct?' It is the latter case where it is easy to hypothesize that children may have (or develop) behavioural strategies before they *know* strategies.

Postscript

After the meeting, *Selman* provided a note on the relations between the papers in this section, of which this is a slightly abbreviated version:

'The Radke-Yarrow and Sherman paper strongly supports the suggestion that for the sake of prediction, information about affective factors, both internal and interpersonal, are important mediators between a child's understanding of prosocial or altruistic behaviour, and the implementation of such understanding in action. Shure, on the other hand, seems to take the position that regardless of mediators, the most effective way to facilitate social development and prosocial behaviour is by providing the opportunity to learn as large a repertoire of social problem-solving skills as (developmentally) possible. Unlike Radke-Yarrow and Sherman, Shure does not emphasize how factors like temperament or affective context may bridge the gap between social thought and action (knowing and doing).

Selman's paper suggests that the emergence (ontogeny) of strategies for interpersonal negotiation (or social problem solving) may have a developmental timetable, related in part to the development of social-cognitive competence. But in accord with Radke-Yarrow and Sherman, Selman emphasizes that social-cognitive developmental models alone are not sufficiently powerful to explain or predict the nature of social behaviour of specific kinds. Unlike the ideas espoused by Radke-Yarrow and Sherman, however, Selman locates certain affective aspects directly "inside" the social behaviour itself, rather than seeing them as independent mediating factors. Unlike Shure, Selman posits that certain strategies or skills are developmentally more advanced than others, and hence the quality, as well as the quantity, of the repertoire must be considered. In each of these three research enterprises, a second question must be addressed: how does social experience and interaction generate greater social understanding? Being a preventive intervention, Shure's programme of work tends to assume that deficits in social cognition, once removed, may lead to "better" behaviour. Selman suggests that disequilibrium between persons leads to disequilibrium within persons that may generate "higher level" strategies and understanding. And together with Radke-Yarrow and Sherman, Selman suggests the interactive nature of social interaction and social understanding.'

Section E

16

Age changes in cognitive abilities
EDITORIAL

Our knowledge of how both cognitive and social performance change with age is now considerable. Understanding of the causal relations between cognitive, emotional, and social development is however a much more difficult task. In general, work in this area ranges from the examination of relations between indices of cognitive development and aspects of social perception or social behaviour, through attempts to specify contextual variables important for development, to attempts to come to grips with process, employing more theoretical concepts. We may consider some examples.

One index of cognitive functioning which has been compared with aspects of social perception is mental age. Because of its high correlation with chronological age (.85), the latter is also sometimes used as an index of cognitive ability. For example, in one study, over the $5\frac{1}{2}$ to $11\frac{1}{2}$ year age range, behaviour in perspective-taking tasks was found to be related to both chronological age (.47 to .78) and mental age (.43 to .77). However, another index of cognitive ability, IQ, was only weakly related to perspective taking, with the relation depending on the particular intelligence test used, since some intelligence tests tap only one ability while others combine many. Furthermore the correlation between IQ and perspective-taking varies with age level, being highest at the pre-school level (.39) and lowest at 5th grade (about 10 years old) (.10; Rubin 1978). Thus the relations between indices of cognitive functioning and indices of social development depend on many factors, including the precise indices used and the age of the subject.

More theoretically based cognitive indices involve tests of such aspects of cognitive functioning as conservation (e.g. the ability to recognize that the number of items in an array remains the same in spite of a change in configuration, or that the amount of liquid in a tall thin jar may be the same as that in a short wide one). These have also been related to various perspective-taking tasks (e.g. Rubin 1973). But whilst there is some evidence for relations between logical thinking and social inference, parallels between social inference and behaviour such as attachment and altruism are not always clear (though see Chapter 12), and parallels with popularity are even less apparent (review Serafica 1983). Furthermore, as Shantz (1983) points out, too often the level of cognitive ability has been merely inferred from a presumed relation between age and theoretical stages of mental development, and then correlated with the ability for social inference.

Identification of the precise causal agents for each aspect of developmental change is again no easy task. In the next chapter Bryant discusses this problem. He argues that studies based on correlations between environmental measures and cognitive performance are in themselves quite inadequate to demonstrate causation. Correlation can give no information about the direction of causal effects, and may in any case arise from the action of a third factor which affects both of the measured variables (the 'tertium quid') (see also Jacquard 1982).

In addition to these issues mentioned by Bryant, there are other problems with correlational studies. Thus correlations imply that if high values of a given environmental variable were associated with high values of a measure of cognitive functioning, the opposite would also be the case. But this is not necessarily the case. It may be that only very high, or very low, values of a particular environmental variable are relevant to cognitive functioning. Again, the correlates of high levels of cognitive functioning may not be the opposite to those of low levels. And cognitive functioning may be affected by diverse environmental variables which are alternatives to each other. For these and other reasons correlational techniques are often poor tools for demonstrating relations between one set of measures (here, the environmental factors) and supposed outcomes (aspects of cognitive functioning), especially with the small samples of subjects that are necessary for intensive study (Hinde and Dennis, in prep.).

Bryant also has misgivings about the conclusions that can be drawn from intervention studies. Usually they involve so many environmental changes that it is difficult to know which matter. Laboratory experiments, involving intervention with careful control of the variables can demonstrate cause and effect, but then it is always possible that the effects observed are irrelevant in the real world. However, Bryant argues that correlational studies with a longitudinal element, *coupled with* intervention studies, can produce hard evidence about cause and effect: he illustrates his thesis from his own work.

In the course of his discussion, Bryant emphasizes the importance of the social context in which the classic Piagetian tests of cognitive development are presented—an issue increasingly important in later chapters and related to Chandler's emphasis on the interaction between social circumstances, cognition, and the understanding of the environment as complexly structured.

The study of the developmental process is often complicated by the difficulty of specifying the nature of the cognitive change underlying a given developmental change in behaviour. For example, developmental changes in moral judgment could reflect not a gradual increase in cognitive complexity, but the operation of a third variable, namely the child's sensitivity to adult behaviour. A relevant example is provided by a study of the development of moral judgment which questions Piaget's proposition that the child moves from a concrete focus on the *consequences* of behaviour to a more subjective focus on the actor's *intentions* for performing the behaviour. Costanzo *et al.* (1973) found that if an act produced a negative consequence, then young children failed to consider the actor's intentions, as predicted by Piaget and others. However, when consequences were positive, children as young as five years rated actors with good intentions more positively than actors with bad intentions. The probable explanation of the failure to consider intention following a negative act lies in a difference in adults' responsiveness to negative and positive behaviour by the child. Adults are prone to condemn negative actions (e.g. the accidental breaking of crockery) regardless of the child's intentions, but are more likely to take intention into account when judging positive actions: children become more independent of this parental bias with age.

Further complexities arise from the fact that the consequences of experience are not to be seen as merely impinging on a passive child. For full understanding it is necessary to see the child as an 'active cognizer', whose mental processes structure his or her social world (Shantz 1983, p. 504). The importance of the role of the perceiver was demonstrated in a study of the types of descriptive categories used by 11 to 13-year-olds at a summer camp. When two children were asked to describe the *same* other child, there was 45 per cent overlap in the categories used, but when one child was asked to describe two *different* others, the overlap rose to 57 per cent (Dornbusch *et al.* 1965). The greater overlap in the latter case demonstrates that what we perceive is a function not only of the characteristics of the perceived, but also of the characteristics of the perceiver. The work on 'implicit personality theory' is based on the view that cognitive factors influence perceptions of others (e.g. Bruner and Tagiuri 1954; Olshan 1970).

Furthermore, these perceptions (at least as revealed verbally) show developmental changes. Around the eighth year, there is a major shift in the ways in which children describe others from highly concrete and global categories (e.g. age, sex, physical appearance, routine habits) to more abstract and inferential categories (e.g. personality traits, motives, attitudes). The 12 to 16 year period shows increases in qualifying terms (e.g. sometimes, quite) and organizing terms (e.g. explanations of behaviour and situational factors) (Shantz 1983, pp. 505–508). Similar developmental changes occur in making inferences about the causes of behaviour. Whereas six-year-old girls merely described behaviour they saw in films, without giving any explanations, nine and twelve-year-olds spontaneously attempted explanations (Livesley and Bromley 1973).

Additional evidence for developmental changes in inferring why a person behaved as he did has been provided within the framework of social attribution theory, with distinctions made between internal (within person) causes and external (outside the person) causes (see Kelley 1971; Shantz 1983, pp. 518–522). In the present section, Chandler emphasizes the interaction between these two sets of variables. He demonstrates experimentally that in order to predict outcome measures, one must specify both the person variables (in this case cognitive ones) and the situation variables—neither on its own will do (see also Moscovici and Andpaicheler 1973). It should be noted that, in keeping with the view that social phenomena involve dialectics between successive levels (see Introduction to this volume), Chandler's view requires neither that social reality determines the person's cognitive structure or vice-versa (an issue we shall return to later) but points to the need to characterize the environment (and the individual's understanding of this environment) as complexly structured. This social understanding, the cognitive developmental level of the subjects and their capacity to take into account different points of view within a group discussion jointly contribute to determine their stage of moral judgment (Oser 1981).

17

Parents, children, and cognitive development
PETER BRYANT

Some questions in developmental psychology are causal questions, others are not. As it happens the ones that deal with causes are the hardest to answer. They are also, at least from the practical point of view, usually the most important and the most interesting. If you know what causes development you are in a much better position to change it. To ask what effect a child's parents have on her cognitive development is to pose a causal question about development and a particularly intriguing one at that. It is intriguing because of the possibility that there are things that parents do together with their child which in the long run have a far-reaching effect on the child's understanding of his environment, and yet are done not to foster the child's intellectual growth, but simply for fun. Games, chit-chat, stories, walks in the park, even arguments and downright confrontations—these are things which are often done only because the child and her parents either enjoy them or cannot avoid doing them, and yet they may be an essential part of that child's intellectual growth.

It seems plausible enough. The child must need some experiences and some stimulation if she is to learn about her surroundings, and in her first few years at least it is her parents who are most likely to provide these things. Their immediate aim will not always be to foster their child's intellect and yet what they do might have far-reaching effects. It is only when one tries empirically to establish that there is a definite link between these two things—the interactions between parent and child and that child's intellectual development—that one begins to see what the difficulties are.

There are two problems, one general and the other rather specific. The general problem is how to establish causes in development. Psychologists have on the whole ignored the limitations of the empirical tools they they use to isolate cause and effect. The specific problem really concerns the strikingly vague nature of the hypotheses and of the measures involved in attempts to show that parents affect children's cognitive development. Although we have detailed hypotheses and very detailed measures of particular aspects of children's cognitive growth the attempts to track some connection between that growth and parents' behaviour rarely take advantage of this kind of detailed analysis. The people who try to make this kind of connection seem much more concerned with demonstrating that parents have an effect rather than with showing exactly what this effect is, and the measures used in such studies usually stay at the very general level of the children's intelligence or developmental quotient. These, as any cognitive psychologist will tell you, are not measures of what is going on in cognitive development. They may show that parents do have some influence, but they do not tell you what precisely that influence is.

Causes and parents

Most of the work that is set up to discover what effects parents have on their children's intellectual growth is correlational. Various aspects of what the parents do and of the environment which they provide for their children are measured. These measures are then related to the child's intellectual powers, and if there is some relationship then the conclusion is usually drawn, albeit with a certain degree of modesty, that the parents have had an effect. Modesty is usually warranted, especially in the case of studies which relate the two things at the same point in the children's lives. The reason for caution about this sort of straight correlation is almost too well known to need rehearsing. The point of course is that the connection may take the form of the child influencing the parent rather than the other way round. The possibility of the child affecting the parent's behaviour is particularly plausible when one is dealing with questions of intellectual development. It is easy for example to think of a parent being stimulated to more conversation and to reading more stories by a child who is more than usually verbal. So to find that parents who talk to their children a great deal tend to have particularly verbal children is to produce a result which, as far as the question of causal direction is concerned, is wholly ambiguous.

Whenever the point about the ambiguity of the direction of correlations is made it is universally admitted. The trouble is that it is often forgotten. Some of the best known attempts to establish links between what the parents do and how their children develop, intellectually and in other ways as well, are designed in such a way that their results are bound to be ambiguous. A case in point is the well known and often quoted book by Yarrow *et al.*, *Infant and Environment* (1975). These research workers looked at a group of mothers and their 5-month-old children, and they applied an impressive number of measures. Their study showed quite strong relations between a number of things that the mothers did and various measures of the child's development. For example, the amount of time the mother spent playing with her infant was positively related to the child's 'mental development' as measured by the Bayley scales. This result has had a great effect, but no one can say what it means.

That study was done 9 years ago, but the problem still persists. Take as an example a study very recently reported by Easterbrooks and Goldberg on *The impact of father involvement and parenting characteristics* (1984). They found a positive relationship between the fathers' (and mothers') attitudes to bringing up their 20-month-old children and those children's intellectual capacities as measured by their ability to solve particular problems. Once again we cannot be sure whether it is the parent or the child who is having the 'impact'. There simply is no way of sorting out the direction, even on the grounds of plausibility. Either direction is plausible: neither can be ruled out. Why then, do studies of this type continue to be done?

I can find no answer to this question and my surprise increases when I reflect that this is not the only reason for worrying about studies like these. There is another crushing problem, which is that of the tertium quid. Mothers who play with their children a lot or fathers who have positive attitudes to the question of their children's upbringing may have children whose intellectual development is unusually rapid for a reason which has nothing to do with their influence on the children's environment. There is the possibility of the genetic tertium quid: both the amount of care which parents devote to children and the rate of intellectual development may have genetic roots and common ones at that. The sort of parent who is scrupulous about bringing up children may also be the sort of person whose genes favour rapid intellectual growth in his children as well as in himself. Of course a genetic influence is but one of many possible tertium quids. It is also quite on the cards that there may be an environmental tertium quid. The parent and the child may both be behaving as they do in these studies because of some factor in the environment which affects them both. In this case neither is affecting the other: both are under the influence of some unknown and therefore unmeasured external factor.

I mention the tertium quid at this point because it is a problem which still afflicts the most common method people have used in their attempts to get round the difficulty of using correlations to determine what is causing what. Most people who have acknowledged and have tried to get round the problem of direction of cause and effect in correlational studies have quite rightly turned to longitudinal correlations. The rationale is simple. If you think that A causes B, measure A before B occurs in development and then follow the children through to see how well A relates to B. If it does you can be reasonably sure that since A in this case precedes B the relation is much more likely to have something to do with A causing B than B causing A. That much is right, but one still has to worry about that tertium quid. A may predict B simply because both are determined by the same underlying and unrecognized third factor.

Let us turn to such studies to see how well they have coped with this problem. Though the number of longitudinal studies that deal with the parents' attitudes and behaviour in relation to their child's cognitive development is now large indeed, their basic results are quite easy to summarize.

First let us take the more general measures of cognitive development which have also been the most popular. Many studies look at the 'effects' of the children's environment on their intelligence or their 'mental development' which is usually measured by the Bayley scales. There are many positive results. The most comprehensive seem to me to be those which involve the use of the HOME measure, developed by Caldwell and her colleagues (Bradley and Caldwell 1984; Elardo *et al.* 1975; Elardo *et al.* 1977). This is a way of measuring the environment which parents give to their infants and children, and it depends partly on observation and partly on interview. It contains subscales such as the emotional and verbal responsivity of mother, the provision of appropriate play materials, and maternal involvement with the child. There are in fact two versions of the scale, one which deals with the environment of children up to three years and the other with that of older children.

The overall score on such a scale is necessarily a general one and it makes some general predictions. There is now an impressive number of studies which show that early scores on the HOME inventory do relate quite well to later intellectual development as measured by IQ tests (Bradley and Caldwell 1984a,b; Elardo *et al.* 1975; Elardo *et al.* 1977; Gottfried and Gottfried 1984; Barnard *et al.* 1984; Johnson *et al.* 1984; McGowan and Johnson 1984). Indeed the relationship with these general indicators of intellectual capacity is consistently closer than with particular aspects of children's cognitive development (Gottfried and Gottfried 1984).

Nonetheless there is some evidence that such measures of a child's early environment do predict some specific cognitive developments. An early candidate in such studies was the grasp of the object measurement, and several research groups used Piaget's measures or variants of them (e.g. the study by Yarrow *et al.* (1975)). Recently Gottfried and Gottfried (1984) tried in a longitudinal study to relate the results of the initial HOME measures to the growth of the object concept. They had little success. The lack of a strong relation between these two things in their study is hard to interpret. Perhaps there is no connection, or perhaps the measures of the object concept were not sensitive enough. Negative results are bound to be ambiguous.

Negative results, however, are not the problem of another equally specific aspect of development—the acquisition of linguistic skills. It is quite clear that these are related to environmental variables (Richman *et al.* 1982; Osborn *et al.* 1984), and it also seems that in longitudinal studies measures of the home environment taken before the child begins to speak are related to his linguistic progress later on (Tulkin and Covitz 1975; Parke 1978; Gottfried and Gottfried 1984). Furthermore there is some evidence from Clarke-Stewart's study (1973) that the rate of the child's linguistic development is specifically related to the amount of verbal stimulation she has in the past received from her mother.

All in all there are connections to be made between what goes on at home and how the child develops. But what form do these connections take? It seems to me that there is no guarantee that they are genuinely causal ones. For one thing there are reasons for doubting their longitudinal rationale. These are often not pure longitudinal predictions. Remember that a longitudinal prediction helps sort out the direction of cause and effect because if A genuinely precedes B it is not possible to maintain that B causes A. But in many of the cases described above, although A predicts B in the future, A does not precede B. B exists part of the time when A is first measured. Even though early measures of the child's environment do predict his intelligence later on, he does have a certain level of intelligence when the original measures of his environment are taken. The usual way round this problem is to adopt some form of the 'cross lagged design' which involves showing that the measure of A at time 1 (the time of the first measures) predicts B at time 2, better than B at time 1 predicts A at time 2. This has been done by, among others, Clarke Stewart (1973) and Bradley *et al.* (1979). Studies of this sort establish that A follows B rather than the other way round, and in my view should be treated with great sympathy. They are beset by statistical perils (Rogosa 1979), but they are still the next best thing we have to showing that A not only predicts B but genuinely precedes it as well.

This leaves the question of the tertium quid, and it is here that all longitudinal studies are at their weakest, for there is no possible way in which they can eliminate the possibility that the relationship between two variables is due to the fact that both are controlled by some other unknown and unmeasured third variable. Of course one should do one's best to think what such third factors might be and then partial them out. One obvious possibility—the genetic one—has already been mentioned, and it certainly seems prudent to do one's best to eliminate it in studies which relate the child's home circumstances to his later intellectual development. In fact this is technically difficult to do without the help of, for example, twin studies or information about adoption (see Rutter, Chapter 7). The nearest that people have come to eliminating genetic influences in studies of the HOME measures has been to measure the parents' IQ and partial that out. The reason for this precaution is simply that the positive relationships reported might be due to the fact that intelligent parents who pass on intelligent genes might also happen to provide environments which score well on such inventories (Scarr and Weinberg 1978). In fact the data on the effect of partialling out the parents' IQ on the relations between the HOME measures and the child's later intellectual development are inconsistent. There are two reports that this relationship disappears when the parents' IQ is taken into account (Campbell 1979; Longstreth *et al.* 1981) and one that it comes through relatively unscathed (Gottfried and Gottfried 1984). We shall have to see, but it is worth pointing out that even if the earlier reports that partialling out the parents' IQ removes the typical relationship between the child's environment at home and his later intellectual development are correct, that does not mean that that whole relation was genetic. It could be that intelligent parents are the ones who provide good environments, and that it is these environments which affect the child's cognitive growth (cf. p. 49).

Whatever the answer to the genetic question, one has still got to settle the problem of the possibility of third factors which are completely unknown and thus impossible to partial out, and this means that longitudinal studies on their own cannot determine cause and effect. It is an intractable problem, and one way round it might be to turn to a completely different type of study. There is one at hand. Intervention or training studies also offer a way to determine what causes what in children's development. The rationale is straightforward. You think that A causes B and so you give some children large doses of A and at the same time you take care to have a control group who have exactly the same experiences as the first lot of children except for the crucial ingredient—A—which interests you. Your hypothesis is that in the end the first lot of children will be endowed with B in far greater strength than the second.

In fact it seems to me that there is precious little that one can say at the moment about parental effects on children's development from training studies. The Headstart studies and their successors tended to tackle such a broad range of parental behaviour, when they had anything to do with parents at all, that it is usually impossible to sort out what is connected with what. There are however some studies which have shown rather more specific effects. Metzl (1980) for example showed that encouraging parents to stimulate their children verbally did have a marked effect on their children's developmental quotient. But there are problems here about controls for attention. We cannot be sure that this was a specifically verbal effect.

It is always difficult to work out the right control, but in principle there is no reason why it should not be found. When it is you can be sure that an intervention experiment which produces positive results really has established a cause. That is what experiments do. But there is a more basic problem about such studies and that is the danger of artificiality. The effects take place in a thoroughly contrived way. There is no guarantee that they have anything to do with what goes on in the real world. The experiment may work as an experiment, but be completely unrelated to cause and effect in the child's world.

The upshot is that the two main methods that have been used to establish the effect of the parents' behaviour on their children's cognitive development (or indeed the effect of any kind of experience on any form of development) are on their own inadequate. Does this mean that one should give up in despair? In my view one should not, because there is a ready solution to these problems.

It is that the strengths and weaknesses of these two very different methods are in fact complementary, and that put together they form a highly effective way of establishing causes in development. The longitudinal prediction establishes a genuine connection between two things: but there is no way of being sure that the connection is causal. The training experiment with proper controls establishes cause and effect, but one cannot be confident that the causal connection genuinely represents what happens outside the laboratory. Thus each method makes up for the weakness of the other, and the solution is to combine prediction with training. When this is done, one really can make a convincing case about cause and effect.

Is this all that needs to be done? There is one other major point that needs to be made about studies which attempt either through longitudinal predictions or by intervention to look at the links between the child's family circumstances and his intellectual skills. It is that such studies tend to ignore what is now known about cognitive development. During the last ten years or so studies of cognitive processes in children have flourished and have led to some very specific and on the whole highly successful hypotheses about the ways in which children solve problems and learn about their environment. There is little sign of this in the studies which I have described. They tend to stick to global measures. 'Mental development', development quotients, IQs, and crude measures of the child's linguistic skills—these are the stuff of this branch of research. All in all, they seem guided by a picture of the child gradually improving his skills and acquiring new ones as he grows older. Yet there are definite signs that this picture is misguided and any way not nearly specific enough.

What has to be explained about cognitive development

One thing that is abundantly clear about tests of cognitive skills is that the same child who fails one task which makes a particular cognitive demand will often succeed in a closely similar task which depends on exactly the same intellectual moves. Why children demonstrate their possession of an intellectual capacity and yet fail to use it on some occasions is now the central question in studies of cognitive development. This pattern is particularly clear in attempts to test Piaget's theory. Piaget, it is well known, thought that children spend a large proportion of their childhood bereft of logical skills, and acquire them very gradually. But subsequent work has tended to show that though children do fail, as Piaget demonstrated in his own highly ingenious tests of logical ability, nevertheless they often do very well in other tasks which apparently make exactly the same logical demands.

Piaget's well known conservation task (1952) is a case in point. This comes in many forms, but I shall stick to the conservation experiments which deal with number. Here the child is shown two identical rows of counters arranged side by side like two ranks of soldiers. He is asked to compare them and usually judges that they are equal in number. Next he sees the experimenter changing one row either by spreading it out or by bunching it up. Then he is asked to compare the two rows again. At this stage children of six years or less usually say that the two rows are not equal and that there are more counters in the longer of the two rows. Piaget concluded from this that they do not have a proper understanding of number. They do not realize that only adding and subtracting will alter a row's number and they actually think that lengthening it or shortening it changes its number too.

This is a result with serious implications for the study of parent-child interactions. It suggests that children who can count often do not know what number means. Counting is something parents and children do together long before school starts. Is it a useless experience as Piaget's ideas suggest it is?

In fact these ideas have been criticized by many people. We can start with the most famous of all the criticisms, made by Margaret Donaldson (1978). She argued that the whole experiment was an unfortunate misunderstanding between the adult experimenter and child. The child, she claimed, really does understand the principle of the invariance of number, but is persuaded to give the wrong answer by the actions of the experimenter. The child sees a strange and important adult who first asks a question about the two numbers, then deliberately makes what seems an important change to one of the quantities and finally poses the question once again. It was Donaldson's view that this strange routine, which centred on the deliberate transformation of one of the quantities, led the child to think that since the experimenter had made an important change and had then immediately posed another question, his (the child's) answer should change too. Hence the different answer after the transformation than before it.

She had the results of her experiment with Jim McGarrigle (McGarrigle and Donaldson 1974) to support her argument. Instead of a serious adult changing one of the two rows, the change was made, apparently as an accident, by a marauding and out of control teddy bear. When this happened the children—even four-year-old children—were much less likely to change their judgements after the transformation; they tended to maintain that the two rows had the same quantities despite the fact that one was now much longer than the other.

Rose and Blank (1974) took much the same line as Donaldson about children being misled, but tested this idea in an entirely different way. They simply decided to omit the first question in the conservation task. They, like Donaldson, argued that the child might be misled by the experimenter solemnly asking the same question twice, and making an apparently important change in between the two questions, and they rightly argued that the simplest way to test this was to stop including both questions in the experiment. In fact the first question—the one before the transformation—is redundant, and so in one condition they left it out. They simply showed the children the two rows, and then transformed one, and then for the first time asked the children to compare the two rows. This turned out to be the more successful condition: children made many less errors in it than they did in the normal conservation procedure. Rose and Blank's experiment was with 6-year-olds, but we have shown recently that their results hold true with five and seven-year-olds as well, and with other materials too (Samuel and Bryant 1984).

We are faced with the phenomenon of children being able to produce the correct answer in one situation, but not in closely similar circumstances. It looks as though the repeated questioning in the traditional form of the conservation experiment does mislead children who understand the rules of invariance in principle, but are in practice easily diverted from using them.

These are examples of what has been a general trend in attempts to pursue Piaget's theories about young children's logical incapacities. Again and again it has turned out that he told only half the story. He was right to point out that young children make some surprising mistakes, but wrong to conclude that each mistake signalled some great logical gap. He did not seem to accept that it is possible to be illogical sometimes and yet to be able in principle to make the necessary logical moves.

When one comes back to the question of parents and their effects on children's cognitive development, the point of all this is that it means that we have to consider not just the effects on the acquisition of skills but also on the way the child deploys these skills. If, as now seems likely, a great deal of cognitive development consists in the child learning how to take advantage of skills which he has had for a long time and which may in many cases have been innate, we should now consider how parents might help a child to do this. As far as I know nobody has tried to do this, and I think that this is because the people who look at parents are not very much concerned with what has been happening in research on cognitive development.

But it is not just a matter of the form of the questions about cognitive development changing. They have become much more specific as well, and that seems to me to mean great hope for future attempts to establish cause and effect, because the more specific a connection the easier it should be to track down. The study of the processes involved in reading gives us a very good example. It is now clear that children's awareness of the way in which words can be broken down into constituent sounds plays a major part in their relative success in learning how to read and write. Poor readers are often also rather poor at working out what these sounds are (Baddeley *et al.* 1982; Frith and Snowling 1983; Bradley and Bryant 1978), and measures of children's awareness of sounds taken before children begin to learn to read are good predictors of the children's eventual success with written language (Bradley and Bryant 1983). In this latter study we adopted a design which combined longitudinal prediction with intervention. Our original measures were of the children's ability to detect rhyme and alliteration—something which depends heavily on the child's awareness of words' constituent sounds—and this was what we trained an experimental group to do better. Both sides of the project produced positive results. Our original measures did predict the children's eventual success in learning to read very well and the children in the experimental group did eventually read better than those in the control groups. So here is an example of the kind of design that I have been advocating demonstrating a genuine cause in development. A child's awareness of sounds in words—acquired before he begins to read—does partly determine how well he reads, and it does so even when IQ and verbal ability have been partialled out.

The obvious next question is what determines this highly specific skill, and one obvious place to turn to for an answer is the child's early experiences at home. Rhymes, word games, and verbal routines are an everyday part of many pre-school children's lives. The time devoted to them probably varies a great deal from one family to another. It seems quite possible that this quite natural type of interaction between parent and child lays the basis in a highly specific way for a specific educational skill.

This seems to me to be one example of the way in which the very specific discoveries that now abound in work on cognitive development can be used to suggest causal connections which start with interactions between parents and their children and which, because of their very specificity, are much easier to test than the traditional and much more global ideas investigated in the studies described in the earlier part of my paper.

I should like to conclude this paper with something of an apology. I have had some harsh things to say about the two main methods of investigating cause and effect in development—correlations and intervention—but only because I think that these are the methods which will eventually solve the problem. With all their weaknesses the studies that use these methods seem to me to be vastly superior to some current work which makes statements about the parent's effects on their children's linguistic development or on their knowledge of the world about them simply on the basis of observations of what happens between parent and child. That is adequate as a beginning and as a basis for hypotheses, but these hypotheses still need to be tested empirically. That means studies which attempt to isolate cause and effect. In my view the best way to do that is (a) to combine longitudinal prediction and intervention in the same study, (b) to base hypotheses not just on what happens between parent and child but also on the details of what is known about cognitive development, and (c) to test specific connections between specific causes and specific effects.

Discussion

Simultaneity of correlation and intervention

Bryant's paper brought a number of methodological questions sharply into focus. *Chandler* asked:

'In your presentation you proposed that an ideal research method would be one which simultaneously undertook longitudinal correlational studies with intervention studies and that work which attempts this two-pronged approach is in short supply. What I don't understand is the requirement for simultaneity. There are, I think, many programmes of research which begin with correlational studies and then turn later to intervention studies. If one permits this two-step approach then research of the sort you are interested in is not as rare as you suggest. Why is simultaneity necessary?'

Bryant agreed that 'Simultaneity is not necessary, but that it does make a *real* connection between correlation and intervention more likely. When people use correlational methods in one study and intervention in another much later study, they tend to change the variables. They use one measure as a predictor and train with a rather different one.'

False security in intervention

Radke-Yarrow sounded a note of caution over an experimental approach:

'It is not a question of whether experiments or interventions are a necessary part of research. They obviously are. However, I do not think we can make quite as strong claims for them as we sometimes do. I'd like to make two points. First, how do we know when we have enough understanding of the phenomenon or process to be able to design the appropriate experiment? What we often fail to attend to sufficiently is how faithfully an experiment represents the elements that are essential in the process we are studying. The second point is that setting up a situation into which one introduces an experimental intervention can lead to false security about the conclusions one may draw. For example, suppose in an experiment on the effects of reasoning on a child's compliance, we have the mothers respond uniformly in a prescribed way. But when this is done, other things change differently for different children: Child A—"My mother is acting strangely; she's never said things like this before." Child B—"There she goes preaching at me again, I'm not going to listen." What can one conclude about the effects of this "critical" variable?'

Bryant took the point, but claimed it suggested not that intervention should be dropped, but that careful monitoring was needed as well.

In defense of Piaget

Doise noted that Piaget did do what Bryant is proposing. Piaget *did* look outside the laboratory situation. He formed a model and then tested the model.

Bryant replied: 'Piaget ignored causal questions in his empirical research, and the intervention studies by his colleagues lacked adequate controls.'

Laboratory vs. real-life

Doise noted that:

'Obviously Bryant is right about the complementarity between experimental studies and field interventions. But in our specific area we had to start with experimental studies. Indeed, our aim was to produce a development of the Piagetian approach, enlarging its paradigm, both on the empirical and on the theoretical level. The first step needed was a demonstration of what kind of social variables could intervene in cognitive development; in other words to give an empirical basis to the theoretical notion of "co-operation". Experimental designs with pre- and post-tests could be inserted directly into the Piagetian framework. Anyhow it is important to note that the Piagetian approach has had an important impact beyond the laboratory, in particular in the education milieu. It was urgent to introduce social variables into this powerful line of thinking as directly as possible. An efficient method to study mechanisms at work outside the laboratory consists in artificially reconstructing them within the laboratory, in a sort of hand-made synthesis. This is the aim of our experiments, but of course we make *no* claims that they cover all the possible social variables that are likely to facilitate operational thinking. But neither can an intervention study pretend to such an ambition. Finally let me add an empirical point already mentioned by Doise and Mugny (1984): several dozen experimental observations show that apparently nothing occurs during a situation of socio-cognitive conflict; effects only appear when systematic comparisons are made between the post-tests of different experimental conditions. Hence it would be very difficult to detect socio-cognitive conflicts and their differential effects by non-experimental procedures.'

Perret-Clermont added that it is difficult to isolate variables in nature, that the one pertinent occasion may be missed, and that a laboratory situation is necessary:

'I don't think that the best way to validate our model of socio-cognitive conflicts is to do longitudinal studies to see what goes on in the environment. First, I might never find a child quarrelling with another one for the same amount of juice because in real life mothers are wise enough to give children identical glasses! Or because social reasons ("You are older." "You are more thirsty.") or moral reasons ("You are not going to quarrel for a drop of juice!") could provide direct explanations for an unequal share. Conservation of quantity is not necessarily of interest to children, they don't necessarily care!

Secondly, we have studied classroom teaching and tutorial teaching, and believe that we can see some important, but rare, socio-cognitive conflicts. That they be rare is not a problem because we do find some, and they are so "powerful" they don't need to be numerous. But the problem is then one of interpretation and of validation of our interpretations: are we right to consider some of these uncommon events as socio-cognitive conflicts and to infer that the behavioural changes that seem to follow are causal consequences? In other words, I am afraid that if I go to the field (as I did: Perret-Clermont and Schubauer-Leoni 1981) to do observational studies, I would come back with descriptions of what I consider to be socio-cognitive conflicts leading to cognitive change, and your causal dilemma will have remained unsolved: indeed it is not possible to infer causality from the observation of co-occurrences. But it is gratifying to predict an occurrence in an experiment constructed in the laboratory, and a rewarding experience to observe a classroom activity with the feeling that some sense can be made of why some interactions (and not always those expected to be "instructive" by the teachers) seem to have more success than others in inducing cognitive growth.

My aim is not to find the Great Cause but to better my understanding of the processes involved in cognition. The laboratory is a workshop for conceptual tools. I then use them in the field (as glasses!) to look at the phenomena. They sometimes make some things clearer—not all, not always. Field work or applied research can in turn raise questions that go back to the laboratory to be worked through (Perret-Clermont 1981).

Artificiality is not a problem: the interesting feature of a laboratory is that is permits us to *construct* phenomena, to control (differently from in the field), at least partially, for their occurrence. The knowledge gained from this (even if limited) control can at least solicit creative imagination for actions in "real life" situations. But of course it never can justify or explain them totally. The conceptual tools are always simplifications of reality.'

Bryant agreed in the value of laboratory studies, and agreed that one cannot assume a relevance to real life. 'In addition, if a real-life event is so rare that it may be missed, is it so important? Most of this meeting has been concerned with more common interactions.'

Later in the conference *Rutter* pointed out that assessing social skills in naturally-occurring contexts carries a particular difficulty: since an important aspect of being socially skilled involves an ability to *avoid* confrontation and conflict, it is necessary to measure *non*-events.

18

Social structures and social cognitions
MICHAEL J. CHANDLER

This chapter is about the confluence of social and cognitive structures. Shortly, attention will be turned to a review of studies which explore such relationships by charting the developmental routes along which children pass in coming to match their own cognitions to the complexities of their social environments. Before turning to the details of this research, however, some context setting is required in order to make clear to what such investigations are relevant, and why, despite the importance I will attach to them, they remain in such short supply. To set this interpretive framework it will be necessary to reconsider some of psychology's most shop-worn controversies including the nature of subject-object relations, the meanings assigned to the term 'interaction', and the much overworked nature-nurture debate. It is to these background considerations that the following section turns.

Person–environment interactions

If, as I will undertake to show, the comprehension and successful navigation of certain social complexities requires as a prerequisite the development of counterpart social-cognitive structures, then why does the point have such an unfamiliar ring, and why are studies which seek to explore the match between social and cognitive structures in such short supply? Why, in short, do those who concern themselves with cognitive structures so rarely speak of environments at all, and why are situational psychologies so typically mute about the details of persons' minds? A partial answer to these questions, I will suggest, is to be found in the curious way in which our intellectual history has led us to drive a wedge between persons, on the one hand, and environments, on the other.

If a particular culprit is required, the candidate villain upon whose head it is currently most fashionable to heap responsibility for this state of affairs is clearly René Descartes (Harré 1984). By doubting everything but his own existence, Descartes established the conceptual primacy of the self, and left for future generations the task of figuring out how one might ever come to know about the existence of the environment and relate one's self to it. Somehow joining together what this Cartesian cleavage put asunder consequently became one of psychology's unworkable tasks.

The usual response to this Cartesian divorce of subjects and objects has been to accept the separation as final, to choose up sides, and to assume that, in terms of psychological conduct, either the person or the environment is alone in the driver's seat. There are then two competing visions. According to the first, it is recognized that persons are inescapably moulded by the circumstances under which they live. According to the second, persons are acknowledged to deform their experience and to construe their environment in accordance with their own nature. In keeping with these contrasting, but equally compelling truisms, persons are held to be either pawns or perpetrators of their environments.

While the sharp dichotomy which I have drawn between situation and subject centred psychologies is often acknowledged to be historically accurate, it is nevertheless seen to bear little resemblance to more modern interactive views. Among contemporary psychologists, doctrinaire adherence to anything remotely resembling such untempered 'nature' or 'nurture' accounts is regarded as hopelessly outmoded and especially déclassé. Instead various pat phrases such as 'dynamic, interactive, continuously ongoing reciprocal person-situation interactive processes . . . ' (Magnusson 1981, p. 7) are very much *de rigueur*, and are commonly offered as a gloss upon the fact that interactionism often means no more than a pinch of this and a pinch of that. Most prevalent among such appeasement views are those that Reese and Overton (1970) refer to as 'weak interactional positions', according to which some aspect of personhood (defined apart from the environment) is set in multiplicative relation to some free-standing aspect of the environment (conceived without reference to the persons who make it up). As expressed in controversies concerning the heritability of intelligence, or the trait versus situation debate in personality theory, for example, most such weak interactional accounts presuppose the independent existence of various contributing parts, that are seen to conjoin statistically as a confused afterthought.

Interactionism in developmental theory

The history of developmental psychology, like that of psychology more generally, has been one of fickled allegiances and inconsistent championing of first one and then the other side of this person-environment split. Like others of their turn of the century contemporaries, the first self-avowed developmentalists were struck by evolutionary imperatives and the ways in which 'nature' overrides 'nurture'. Following the First World War and into the depression years, second generation developmentalists became impressed instead by the power of the environment, and came to portray the socialization process as one in which strong situational forces dictate the course of childhood. For more recent generations of developmentalists, however, these earlier single vision accounts have come to be viewed as outmoded. At least since the advent of the Piagetian revolution in the early sixties, such either-or accounts have been eclipsed by a new and stronger form of interactionism, according to which adaptation is defined as a joint product of the equal and opposite processes of assimilation and accommodation. For many this solution has laid to rest the need for further agonizing over the shop-worn question of person–environment interaction, and has freed previously conflict-bound energies for the demanding task of specifying the changing shape of children's cognitive structures. More recently, however, for developmentalists in growing numbers, the newly arrived post-Piagetian era has brought with it a wave of renewed doubts. The question in the minds of many has become whether, all good intentions aside, Piaget's theory is more than an 'assimilation side' psychology (Chandler 1982) which sacrifices the environment on the altar of children's strongly appropriating cognitive structures. This crisis of faith is due in no small measure to the persuasiveness of Piaget's own empirical demonstrations. Setting aside their widely different contents, what his studies repeatedly demonstrated was that at various points in their developmental histories the same children would respond very differently to objectively identical stimuli. The effect of all of this has been that the environment, at least when it remained within some average-expectable range, could be seen to function as a kind of undifferentiated backdrop against which only the figure of cognitive structures could be seen to emerge.

In order to be fair to the tradition of Piagetian research it is important to stress that structuralists' views need contain no natural antipathy toward the environment, and commonly assume instead that structure is a property of both subjects and objects. Adaptation, in fact, is seen by Piaget to hinge on this fact, with mental structures gradually conforming to, or developmentally approximating, structures of reality (Broughton 1981). Without such an assumption the course of structural change would not qualify as a means of adaptation, and simple random change could not be distinguished from true developmental progress or increasing maturity (Blasi 1980). Despite the essential role that the environment is assigned by Piaget in accounting for the diachronic thrust of development, however, the more synchronic accounts that his theory provides of the interfacing of persons and their environments nevertheless often fall short of what might be regarded as an authentic dialogue. Because of the constructivistic character of the theory, and the epistemology of active transformation that this entails, the organism is most often portrayed within Piaget's model as appropriating the environment by way of strong structures capable of deforming the stimulus world beyond recognition.

These problems, of course, did not go unnoticed by Piaget, and the distorting effects of assimilation were intended by him to be matched by the countervailing process of accommodation, by means of which cognitive structures were said to be transformed through the force of environmental resistances. Despite this system of intended checks and balances however, the mechanisms of assimilation and accommodation, as portrayed by Piaget, do not seem fairly matched. As it has been presented, assimilation appears to profit from a kind of primacy or right of the first blow, which leaves to accommodation only the more menial task of cleaning up after assimilation's failed attempts to force round environmental pegs into square cognitive holes. Put somewhat differently, it is hard to see how, from a Piagetian perspective, a privately constructed world could possibly offer much resistance to the forces of assimilation (Broughton 1981). In a similar way, history would appear to be on the side of assimilation in that increasingly elaborate structural systems would seem to come quickly to outweigh challenges from a moment to moment stimulus world. As a consequence the individual, as portrayed within a Piagetian framework, becomes increasingly immune to and sealed off from the environment (Moessinger 1978). The effect of all of this seems to be that, in the post-infancy period, the assimilation–accommodation balance described by Piaget tends to break down, with the result that there is an involuntary curtailment and eventual eclipse of the accommodation pole of experience (Russell 1979).

Complaints about this state of affairs have begun to arise from many quarters. As has been the case in personality theory where situationalist critiques of the previously dominant trait theories have become commonplace, developmental psychologists in increasing numbers have become suspicious of models that seemingly sacrifice the particulars of experience to abstract, universalistic structures. Neo-Marxian scholars have faulted Piagetian Theory for its apparent lack of concern with social-historical context, language theorists have begun to abandon universalistic syntactical models in favour of more experientially grounded semantic accounts, life-span developmentalists have emphasized the increasing role of situational determinants in the adult years, and information processing models have demanded greater specificity in detailing the nature of stimulus input.

In all of these critiques the common theme is a commitment to the restoration of the importance of the stimulus environment. The potential problem presented by such restoration projects, however, is that they often threaten simply to exchange new lamps for old, and to reinitiate a perseverative cycle that abandons hard won insights concerning cognitive structure in favour of some new wave of single-minded environmentalism.

What is needed, it will be argued here, is not some frank neo-situationalism, which simply substitutes a new brand of one-sided interest in the environment for current preoccupations with subjectivism, but a new integration which inquires in serious ways about the intersection of structured persons with their structured environments.

The problems with any such interactionist ambitions is that one may not reasonably retain a Piagetian-like view of actively organized and structuring individuals, and simply nail back onto this image the lost role of the environment. The fragmented mosaic of contingent stimulus events that preoccupied the psychology of the first half of this century can not simply be abutted to an active self-organizing subject. Any such effort to affix artificially an unstructured environment to a structured construer of random details, will simply collapse into an alternative form of insulated structuralism. What is needed instead is an organized and structured view of the environment that offers active resistance to the deforming influences of subjects with a constructivistic or assimilatory intent.

While such an independently organized and separately structured account of social reality certainly is possible, the problem is that, at least to date, most efforts to produce such independent accounts of the environment have been rendered in languages which prevents their easy interdigitation with counterpart accounts of cognitive structure. Sociologists and environmentalists commonly speak a different language from psychologists, and the resultant translation problems block easy movement between one of these descriptive systems and the other.

An obvious solution to this dilemma would be for psychologists to produce their own structural accounts of the stimulus environment and to do so in a vernacular that facilitates the drawing of connections to counterpart descriptions of subject-centred cognitive structures. While such an undertaking is certainly possible in principle, it promises to be an extremely formidable task in practice. Piaget spent an unusually long and productive lifetime generating an account of evolving cognitive structure, and several average lifetimes could easily be spent in articulating any counterpart description of the environment. Beyond the sheer workload, there are seemingly insurmountable epistemological impediments to such an undertaking. Trying to provide a conceptual account of preconceptual or premasticated environmental structures is a bit like trying to lift oneself by his or her own bootstraps. Even if it could be done the independence of the effort would be hard to argue for.

A potential means of short-circuiting this problem is presented, however, by that sector of reality that specifically constitutes the social environment. Whereas, in the instance of interactions between persons and their natural or non-human environments, possible correspondences between cognitive and environmental structure are at best problematic, social reality is constituted by the same kind of processes as are at work in the individuals engaged in comprehending it. In other words, because social life is produced and reproduced through the operation of individual cognitions, both constitute different embodiments of the same underlying structure and there is between them a kind of complicity that establishes the conditions for possible social understanding (Chandler 1982).

What this translates into is the opportunity to draw upon the language of cognitive developmental structures as a descriptive vehicle for characterizing the structure of the social environment. If, as developmental psychologists, we are prepared to characterize our subjects as organized in understandable ways, then it ought to follow that such individuals should retain that structure as they become the objects of someone else's cognition. By adopting this 'what is good for the goose is good for the gander' orientation, which allows persons a structural description whether they are thinking or being thought about, the problem of devising a natural language within which to discuss social-environmental organization is finessed and the way is cleared for asking truly interactive questions about the degree of match between the structures of subjects and their social objects. Having piggy-backed a means of describing the complexity of social objects, on the strength of the fact that they are also subjects, the interpersonal environment need no longer be assumed to lie in wait like a random heap of stones, or some interminable Rorschach test for which there are no right or wrong answers. Instead, one could anticipate that persons whose cognitive structures are less complex than the structure of the slice of social reality which they set out to understand would meet with resistance in their constructive efforts and would fail to grasp the essentials of the target of their understanding.

The studies summarized below pursue the binocular focus just outlined by examining the efforts of children, representative of different levels of cognitive developmental maturity, to come to grips with comparably structural segments of social reality. While the contents of the two studies to be reported are unrelated, both share the common feature that they employ as stimulus materials social circumstances structured and open to interpretation at a range of levels of complexity. The common expectation across these demonstration studies was that subjects would succeed in coming to terms with only those situations the structural complexity of which was less than or equal to their own, and that they would fail to comprehend and deal with social situations more complexly organized than themselves. Included in this summary are two studies: one which explored the interfacing of social and cognitive structures as an explanation for the experience of crowding, and a second which examined the confluence of cognitive and moral structures.

Study one: cognitive complexity, social seriation, and the experience of crowding

This first study (Chandler *et al*. 1977) constitutes something of a transition case between more traditional research into children's changing conceptions of their impersonal environments, and other less arguably social investigations into their developing abilities to grasp counterpart structures inherent in the interpersonal environment. The initial impetus for this project grew out of the anticipation that persons probably experience as crowded or congested only those social gatherings which are for them lacking in any degree of understandable social structure or organization. This interactive hypothesis, which evokes by way of explanation a presumed failure of concordance between cognitive and social structures, is at sharp odds with more conventional interpretations which tend to view crowding in exclusively environmental or intrapsychic terms. The first of these competing views (e.g. Calhoun 1971; Christian *et al*. 1960), promoted primarily by comparative psychologists concerned with the study of overcrowding in animal colonies, holds crowding to be the direct consequence of missing leg room, indexed in exclusively situational terms by way of simple body counts per cubit. The alternative conception has been an exclusively person centred view which interprets crowding in those more subjective terms which permit even three to be a crowd (e.g. Sommer 1969; Esser 1972). In lieu of either of these single vision conceptions, the research to be described here reads the experience of crowding as the byproduct of a failed match between the kinds of conceptual order of which one is capable, and the complexity of the prevailing social structure within which one is immersed.

A test of this transactional hypothesis required a means of systematically manipulating the structural complexity of either a group of experimental subjects, or their immediate social environment, or both. In this study the selection of subjects representative of different levels of cognitive developmental maturity allowed manipulative control of the first of these variables, while alterations in the experimental setting permitted manipulation of the second. The design of the experiment drew upon the fact that social groups are frequently organized as simple seriated sequences, with some group members outstripping others along a given dimension of comparison such as status, power, etc. The ability to appreciate the structure latent in any such sequence obviously requires both an understanding of the relevant dimension of difference and, more fundamentally, the capacity to appreciate the transitive relations upon which any such seriated sequence necessarily rests. The comparative dimension of relevance chosen for use in this laboratory analogue of social crowding was the physical height of the participants, and the experimental setting and task consisted of a small 10 by 14 foot room (actually a classroom coat closet) in which groups of children were required to arrange themselves from shortest to tallest.

The metric of individual cognitive complexity which served as the basis for subject selection and group assignment was the degree to which the participants approached an understanding of the concept of unit measurement. As Piaget *et al.* (1960) have shown, most young preschool children are puzzled by the notion of relative size, apparently because of their inability to attend simultaneously to all of the extremities of objects at once. In looking at the tops of things they fail to keep in mind the bottoms. By early school age, however, most children acquire some skills at ordering by magnitude things which can be directly compared, and can successfully seriate by height things which can be stood back-to-back, or conveniently rest upon a common surface. Only some years later however, do these same children typically come to the notion of unit measurement and appreciate that, with the aid of portable measurement tools, it is possible to determine the relative size of things which can not be directly compared. Utilizing a set of standardized assessment procedures (Elkind 1964) 36 boys between the ages of five and nine were identified, 12 of whom were clear representatives of each of the three ability levels just outlined. Three boys from each of these ability levels were constituted into four working groups of nine members. Poorly prepared as some of them were for the task in hand, these teams were then armed with yard sticks, sent into their classroom coat closet, and told to line up according to height.

As a means of further manipulating the complexity of this experimental task a final indignity was perpetrated upon half of these subjects. As previously described, two of the four teams undertook their assignment in a standard cloakroom. The remaining subjects, however, were required to attempt the same task only after the floor of the cloakroom had been contoured by the installation of a network of canvas-covered theatre risers which left the room a broken field of highs and lows.

What all of these experimental machinations were meant to accomplish was a social occasion upon which persons of known levels of cognitive complexity could be observed in the act of constituting a more or less simple social ordering. Given the details of the two experimental settings and the three ability groups represented, two contrasting sorts of outcomes could be anticipated, neither of which could have been predicted from a knowledge of the complexities of either the participants or of the settings alone. For those subjects totally lacking in seriation skills, and for those assigned to the room with the contoured floor who possessed only comparative but not unit measurement skills, the structural demands of the experimental task were anticipated to outstrip the complexity of their own cognitive structures and, as a consequence, they were expected to fail in their efforts to find or recognize their place in the emerging social organization. By the standards outlined above, these failure or 'mis-match' subjects were those who were expected to experience the experimental setting as crowded. By contrast, for all of those subjects with unit measurement skills, and for all those individuals assigned to the room with the conventional floor who were capable of back-to-back comparative measurement, the complexity of the task demands were expected to remain within their competency range, making it possible for them to locate themselves correctly in relation to other group members. Relative to the subjects of the previous 'mis-match' group, these individuals whose cognitive complexity matched that of their social setting were expected to show less evidence of experiencing the experimental setting as crowded.

The dependent variable of experiential crowding was indexed in this study by a triad of physiological, behavioural, and cognitive measures. Pre- and post-experimental indices of palmar sweating served as a measure of stress, experimental sessions were rated for individual evidence of behavioural disorganization, and a structured interview administered by each child's parents were used to obtain recollected estimates of the length and width of the experimental room, and the number of participants in the working groups. Support for the initial hypotheses was found with all four response measures. Relative to other participants, children in the 'mis-match' group, whose level of cognitive complexity was not equal to the complexity of their social task and setting, showed greater levels of physio-logical stress, displayed more behavioural disorganization, systematically underestimated the size of the experimental room, and overestimated the number of its occupants. By contrast, neither the variable of cognitive complexity, nor situational complexity, when considered alone, successfuly accounted for the variability observed on any of these outcome measures. Taken together these results lend strong support to the hypothesis that the conditions responsible for the experience of crowding are neither personal nor situational, but occur instead at the interface of personal and social structures.

Study two: cognitive and moral complexity

Within the cognitive developmental tradition initiated by Piaget and expanded upon by Kohlberg, the moral reasoning process has come to be studied in ways which are typically quite remote from the actual morally relevant personal experiences of the individuals who usually serve as research subjects. Instead, most of the work in this field has concerned the interpretive commentaries which subjects offer about moral choices arising in the interpersonal lives of others. While the once-removed character of such moral dilemmas has been the subject of frequent criticisms, this other-wise questionable measurement practice should at least have served to guarantee that what is known about moral reasoning would include knowl-edge about the impact of diverse social contexts. Put differently, it would seem reasonable that, given the opportunities afforded by the license to invent moral dilemmas of every conceivable stripe, one might find a litera-ture filled with studies which systematically explore the numerous ways in which various sorts of prescriptive obligations come to loggerheads. Such is not the case, however. Instead, standard dilemmas, such as those involving Heinz and his dying wife or Joe and his vacation hungry father, have become an increasingly familiar part of psychology's folk tradition, endlessly repeated with little heed to what precisely is being served up. Responsibility for this seemingly paradoxical repetition compulsion would appear not to lie in any peculiar lack of inventiveness on the part of investi-gators of moral development, but can instead be laid at the door of what was described earlier as contemporary cognitive-developmental psy-chology's widespread assimilation-side bias (Chandler 1982), a view according to which what one does by way of choosing stimulus materials hardly matters.

The apparent logic behind such studied disinterest in who is doing pre-cisely what to whom is that, since subjects are thought to appropriate naturally such stimuli in any case, and quickly distort them in ways which are in keeping with their own cognitive organization, all of the variability worthy of note is expected to prove to be a feature of subjects rather than social settings. To the extent that something like this assimilation over-kill is thought to be unavoidable, it follows that one might just as well settle for whatever stimulus opportunities are most easily at hand.

Following the same research strategy just illustrated by the work on crowding, the second study to be reported (Chandler et al. 1980) under-took to challenge such traditional assimilation-side accounts of the moral reasoning process by focusing research attention upon the formal struc-tures latent within the dilemmas upon which moral judgments operate. The central premise upon which this work was grounded is that the com-peting horns of all moral dilemmas necessarily represent one or another of several distinguishable levels of prescriptive obligation, which in combi-nation yield a limited set of structurally distinct dilemma types. It was the thesis of this study that the outcomes of such internal moral debates are the predictable consequence of interactions between the structural level of the dilemma in question and the cognitive structures of those who deliberate about them. Tradition has it (Kohlberg 1968) that studies concerned with the form of moral reasoning can yield generalizable information concern-ing certain stylistics of judgment, but are necessarily mute about the par-ticular moral conclusions that a given subject might draw. In short, such accounts abandon us just when we need them the most—at a point where we thought we were about to learn how things were going to turn out. The thrust of the present research is that such failures of theoretical nerve are in no sense obligatory and come about as a consequence of an earlier failure to consider, along with the cognitive structure of their subjects, the underlying structure of the moral conflicts in question.

As with the previous study of children's responses to social crowding, this research began with a formal job analysis of the stimulus environment; in this case an attempt to specify the latent structure of alternative dilemma types. Drawing upon the work of Hare (1964), Taylor (1961), and others, it was argued that as normative prescriptions, the conflicting obligations out of which moral dilemmas are composed bear the structural stamp of alternative moral orders and, consequently, may be understood to vary across levels of generality or scope of intended application. Consistent with this view, qualitative distinctions were drawn between prescriptions of three sorts, referred to here as *commands*, *rules*, and *principles*. At the lowest registry of this scale of generality are simple commands, such as, 'Keep quiet' or, 'Take off your shoes', which are intended as restricted, one-shot injunctions meant to apply to particular persons, in particular places, for particular periods of time. Principles, in direct contrast, represent prescriptive obligations of a less ambiguously 'moral' sort which achieve their special status precisely because no hedge is placed against their unremitting generality. As such, principles are meant to apply to all comers, at all times, under every conceivable circumstance. Intermediate between the extremes of particularistic commands and universalized principles are concrete rules, which are intended as prescriptive obligations of real but intentionally limited generality. Rules are meant to apply only under certain conditions, and prescribe for only limited classes of events. For these reasons rules, in contrast to principles, can have ambiguous and overlapping domains of application and, like commands, may consequently contradict one another.

In the context of this research, moral dilemmas were understood to be the product of a clash between commands, rules, or principles. When set in opposition to one another, in all possible combinations, these three prescriptive forms generate a typology of five structurally distinct moral dilemma forms. Commands can be made to contradict other commands, rules, or principles, and rules can clash with principles or other rules. By definition, principles may not contradict other principles without jeopardizing their claims to universality, and thus degenerating into concrete rules. Using this formal typology as a guide, a series of short, child-centred story problems were written, exemplifying each of these dilemma types. These story problems served as the structurally specified stimulus materials in the study of moral deliberations to be described.

As in the previous study, the subject dimension of structural complexity adopted was again Piaget's familiar distinction between preoperational, concrete operational and formal operational modes of thought. In order to obtain samples of children representative of these three structural types, 150 boys and girls between the ages of five and fourteen were screened, utilizing a standard battery of Piagetian-inspired assessment procedures (Kuhn and Ho 1977). On these grounds, 20 children were chosen as clear representatives of each of Piaget's three levels of cognitive-developmental maturity. Following this screening procedure the subjects listened to tape-recorded moral dilemmas representative of the five logical types previously distinguished, and were asked to decide how the central story character should act and to justify their decisions.

The consequent design of this study can be mapped as a person-by-situation matrix representing three levels of cognitive complexity and five structurally distinct moral dilemma types. The expected outcomes for each of the 15 cells of this matrix were decided upon on the basis of a hypothesized two step deliberation process. First, it was assumed that, given their obligation to assimilate available inputs to existing cognitive structures, individual subjects would necessarily fail to comprehend prescriptive obligations which rested upon universality claims more general or abstract than their own current level of cognitive operational competence, and would instead read such situations in ways consistent with their own cognitive structures. By this transformational rule it was reasoned that:

(i) preoperational children, said to lack reliable systems for dealing with class concepts, would reinterpret all rules and principles as instances of simple commands; and

(ii) that concrete operational subjects, understood to lack those propositional forms of thought required for an appreciation of universals, would misconstrue principles as instances of concrete rules.

Second, it was hypothesized that individuals of all cognitive developmental levels would consistently prefer and choose whichever of two available morally relevant alternatives they judged to be the more universally prescriptive. By this standard, principles would be preferred to rules, and rules to commands by all who were in a cognitive position to make such distinctions. Alternatively, whenever conflicting courses of required action were, or were read as being, prescribed by systems of obligations couched at the same level of prescriptive universality, it was anticipated that the course of action settled upon would cease to be a predictable function of either cognitive complexity or dilemma type. In short, when none of one's morally relevant options can be seen to be more lofty than any other, then the decisions taken will no longer hinge upon one's level of cognitive maturity.

The joint effect of these two guiding expectations was to parcel the resulting person by situation matrix into expected outcomes of two contrasting types:

(i) those with regard to which subjects are capable of recognizing and preserving levels-distinctions between the two horns of the dilemma in question; and

(ii) those for which no such levels-distinctions were present or noted.

In the instance of the first of these response categories all subjects were expected to take the high road and uniformly choose the loftier of the alternatives presented. In the remaining instances, which together form a legitimate space of moral indeterminacy, it was anticipated that no systematic preferences would be in evidence. Stated somewhat more formally, in all cases in which the subjects in question were hypothesized to judge the available alternatives to be prescriptions of the same logical type, no structural grounds were available for dictating strong preferences and the proportion of subjects anticipated to elect one course of action or another was expected to approach 0.5. When, by contrast, the moral alternatives presented were, and were understood to be, of a different logical type, strong preferences were anticipated and the proportion of subjects expected to choose in favour of the higher order alternative was set at 1.0. What all of this amounts to is the decidedly Socratic prediction that 'the "man" who knows the good will do the good'.

The results obtained from the 60 subjects tested were analyzed by charting the proportion of their choices which favoured one or the other of the morally relevant alternatives presented by each dilemma, and comparing these data with the theoretically generated expected values. The overall descriptive adequacy of this proposed transactional model was tested and supported by a 'pattern hypothesis' analysis (Steiger 1980), which indicated that the obtained results closely followed and were only minimally different from those predicted. These results argue strongly that, given information concerning the cognitive structures of one's subjects and an appreciation of the structural complexities of the moral conflicts which they face, the particular outcomes of moral deliberation can in fact be directly predicted.

Conclusion

The studies summarized in the preceding sections are meant to be illustrative of a research strategy aimed at restoring a place for the lost role of the social stimulus environment in the study of cognitive-developmental processes. Capitalizing upon the presumption that cognitive and social organizations may be usefully understood as expressions of common underlying structures, these research efforts undertook to show that our understanding of various social-cognitive phenomena can be enhanced by exchanging a somewhat tired assimilation-side psychology for a viewpoint and investigatory strategy which attempts a more balanced consideration of the structural features of both subjects and their social environments.

Discussion

Relating theory to data

Selman stressed the need to distinguish between subjective complexity and environmental social world complexity. Similarly, *Hinde* commented, 'Your introduction reminded me of Verplanck's (1954) discussion of what we mean by a stimulus—is it something in the physical out-there or something to which the individual responds. Your experiment could be described in part in terms of individuals' differential responsiveness to order out there and in part in terms of complexity out there. Is your problem a storm in a Piagetian tea cup or is it something more general?'

Chandler replied:

'It is true, as you suggest, that the subjects of the research which I reported were influenced by the complexity of the socially ordered events to which they were exposed. Exclusive knowledge of this event complexity, without a counterpart understanding of the cognitive complexity of the subjects themselves, would not be adequate however, to account for the data presented. It is exactly my point that the separate consideration of personal or situational variables is not sufficient, and some joint attention to the complexity of persons in relation to the separate complexities of their environment is required.'

The meaning of the measure

Kummer questioned the relevance of some of the measures used:

'Could not palmar sweating in the non-seriating boys derive from their sense of incompetence for the task and their recollection of crowding be produced by being shoved around, and touched more frequently by boys during the lining up?'

Chandler answered:

'I agree that the stress which the non-seriating subjects demonstrated can be understood as an expression of perceived incompetence. The point, however, is that the differential stress reactions demonstrated by the subjects of this study can not be predicted on the basis of information about either their cognitive maturity or the complexity of the task alone, but requires a joint consideration of both person and situation variables. I am not persuaded that, as you suggest, the differences observed can be attributed to the amount of pushing and shoving going on, however. All of the experimental groups contained equal numbers of subjects at each cognitive level, and both settings seemed to have their equal share of such contact.'

Ignjatovic-Savic asked a similar question:

'Why do you think this situation of crowding is adequate for the investigation of social cognition? That is, here you are treating persons as physical objects. The question of what is children's social cognition in this situation might be: why does the experimenter put us in this stupid place?'

Chandler agreed:

'Of the various studies described in my written paper, the work on crowding is in some sense the least social and, as you suggest, permits the participants to regard one another as physical objects. It was my intention to present this work as something of a transition case between studies which concern children's understanding of exclusively non-social events, and those, like the others which I describe, which are more clearly social in character. Standing on its own, I would agree with you that the crowding study could be interpreted as only partially focused upon what we usually have in mind when we discuss social cognition.'

Interaction or main effects?

Rutter queried the interpretation of the data:

'It seems to me that the data from your crowding study do not support your conclusion regarding an interface between personal and social structures. Surely the finding that the effect of the situation was closely similar in both groups of children argues that there was an effect of both person and social situation (i.e. two main effects in statistical terms), but no interface or interaction between them in either a statistical or conceptual sense?'

Chandler replied:

'I can appreciate how the summary tables which I presented might lead you to the conclusion that the data supports an interpretation of a main effect for both cognitive ability level and experimental situation. In fact, when these data, which represent ranked difference scores computed in terms of ohms of resistance, are evaluated by means of an appropriate non-parametric statistic (in this case the Mann-Whitney U test) no significant effects for either condition or ability level was observed. When however, subjects whose cognitive abilities were structurally less complex are compared with those for whom the opposite is true, then significant differences are observed. You are correct in assuming however, that with larger samples there may very well have been evidence for the main effects you propose. This is not the case however, with the other study described in the paper.'

Bryant asked:

'I wonder whether it wouldn't be interesting to add to your experiment groups who were all at the same cognitive level. This would sort out whether the cognitive differences in the groups had an effect.'

Chandler replied that he has done this, with no effect.

Section F

19

Socio-cognitive dialectics
EDITORIAL

In the present section the authors are largely concerned with the difficulty of clearly delineating cognitive factors from social factors. To some extent the material presented in previous chapters has shown that this distinction is possible and fruitful for gaining information from certain types of observations, surveys, and experiments. But within these lines of research the question of the precise nature of 'cognition' constantly emerges, and the nature of its observed dependency or interrelation with environmental social factors is from from clear. The notion of 'social cognition' seems to form a frontier zone where the independence of cognitive growth (or even its primacy) is particularly difficult to assert: the 'social' could be considered as the mere content of basic 'cognitive' processes, but these still appear to be elicited by certain types of situation that bring the individual to discover the existence of other persons, their importance, their individuality as independent centres of causality, emotions, appeals, threats, etc.

But whilst social cognition obviously has this special status within the realm of psychological processes, it is not the only area where social and cognitive processes meet: cognitive processes *never* take place within a social vacuum, nor indeed are the very forms that they take free of socio-cultural influence. The development of language, reasoning, drawing, scientific competence, and other communicative skills rely—at least in their expression and actualization—upon culturally constructed and learned forms, models, and schematas. Thus vocabulary, grammar, counting skills, theories, and concepts are not completely reinvented from the start by each individual, and it is not pure chance that individuals of different cultures use many different conceptual and communicative tools. Studying changes in religious beliefs, Deconchy (1971, 1980) has shown how cognitive and social factors are likely to interfere strongly in some circumstances (see also Hinde 1984).

But the social factors themselves also do not have an independent and distinct existence of their own. Social circumstances, as shown previously (Chandler, this volume), have different impacts on individuals according to how they are perceived and interpreted by them. Indeed social factors are perceived along dimensions that are cognitively structured; known and unknown, threatening or comforting, useful or dangerous, desirable or dreaded, etc. Humans transmit to each other and from generation to generation various ways of interpreting the social world: as a consequence the individual does not merely react directly to the environment, but develops and acquires socially constructed means to regulate his behaviour and thereby gain in control of himself and of the situation. These means can be described as cultural schemes of interpretation (encoding for expression, decoding for understanding) and cultural anthropologists describe them as at work in all spheres of daily life (affective, relational, cognitive, social, material, working, etc.) as well as in more exceptional events (e.g. rîtes de passage) (Centlivres and Hainard in press). These schemes of interpretation, which are socially transmitted, usually convey meaning by relating (through the pointing out of similarities or divergences, analogies, metaphors, social recognition or avoidance, explication of norms, and punishment, etc.) one or several other events in the same or different areas of behaviour (for instance when parents try to reinforce sex-typed behaviour, or comment on disobedience: automony may then be perceived as incompatible with the affective parent-child bond). Some of these cultural schemes seem even to weave together the meaning of events that take place in different psychological areas. Doise *et al.* (1976) and Bell and Perret-Clermont (1985) report observations illustrating how adolescents with poor school careers tend subsequently to behave as if they had a deeply interiorized sense of failure, presenting a negative social identity, and limited and rigid social cognitions in the culturally valued areas of social life. It seems as if failing on certain cognitive school tasks had influenced meanings in a wide range of contexts. Such an interpretation is corroborated in part by the clinical observations reported by Pain (1981).

We are thus concerned with a triple process involving development itself, socialization (i.e. learning to depart from an egocentric point of view or from an ethnocentric conception of reality, and to live in coordination with others) and acculturation (the learning of the specific symbolic mediators of the culture in which the individual is immersed). How does this take place? Among many others two authors have occupied an outstanding role in promoting research around this question: Piaget and Vygotsky. For the first, development is seen as being fundamentally of biological origin, situated within the individual organism who develops actively his potentialities, organizes them and structures their whole mode of equilibrium according to the feed-back he receives from interacting with the environment. The source of the development described by Piaget is the equilibration process located within the living activity of the individual. Of course this process is likely to be facilitated or inhibited by biological maturation processes, and by external factors such as ecological conditions, and experience with objects and persons, but these factors that Piaget considers (in accordance with his own focus as a biologist on the individual body as the unit of analysis) as completely external to the inner development of the individual and thus unable to affect the deep structures and forms of this inner development. The cognitive structures that Piaget describes are supposed to be independent of cultural differences, and this is supposed to explain why all human beings have them in common provided, and this is very important, that they have encountered the environmental conditions necessary for them to develop normally. What is a proper social environment for Piaget? One that is free of social constraints and permits the child to explore autonomously the physical environment, discovering it by personally acting upon it. For Piaget only situations of co-operation among equal peers are favourable to development. By contrast those social situations in which beliefs or norms are transmitted culturally (i.e. by the constraints of older generations, authorities, powerful or prestigious social agents) actually inhibit normal development because they impose on the individual particular modes of thinking and thereby hinder him from constructing them for himself as autonomous inner knowledge. In accordance with this view, Piaget was not interested in studying knowledge developed in children via instruction in schools: he carefully selected his tasks as unrelated as possible to classroom activities and formal instruction. (Yet this did not prevent later cross-cultural studies [by Bruner *et al.* 1966 for instance] from showing that the children who had benefitted from longer formal schooling performed better on Piagetian tasks. Piaget himself referred to the cognitive décalage between upper school students and young workers but attributed the difference to unstudied factors).

Seen from a Vygotskian perspective, Piaget's understanding of development does not take sufficient account of how imbedded in a social environment the individual is from birth, and how his growth depends on stimulation by the older persons around him. For Vygotsky the child's discovery of his environment (even the physical one) is socially mediated. It is through his interactions with other persons that the individual learns to act on objects and to use signs to regulate interpersonal (and later, by interiorization, also personal) behaviour. First the child is led by other individuals into certain procedures of actions and signs at the interpersonal level. This opens the way for his later more independent (with more initiative) and finally autonomous action and thinking. For Vygotsky autonomous thinking is not inhibited by external social agents. Rather it is first by acting with the individuals on whom the growing child is dependent—or, in other words, within the relationships that he has established—that the child is led to relate certain acts with outcomes, certain procedures with results, certain signs with actions, and to acquire language as a major mediating tool in regulating behaviour. Only afterwards can he enact these behaviours alone. Culture provides tool-mediated actions and these the child can acquire only through the specific type of instruction that adults can provide when they interact with the children within their zone of proximal development.*

Wertsch and Sammarco's contribution is directly inspired by these considerations. By comparing mother-child interaction between language impaired and normal children, these authors illustrate how in interaction symbolic tools, if not impaired, are likely to support joint action and to permit autonomous thinking as revealed by the subjects' taking of initiative, for instance. Of all the culturally transmitted mediating tools language appears to be the most important one. And this is perhaps the domain in which the reciprocal effects of cognition on social experience and social experience on cognition within learning are the most obvious.

Language development has been much discussed recently, and has been largely omitted from this volume. However, whilst the influence of linguistic abilities on relationships can perhaps be granted, at least in a general sense, the crucial importance of social experience within long-term relationships on linguistic development should perhaps be emphasized. Whilst at least some early signal movements are part of the repertoire of the newborn, it is reasonable to regard them as merely expressive of physical states rather than communicative: the development of intentional communication depends on the responses of caregivers, and perhaps on caregivers treating the signal movements of young infants as if they were intentional. Much evidence has accumulated that such activities as mutual gaze, pointing as an indication of interest with the accompanying vocalizations, and turn-taking in gaze and vocalizations, *could* constitute steps relevant to the development of language (e.g. Stern *et al.* 1975; Collis 1977), and that caregivers behave in ways that give infants clues that might be necessary for the development of conversational skills (e.g. Snow 1977). However attempts to relate individual differences in such pre-linguistic measures to individual differences in linguistic abilities have in general not been outstandingly successful (e.g. Kaye 1979), probably largely because the measures of early interaction have no logical relation to the linguistic parameters assessed (Camaioni *et al.* 1984). Bruner (1975), who tried to relate the mother-infant routines of joint attention and activity to the attention/action structures present in language, ran into similar difficulties. However, recently Camaioni *et al.* (loc. cit.) have emphasized three basic principles necessary for success in such endeavours: social interaction and language must be seen as constructing cognitive categories and not as rules operating on previously defined categories, since only

---

* The 'zone of proximal development' refers to the view that learning can occur only within limits set by the stage the child has already reached. The child can enter into an action, discourse or perspective suggested by an adult only if it is not too different from what it can already cope with. Thus learning does not occur on a *tabula rasa*, but is anchored at every stage to the current capacity of the child.

in that way can their relations to each other be studied; language must be seen as operating on reality to make it accessible to knowledge, the capacity of language for structuring the environment depending on the fact that it is, right from the start, intersubjective; and the basic unit of language should be not one utterance but at least two—in other words, the dialogue. Their socio-interactionist approach accepts that a dialogical structure is present in pre-linguistic routines for assuring joint attention, but argue that the non-linguistic intersubjective processes of singling objects out from the flux of stimulation differ from establishing perspectives through the use of nouns in that the latter already incorporate shared perspectives on the world. Thus a link between the two depends upon demonstrating that intersubjective processes determine the way children learn to structure the world by the use of nouns. By a detailed analysis of observations on the early use of nouns, Camaioni and colleagues produce much evidence that this is in fact the case. For example, even the first use by the child of labels for objects in the external world could be traced to previous episodes of interaction with a significant adult. Their work thus links with Vygotsky's (1962) view that intrapsychic functioning is derived from interpsychic (Hartup; Wertsch and Sammarco, this volume). Camaioni and her colleagues emphasize that it is interaction episodes characterized by agreed-upon rules and conventions (and which are thus most likely to occur within long-term relationships) and not social interactions *per se* which are likely to be important for later language development.

It has also been suggested that language development depends on careful adjustment by the caregiver of the complexity of the language used to suit the cognitive capacity of child (Wertsch 1978). If, as Vygotsky would suggest, cognitive development depends on others' linguistic utterances, the child must have the cognitive ability to comprehend those utterances already. Wertsch thus suggested that communicative acts by the caregiver increase in complexity, requiring more linguistic and other knowledge in the child as he ascends the hierarchy. Successful comprehension of the later and more complex utterances would depend on progress made as a result of earlier ones. This parallels the suggestion, noted above, that at an earlier stage the transition from expressive to intentional communicative acts may depend on the caregiver treating them as if they involved more than mere expression. In both cases it is reasonable to suppose that a caregiver with a longstanding relationship with the child would be more effective than a casual acquaintance in tuning his or her discourse to the child's abilities.

Recent summaries of language development have been provided by Shatz (1983), Clark (1983), and Maratsos (1983). Returning to the chapters in this section, Doise's work stems from Piaget's (1929) suggestion that quarrelling between peers might provide an important stimulus for cognitive development. As we have seen, mutual sharing of perspectives between adult and child may be much more difficult than between peers. Doise emphasizes that cognitive progress often comes from cognitive conflict with another individual who can be seen as comparable with the self but who offers a discrepant solution to a problem, and that it comes also from contradictions between responses resulting from the child's own cognitive structures and those based on social rules. In both his theoretical approach and his empirical work, Doise focuses primarily on the influence of social factors on cognitive development, but he nevertheless fully acknowledges the complexity of the interaction between them. Indeed his paper shows that Piaget was fully aware of the difficulty: if cognitive and social development occur together, how can the effects of one on the other be distinguished? An answer, of course, lies in an experimental approach that provides for a reconstruction within the laboratory of the interaction of different levels of analysis. Doise illustrates how the laboratory provides a means to isolate and control for some factors and thereby enables the researcher to construct *models* of reality that can be tested. In his view the laboratory enables the social scientist to test the *model* but not reality. Yet if the model has some reality it should at least function in the laboratory! The laboratory permits the isolation of certain factors, pertaining for instance to different levels of analysis, and the empirical and systematic examination of their interrelations. It does not mean that in 'natural' settings the same phenomena could be observed because other factors there are likely to interfere. The natural setting is a more complex and less controlled scene than the artificial setting of the laboratory. (But one should not forget that even in the laboratory many variables are not controlled: the image that the subjects have of the laboratory, for instance, is rarely controlled.)

This brings us back to a second issue raised by the contributors to this section: the question of the choice of units of analysis relevant for those processes that are uniquely human (i.e. cultural). Cultural processes are performed by individuals situated within groups themselves socio-historically situated, and cannot be construed merely as biological processes within the boundaries of the skin of an individual, or the limits of his nervous system. But neither can individual psychological life be correctly accounted for as the mere interiorization of external social processes. Behind the question of the dialectics between the social and the cognitive arises the question of what is the object under study. As long as it is only one specific type of behaviour one can consider all the others as external variables affecting it. In such a perspective Perret-Clermont and Brossard illustrate in their contribution how deeply a cognitive competence, as assessed by a Piagetian type of test, can depend upon the different variables that characterize its socio-cultural situation. Operational level appears no longer as a characteristic only of the individual but rather as a characteristic of a certain type of socio-cognitive relationship that in given circumstances certain children with the necessary prerequisite social experience can establish with an adult experimenter. Something is known, but as yet not much, on how children gain autonomy for this kind of socio-cognitive competence (Perret-Clermont 1980; Doise and Mugny 1981; Perret-Clermont *et al.* 1984; Mugny 1985).

If the object of study is not one category of behaviour in relation to others, but the *integration* of behaviour in individuals, then the task is completely different: the unit of analysis cannot pertain to only one of the previously defined (see Introduction) levels of analysis but has to be relevant to the integrative processes themselves. And in that perspective it is interesting to note that the 'founding fathers' of modern contemporary psychology often had a more holistic approach to human bio-psycho-social life. The further attempts to improve the description and understanding of the variety of behaviour seem to have had two consequences: the multiplication of levels and units of analysis considered on their own; and the ordering of these on a rather unidimensional scale with age as a reference (and often the main) criterion. The pursuit of psychological research during the last decades has divided the object of study into ever smaller parcels, smaller and more refined units of analysis. It has also resulted in a growing number of distinct levels of analysis (for instance: physiological vs. psychological vs. interindividual vs. intergroups vs. anthropological vs. sociological, etc.) and these have tended either to be considered on their own, or to be ordered on a one-dimensional scale (from the less complex to the more complex, from the less advanced to the more advanced). This has been useful for a better description and awareness of the variety and complexity of behaviour, but this analytical process can be misleading if it loses sight of the whole. The dialectics between the various levels of analysis and between the behavioural units belonging to the different levels, is complex and must be studied as such. The present focus of attention on the dialectics between social and cognitive processes can perhaps be viewed as an effort towards the integration of current understanding of human behaviour (Fig. Int. 1) Yet it will not be sufficient if it does not draw upon knowledge on the other biological and cultural levels and if it does not choose units of analysis that illustrate the mechanisms of the interdependency.

20

Social precursors to individual cognitive functioning: the problem of units of analysis

JAMES V. WERTSCH AND JOAN G. SAMMARCO

As is the case with virtually all disciplines everywhere, cognitive developmental psychology in the West is grounded in certain assumptions that are seldom made explicit. One of the most fundamental of these assumptions is that the boundaries of the individual provide the proper framework within which psychological processes can be adequately analyzed. Recently, this assumption has been brought increasingly into question, but it continues to guide much of the thinking behind studies of cognitive development.

At first glance it may appear strange to question this assumption. After all, in contrast say to sociology, psychology does focus on processes carried out by the individual. However, we believe that to leave things thus is to overlook an essential avenue of explanation that would otherwise not be available to this discipline. Indeed, we would argue that by restricting its focus solely to phenomena that occur within the boundaries of the individual, cognitive developmental psychology cannot properly address some of the very problems it has set itself. In an important sense, we believe that to explain the individual, one must begin by going beyond the individual. The kind of irony involved here was recognized by Luria. One of his beginning axioms was that:

'In order to explain the highly complex forms of human consciousness one must go beyond the human organism. One must seek the origins of conscious activity and "categorical" behaviour not in the recesses of the human brain or in the depths of the spirit, but of the external conditions of life. Above all, this means that one must seek these origins in external processes of social life, in the social and historical forms of human existence.'

(1981, p. 25)

The position Luria took toward the relationship between social and individual phenomena has several implications. First, the explanation of social processes cannot be reduced to principles that apply to individual psychological phenomena. This point has been made in Western psychology by figures such as Adorno (1967) and Mead (1934). In the latter's formulation, individual psychological processes are involved in social interaction, but the relationship between the social and individual planes involves a complex dialectic rather than a form of reductionism. Second, and conversely, Luria's position assumes that the explanation of individual psychological phenomena cannot be reduced to principles that govern social processes. That is, it rules out a form of 'social reductionism'. In accordance with this point, it would be incorrect, for example, to view individual psychological processes as direct, internalized copies of social interactional processes. Rather, there is a complex dialectic here as well, a point to which we shall return in analyzing our empirical example.

Instead of falling into either of the two reductionistic traps just outlined, we believe that disciplines such as developmental psychology must be viewed as part of an overarching theoretical enterprise in which multiple levels of explanation are involved. The ultimate goal of an account of human action and its development is to specify how various levels of explanation are interrelated without reducing one to another.

The way that we propose to work toward this goal is grounded in the ideas of Vygotsky and his followers in the USSR, especially Leont'ev (1959, 1975, 1981) and Zinchenko (1985). There are many ways to explicate the complex general approach proposed by these figures, but here we shall focus on one issue in particular, the issue of an appropriate unit of analysis for understanding higher (i.e. uniquely human, sociocultural) mental processes.

As Zinchenko (1985) has argued, the choice of a unit of analysis in a psychological theory says a great deal about the way in which the rest of the theory can be organized. For example, it makes a great deal of difference whether say, an S–R bond or a figure-ground relationship is assumed as an appropriate unit for analyzing psychological processes. The choice here determines to a great degree what the theory is capable of formulating as questions and 'seeing' as data.

In Soviet psychology a great deal of effort has been expended on identifying a unit of analysis appropriate for modern psychology. The efforts of figures such as Vygotsky, Leont'ev, and Zinchenko have produced several candidates. For example, on the basis of an explicit metatheoretical review of the problem of units of analysis, Vygotsky (1934) suggested that word meaning is a construct that can fill this role. Over the course of several decades, some of his students (e.g. Leont'ev 1959, 1975, 1981; P. I. Zinchenko 1939) conducted critical analyses of this proposed solution and argued that the notion of goal-directed action is a more adequate unit for Soviet psychology. Quite recently, V. P. Zinchenko (1985) (also see Zinchenko and Smirnov 1983; Wertsch, in press) has outlined ways of integrating the ideas of Vygotsky and other Soviet psychologists such that the strengths of various perspectives are preserved.

Although there are important differences among the accounts of analytic units provided by these various authors, they all have certain fundamental assumptions in common. For our purposes, the most important of these is that one and the same unit can serve as an appropriate analytic construct both on the social interactional and on the individual planes of functioning. This is true for Vygotsky's account of word meaning, for Leont'ev's account of action, and for Zinchenko's notion of 'tool-mediated action'. As we shall see, this does not amount to a claim of strict parallelism as outlined in our earlier critical comments on social reductionism. Rather, it constitutes a claim that social and individual processes can be interrelated through a complex process of socialization and internalization (cf. Leont'ev 1981; Wertsch and Stone 1985).

The specific analytic unit that will be our focus here is Zinchenko's construct of a tool-mediated action. Not surprisingly given the term, the two basic notions involved in it are mediation and action. The former derives primarily from Vygotsky's claims about the mediation of human action by 'technical tools' and 'psychological tools', or signs (see Wertsch in press). It is concerned with the idea that the mastery of tools, especially psychological tools such as human language, is a crucial factor that distinguishes humans from animals, and one cultural epoch of humans from another. In this view, signs are not considered to be auxiliary means whose use merely facilitates already existing social or psychological processes. Rather, they are viewed as playing a fundamental role in determining what these processes can be.

In the case of human ontogenesis, Vygotsky argued that the mastery of technical, and especially, psychological tools has the effect of transforming the elementary mental functions of the infant into the higher mental functions of acculturated children and adults. Here again his point was that mediational means do not merely facilitate forms of action that would otherwise exist. Rather, they have a powerful transformatory force. According to Vygotsky (1981a):

'By being included in the process of behaviour, the psychological tool [i.e. sign] alters the entire flow and structure of mental functions. It does this by determining the structure of a new instrumental act just as a technical tool alters the process of natural adaptation by determining the form of labor operations.' (p. 137)

If the notion of mediation in Zinchenko's account of tool-mediated action comes from Vygotsky, the notion of action comes primarily from Leont'ev (1975, 1981). In particular, Zinchenko is indebted to Leont'ev's comments about the level of analysis concerned with action (deistvie) in a more general theory of activity (deyatel 'nost'). Two of Leont'ev's ideas are particularly relevant for our purposes. First, in his account an action is defined largely in terms of a goal. There is no mechanistic limitation on the nature of a goal in human action. It may be getting to point N, writing a book, eating breakfast, laying a pipeline, etc. Furthermore, as we shall see, a goal may be embedded as a subgoal in a larger framework, and it may itself be comprised of subgoals. Because the notions of goal and action are interdefined in Leont'ev's and Zinchenko's approach, we shall treat the terms *action* and *goal-directed action* as equivalent.

The second point to note about action is closely related to Vygotsky's account of mediation. In keeping with their interest in accounting for uniquely human, socio-historically situated psychological processes, Soviet psychologists such as Zinchenko tend to focus on socio-historically specific goal-directed actions. In this view, actions that strike humans in one socio-historical setting as obvious or natural may not strike those from other settings in the same way. Authors such as Cole and Scribner (1974), Scribner and Cole (1981), and Lave (1977) have provided support for this claim in several empirical studies. Their findings indicate that instead of attempting to understand human action in terms of universal, ahistorical categories, we must pay much greater heed than we have in the past to its socio-historical context.

Zinchenko outlined the notion of tool-mediated action to deal with a broad range of issues in psychology. Our focus here will be somewhat narrower; we shall be concerned with its role in an account of the socialization of human cognitive processes. We begin our account of this issue by turning to Vygotsky's 'general genetic law of cultural development'. According to this law:

'Any function in the child's cultural development appears twice, or on two planes. First it appears on the social plane, and then on the psychological plane. First it appears between people as an interpsychological category, and then within the child as an intrapsychological category. This is equally true with regard to voluntary attention, logical memory, the formation of concepts, and the development of volition.' (1981b, p. 163)

Vygotsky viewed this law as applying primarily to mental functions such as thinking and memory, as well as to his primary unit of analysis, word meaning. In our reformulation of his approach, however, it is tool-mediated action that is viewed as appearing on the interpsychological and intrapsychological planes.

To make this change in Vygotsky's approach does not prevent us from retaining one of the fundamental strengths of the general genetic law of cultural development. This strength lies in the relationship Vygotsky saw between social and individual processes. Instead of simply arguing that intra-psychological functioning somehow grows out of participating in social life, he was arguing that the same functional units are involved on both planes. This is the key to the problem of analytic units in an account of social and individual functioning. Among other things it means that in this approach, thinking, remembering, and goal-directed action are terms that can be appropriately predicated of groups as well as individuals.

The general approach used here is one in which an unsocialized child enters into tool-mediated action on the interpsychological plane and then comes to master the action such that he or she can execute it independently on the intrapsychological level. The more specific question we shall examine concerns the course of development when participation in interpsychological functioning is disrupted due to some sort of disorder in language processing. In a sense such disorders (e.g. problems in language comprehension) are intrapsychological, but in the approach being used here they quickly take on an interpsychological aspect as development proceeds.

Tool-mediated action—an example

In order to explicate our argument about the role of tool-mediated action in the emergence of interpsychological and intrapsychological functioning, we turn to a concrete example. The setting that we shall examine here is one in which the overall goal is to use various 'pieces' to construct a 'copy object' in accordance with a 'model object'. In order to reach this goal in this setting, one must carry out a set of subgoals. The number of subgoals that one can identify depends on the materials and on the level of detail one specifies about psychological, motor, and psychophysiological processes. For our purposes, we shall use a fairly general level and focus on the following three subgoals:

(i) consult the model to identify which piece comes next;

(ii) select the piece identified in step (i) from the pieces pile; and

(iii) insert the piece selected in step (ii) into the location in the copy identified in step (i).

Each of these subgoals may be considered to be a goal in itself, but the fact that they are subordinated to a larger goal is reflected in their formulation. Hence, (ii) presupposes information about (i); (iii) presupposes information about (i) and (ii); and if we were more detailed in our description of these subgoals, we would have to recognize that the term *next* in (i) presupposes information about previous executions of subgoal (iii). For our purposes, we shall term subgoals (i), (ii), and (iii) *steps* and reserve the term *goal* for the overall task of placing a piece in the copy in accordance with the model. As should be apparent by this point, the goal-directed action at issue here is an integrated structure that is interdefined with the steps and the very objects (i.e. their functional significance) in the task setting.

In addition to being a goal to which subgoals or steps are subordinated, the goal we have in mind here is subordinated to a more inclusive goal. This is so because the three steps we have outlined apply to a single piece only, whereas the task setting of interest to us requires that several pieces be identified, selected, and placed in the copy. Therefore, what we are call-ing an action in reality represents an intermediate level between subordi-nating and subordinated goal-directed actions. Again however, we shall reserve the term goal-directed action for the intermediate level.

The actual behaviour involved in executing one such action comprises what we shall term an *episode*. In order to carry out an episode successfully (except by chance), each of the three steps listed above must be executed. That is, in order to end up with a piece in its correct position in the copy, someone must consult the model, select an appropriate piece from the pieces pile, and place it correctly in the copy.

In most contemporary studies of cognitive development an analysis of this task would focus on how the individual performs it. While one of the aims of this chapter is to examine the development of individual function-ing, Vygotsky's general genetic law of cultural development suggests that a preliminary step in the analysis is needed to get there. Namely, one must begin with the interpsychological origins of intrapsychological functioning. This is precisely our approach to analyzing the execution of the steps in the goal-directed action of constructing one object in accordance with another.

In order to conduct such an analysis one must invoke the notion of *div-ision of responsibility* for the steps in the action. The point here is that to say that all three steps must be executed in any successful episode of this task setting says nothing about *how* they are executed. The 'how' of a con-crete instantiation concerns what Leont'ev (1981) called the 'operational composition' of action. As noted by Wertsch *et al.* (1984), one aspect of the operational composition of action concerns whether the action is executed on the interpsychological or intrapsychological plane.

In what follows, we shall examine the operational composition of the goal-directed action of constructing a copy object in accordance with the model. The empirical findings reported here are part of a larger study by Sammarco (1984). This study involved two groups of six mother-child dyads. The children were boys, ages 3–7 and 3–11, matched for nonverbal intelligence, age, social-emotional status, and sensory acuity, whose under-standing of language was at least average (the 'normally achieving' group) or significantly below average (the 'language disordered' group). All dyads were from monolingual Anglo-American homes in which English was the primary language spoken. All mothers had at least a high school education.

Since problem-solving is generally assumed to be associated with intelli-gence (Snow 1980), it was necessary to control for it. A nonverbal measure, one that does not require verbal responses, was used so as not to penalize the language disordered group. There are several nonverbal measures (Hiskey 1966; Leiter 1969) appropriate for the age range being studied. The Hiskey-Nebraska Test of Learning Aptitude (Hiskey 1966) was selected because it includes a greater variety of items relevant to this investigation. All of the children demonstrated average intellectual ability by falling within one standard deviation of the mean (84–116) on the Hiskey.

Language disordered subjects were selected from children enrolled in preschool special education programmes in the suburbs of Chicago. These children had previously been identified by their local school district as having special needs. School records of audiological and visual screenings were examined to screen out any children with known sensory impairments. Teachers were asked to recommend children who did not have any *primary* social or emotional disorders. In addition, teachers were asked to recommend children for screening who had receptive language problems and were thought to have average nonverbal abilities.

Children recommended by their teachers were screened for inclusion in the study. The ability of subjects in the language disordered group to comprehend single words was significantly below age level, i.e. one or more standard deviations below the mean on the Peabody Picture Vocabulary Test-Revised (Dunn and Dunn 1981). For all subjects in this group, there was a difference of at least 10 points between their Hiskey and Peabody standard scores.

The materials for the experimental task consisted of two identical three-dimensional toy airport scenes. Each scene contained a terminal plus 12 vehicles arranged on a board. The terminal remained stationary on the copy board when the rest of the copy was disassembled and was not considered in the analysis of the data. The experimental task also included a set of distractor items. Four sets of three test objects each were used in this study: helicopters, baggage cars, airplanes, and cars. The objects within each set varied along two dimensions: size (big and little) and colour (yellow, blue, red). There was one object of each colour in each set.

An informal measure was designed to assess each child's understanding of task-related vocabulary. Items included nouns (*helicopter, car, baggage car, airplane*), adjectives (*big, little, red, yellow, blue*), position expressions (*in front of, behind, next to, first, middle, last*) and adjective sequences (*little red, big red*). Actual objects, identical to those included in the experimental task, were used in the informal measure. Each child was asked to label the objects and colours. If the child responded appropriately, it was assumed that he could also point appropriately, and therefore, he was not asked to do so. If the child failed to provide any of the labels, he was not penalized and was given the opportunity to demonstrate his understanding via a pointing response. There was no overlap in scores between the normally achieving and the language disordered groups (Mann Whitney U test, p < .001, one-tailed).

Mothers of children who met the criteria established for this study were asked to accompany their children to school and engage in a problem-solving task. This session took place no later than three weeks after the child's initial screening session with the experimenter. The problem-solving session lasted approximately 20 minutes. Identical procedures were used with both groups of dyads.

The session took place in an empty classroom of the school. The mother and child were seated side-by-side at a specially designed table. The model and completely assembled copy were already on the table when the dyad was seated. The mother was seated behind the model and the child was seated midway between the copy and the pieces pile. The experimental task was presented in the following way:

Here we have two toy airports (*Point to the model and then point to the copy*). As you can see, the two toy airports are exactly alike. All of the pieces are the same. They are also in exactly the same places (*Point to the cars in the model and then in the copy*) and looking in the same direction (*Point to the airlanes in the model and then in the copy*). We know this piece goes here, because there is a dot here (*Point to the dot*). These toys (*Point to the model*) don't come out because they're glued. In a minute, I will take this airport (*Point to the copy*) apart. I want (child's name) to make this airport (*Point to the copy*) look exactly like this airport (*Point to the model*) with all the pieces in the same places. In the end, both of them should look just like they do now. I will put some extra pieces here also (*Present board with extra pieces*). They do not fit into the airport. So, in the end, they will be left out. If (child's name) doesn't know how to put the airport together, I would like you to help him. Provide any assistance that you think is necessary for (child's name) to complete the airport. There is no hurry. Take as long as you need. Do you have any questions? *Dismantle the copy and randomly place the pieces on the board with the extra pieces*). You may begin.

Videotapes were made of the dyad's behaviours. The audio portion was transcribed, and the nonverbal behaviours such as gazes to the model, copy, pointing, picking up, and placing pieces were co-ordinated with this transcript. The detailed record of interaction that resulted provided information about the operational composition of the actions carried out in this task setting. We shall present our findings in terms of a three level analysis.

The first level of the analysis was designed to determine, for each episode, who was physically responsible, i.e. the *mother* or the *child*, for the three substeps: (a) looking to the model; (b) picking up a piece; and (c) placing the piece in the copy. If (and *only* if) the child was physically responsible for a substep at Level 1, the analysis proceeded to Level 2. At this level the various degrees to which the child was responsible were determined. Given that the child carried out a substep, this second level was concerned with whether the child performed the behaviour on his own, or because his mother directed him to do it. When the child acted independently, his performance was coded as *self-regulated* and when the child's performance was instigated by his mother, it was coded as *other-regulated*.

A substep was coded as other-regulated when one or more of the adult behaviours specified below occurred before the actual behaviour identified with the substep (i.e. looks to the model, picks up piece, places piece in copy) but after the beginning of the episode: (a) pointing to the model, to the pile of pieces, to the place where the piece should go, or in the general direction appropriate for a particular piece; (b) making a complete utterance which explicitly mentioned the model, the piece to be picked up, or the place where the piece should go; (c) making a complete utterance which implicitly involved any of the above.

A substep was coded as self-regulated if the adult did not carry out any of the behaviours specified above before the actual behaviour identified with the substep and after the beginning of the episode. In other words, substeps were coded as self-regulated if they were not preceded by a behaviour on the mother's part which could be construed as an effort to get the child to perform the substep.

At Level 3 the focus was on whether the other-regulation was *direct* or *indirect*. The distinction between direct and indirect focuses on what would be required of the child if he were to respond appropriately. The child can comply appropriately to a direct form of other-regulation without having to carry out any implicit subdirectives involving both airports, i.e. the child can respond appropriately to direct other-regulation on the basis of information about only one of the two airports. Indirect forms of other-regulation, on the other hand, require the child to carry out implicit subdirectives involving both the model and the copy. Indirect other-regulation does not explicitly instruct the child to consult both the model and the copy, but in fact does implicitly require the child to make a comparison between the two if he is to respond appropriately. Consider the following examples (the mother's speech is in italic print and the mother's and child's nonverbal behaviours are in parentheses):

Example 1

(Mother starts to point to big red car in the model.) *Look over here.* (Child looks at the model.) *What's next? huh? Find,* (mother points to the big red car in the pieces pile and at the same time the child looks to the pieces pile), *The next piece.* (Child picks up the big red car and then looks to the copy board.) *Okay now where,* (mother points to the dot on the copy board where the piece is to go). *Does it go?* (Child places piece in correct spot on copy board).

Example 2

Okay now find the next piece (child looks to the model). *What do you need?* (Child looks to the copy, back to the model, and then to the pieces pile, picks up the big red car and then looks to the copy board). *Okay now where does it go?* (Child looks to model, then back to the copy, and places the piece in the correct spot on the copy board).

In the first example, all three of the strategic steps (looks to the model, pick-up, and placement) are preceded by direct other-regulation. The second example, on the other hand, illustrates indirect other-regulation for each of the strategic steps. Consider the substep of looking to the model in more detail.

In the first example, the child's look to the model was preceded by the mother's pointing and the directive, 'Look over here'. In order to comply with this direct other-regulation the child had to follow his mother's finger visually. This regulation explicitly or directly told him what to do. No inferences were required. Contrast this with what was required of the child in the second example. Here the mother said, 'Okay now find the next piece'. Note that there was no pointing to guide the child's behaviour. This indirect other-regulation required the child to carry out a number of implicit subdirectives (involving both the model and the copy) if he was to respond appropriately. In other words, what comes next can only be determined by looking to the model. An understanding of this subdirective was required of the child if he was to respond appropriately.

On the basis of our three-level analysis, one can derive a summary measure of 'direct responsibility' (cf. Wertsch, Minick, and Arns 1984). The mother was considered to have been directly responsible for a substep if she physically carried out a step or provided a form of direct other-regulation to get her child to perform the step. Conversely, the child was considered directly responsible if he carried out a step using self-regulation or a form of indirect other-regulation.

In instances where the child carries out a step through direct other-regulation, the mother is considered directly responsible because the child did not need to understand the strategic significance of the behaviour in order to carry out the step physically. For example, when looking, pick-up, and placement behaviours are directly other-regulated by the mother, the child does not need to understand the significance of the model. On the other hand, when the child carries out a step through indirect other-regulation, he must assume some of the responsibility for the strategy involved in the task. Recall that indirect other-regulation requires the child to carry out implicit subdirectives involving both the model and the copy.

Our coding procedures are represented in summary form in Fig. 20.1. The underlined terms *self-regulation* and *indirect* identify behaviours that constitute the child's being directly responsible for a step's execution. If a step was coded under another heading, the direct responsibility was assigned to the mother.

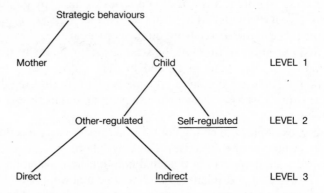

Fig. 20.1 Tree diagram representing relations among the three levels of analysis.

The results of this three-level analysis vary by level and strategic step. Level 1 examined who was physically responsible for each of the three strategic substeps. Significant differences (based on Mann–Whitney U tests) were found between the normally achieving (NA) and the language disordered (LD) for each of the strategic steps. The NA children were physically responsible for each of the three strategic steps virtually 100 per cent of the time. In contrast, the proportion of time the LD children were physically responsible was significantly less for each strategic step. They were responsible for 91 per cent of the pick-ups, 79 per cent of the placements, and 68 per cent of the looks.

For those behaviours which the child physically carried out in the Level 1 analysis, Level 2 examined whether they were self- or other-regulated. Significant differences were found here as well between the two groups for each of the three strategic steps. This difference reflects the fact that the NA children demonstrated more self-regulated behaviour (looks to the model, 63 per cent; pick-ups, 37 per cent; placements, 51 per cent) than did the LD children. In fact, the LD children rarely used self-regulation when carrying out a step (looks to the model, 13 per cent; pick-ups, 9 per cent; placements, 3 per cent).

For those behaviours that were other-regulated in the Level 2 analysis, Level 3 examined the nature of the other-regulation, i.e. direct or indirect. Significant differences were found between the two groups for pick-ups and placements. These differences reflect the fact that the mothers in the NA group used more indirect other-regulation (pick-ups, 100 per cent; placements, 70 per cent) than did the mothers in the LD group (pick-ups, 35 per cent; placements, 31 per cent). There was no significant difference between the two groups with regard to the nature of the other-regulation that preceded the child's look to the model. This finding was related to the fact that by the time one reached Level 3 in analyzing eye gazes, too few instances were available for meaningful comparisons.

As outlined earlier, these three levels of analysis for each strategic step can be summarized in terms of a measure of direct responsibility. Recall that the child was considered directly responsible if he carried out a step using self-regulation or indirect other-regulation. Conversely, the mother was considered directly responsible if she physically carried out a step or provided a form of direct other-regulation to get her child to perform the step.

There was no overlap between the LD and NA dyads in the proportion of episodes in which the child was directly responsible for a look to the model (Mann Whitney U test $p < .001$, one-tailed). Comparison of the two groups with regard to who was directly responsible for the looks to the model revealed that in the LD group the children were directly responsible for 12 of 77 (16 per cent) of the looks as compared to 54 of 70 (77 per cent) in the NA group. In other words, it was the mothers in the LD dyads, as compared to the chidren in the NA dyads, who assumed the bulk of the strategic responsibility for looking to the model.

There was also no overlap between the groups in the proportion of episodes in which the child was directly responsible for a pick-up (Mann Whitney U test, $p < .001$, one-tailed). Comparison of the two groups with regard to who was directly responsible for the pick-ups revealed that in the LD dyads the children were directly responsible for 23 of 61 (38 per cent) of the pick-ups as compared to 69 of 69 (100 per cent) in the NA group. As with looks to the model, it was the mothers in the LD dyads, as compared to the children in the NA dyads, who assumed the bulk of the strategic responsibility for pick-ups. In fact, the children in the NA group were directly responsible for pick-ups 100 per cent of the time.

Finally, a Mann-Whitney U was calculated on the proportion of episodes in which the child was directly responsible for a placement. This test revealed a significant difference between the LD and NA dyads ($U(6,6) = 2$, $p < .004$, one-tailed). Comparison of the two groups with regard to who was directly responsible for the placements revealed that in the LD dyads the children were directly responsible for 20 of 77 (26 per cent) of the placements as compared to 58 of 70 (83 per cent) in the NA dyads. As with looks to the model and pick-ups, it was the mothers in the LD group, as compared to the children in the NA group who assumed the bulk of the strategic responsibility for placements.

In summary, the measure of direct responsibility provided an overall assessment of who was directly responsible (mother or child) for each of the three strategic steps. Significant differences were found between the two groups for each of the strategic steps. The NA children were directly responsible for 77 per cent of the looks, 100 per cent of the pick-ups, and 83 per cent of the placements. In contrast, the proportion of time the LD children were directly responsible was significantly less for each of the strategic steps, i.e. 16 per cent of the looks, 38 per cent of the pick-ups, and 26 per cent of the placements. Thus, across the three steps, the mothers in the LD group, as compared to the children in the NA group, assumed the greater part of the strategic responsibility.

Conclusion

With these results in mind, let us return to the issue of the role of units of analysis in the general genetic law of cultural development. First it is important to recognize that we are dealing with a tool-mediated action of the sort outlined by Zinchenko. The action at issue can be mastered only by participating in socioculturally situated routines. It is not something that is innate or structured by the nonsocial environment. Unlike some other puzzle tasks, there is not even feedback built into the materials in our study that provide clues which would allow a child to master the task without the 'on-line' assistance of an expert. In general, there are countless sets of manipulations that could be carried out with the task materials that we provided to the dyads in this study. The fact that the copy was always completed 'correctly' means that a specific socioculturally defined plan was being imposed on the materials.

Our data show that the two groups of dyads varied in how they structured the operational composition of the actions. Specifically, the 'mix' of interpsychological and intrapsychological functioning differed. Even though this is so, an essential point to our argument remains that in an important sense one and the same goal-directed action was being carried out regardless of its operational composition. This remains the unit of functioning that appears in social interaction and is ultimately mastered in individual functioning.

Obviously, this does not mean that the same psychological processes are involved in the individuals participating in the task. It does mean, however, that the structure of what is to be mastered on the intrapsychological plane is 'out there' in concrete social interaction. As far as the actions involved and their results are concerned, one and the same thing is done on the social and individual planes; what changes is the division of responsibility—the operational composition of the action—and, correspondingly, the intrapsychological processes executed by each participant. This line of reasoning reflects once again the ironic point made by Luria that it is only by going beyond the boundaries of the individual that we can arrive at a full account of the intrapsychological processes carried out within these boundaries.

Saying this does not address the mechanisms that make possible the transition from interpsychological to intrapsychological functioning. As authors such as Wertsch and Hickmann (in press) have noted, there are several possible mechanisms involved in the transition from social to individual functioning, but one seems particularly important in the kind of setting under consideration here. This is the procedure of carrying out a task first and later coming to understand it. This is the reverse of what is typically understood as occurring in cognitive development. It is generally assumed that one must understand a task before one can carry it out. Our suggestion is that in many cases one must carry it out before one can understand it. The process we have in mind here is similar to Cazden's (1983) notion of performance before competence. Of course our notion of carrying out a task means participating in its execution on the interpsychological plane. Hence the notion of carrying out something before understanding it is not as paradoxical as it may first appear.

With this in mind, let us return to the differences we found between the two groups of dyads in our study. Recall that major, statistically significant differences were found between the two groups in the division of direct responsibility for each of the three strategic steps of the goal-directed action under consideration. In well over half of their episodes, normally achieving children were directly responsible for all three steps; precisely the opposite pattern characterized the language disordered children. A cursory glance at the tapes of the two sets of dyads reveals some of the reasons for this difference. In general, the mothers of the LD children had great difficulty in getting their child to understand and follow their verbal other-regulation. This is hardly surprising given that the selection criterion for the LD children was a comprehension deficit. The major problems in verbal communication that arose for these mothers meant that the forms of mediation and representation used by the NA dyads on the interpsychological plane were not employed. The LD children were not presented with the same 'semiotic challenges' (Wertsch in press) that confronted the NA children; as a result, they could hardly be expected to master the representation of the task setting that emerged naturally through the normal achievers' experience in interpsychological functioning.

The integral nature of action as an analytic unit becomes particularly important here. This point is perhaps best made if we return briefly to Level 1 of our analysis for the three strategic steps. There we reported that the NA children physically carried out all three steps in almost 100 per cent of the episodes, whereas the LD children were physically responsible for 91 per cent of the pick-ups, 79 per cent of the placements, and 68 per cent of the looks to the model. The figure for looks to the model is particularly revealing here. In order to understand the functional significance of picking up a certain piece (Step ii) and placing it in a certain location in the copy (Step iii), it is essential to understand the role of information from the model. In the absence of such understanding, the significance of picking up and placing pieces is quite different from what it would be if these behaviours are interpreted as steps in an integrative goal-directed action; indeed the entire 'situation definition' (Wertsch 1984) is quite different in the two cases.

The comprehension deficits that characterized children in the LD group constitute a form of disruption in individual functioning. As noted earlier however, such a disruption soon has a powerful effect on the development of interpsychological functioning. One could say that the LD children never entered into interpsychological functioning in the way characteristic of the NA children. Specifically, they did not participate fully in the goal-directed action on the interpsychological plane. While they carried out some behaviours that were outwardly identical to those executed by the NA children, the fact that they did not execute others (namely, consulting the model) means that their definition of behaviours and objects in the task setting was likely to be very different from that of the NA children.

What we envision here is, in effect, a developmental model comprised of a series of developmental transitions, each of which can have an impact on several subsequent ones in complex and indirect ways. Thus a disruption in individual language processing abilities may not be directly tied to intrapsychological functioning. Rather, for the type of cognitive processes examined in this chapter its influence may be indirect, through the interpsychological plane. In essence, one disruption is viewed as setting off a series of consequences.

If we are to follow the general lines of the reasoning we have outlined here, our central point about an analytic unit must be incorporated into the argument. This is so because it provides the key to understanding the nature of the relationship between interpsychological and intrapsychological functioning. It provides a much needed way of characterizing the social contexts that give rise to individual cognitive skills. For the specific purposes of this paper, invoking the notion of tool-mediated action has made it possible to provide some insight into how disruptions in interpsychological functioning may impede progress in making the transition to intrapsychological functioning. As a more general point, we would assert that many aspects of acculturated, individual, cognitive functioning will never be accessible to interpretation unless we identify analytic units that allow us to specify principled interconnections between social and individual functioning. This is an issue that has been a source of confusion for at least half a century. If we are to make progress in accounting for cognitive development, we must address it more seriously than we have done up until now.

Discussion

Interpretation

Ignatovic-Savic noted:

'I find both similarities and dissimiliarities between your and my interpretation of Vygotsky's basic idea. As for the question of units of analysis your interpretation is that for Vygotsky word meaning is a unit appropriate for modern "psychology". I think that is not correct. Vygotsky's claim was that it is the unit appropriate for the study of relations between thought and language. The essence of what Vygotsky intended to say is: whenever you study complex psychological phenomena you have to find a unit that conserves the meaning of the whole. This methodological principle applied to the study of interaction between adult and child leads us to take a *dyad* as a unit of analysis.

Also, I could not agree with you that the first phase in the development of the child is "individual pre-interpsychological functioning". Vygotsky's view was quite opposite to that. The habitual methodological practice in psychology has been to count what one partner has done separately from what the other has done, and later on to look for the relation between these two sets of data. This new principle requires that you count what the dyad as a whole does at each moment.'

Wertsch replied:

'I believe that both of your comments about my presentation of Vygotsky's ideas raise issues of interpretation of a complex theoretical figure. As is known, Vygotsky wrote a good deal in a relatively short period. This, plus his tendency not to edit many of his writings, has resulted in various interpretations being possible on many particulars in his approach. Hence, while I agree with your first point (that Vygotsky was focusing primarily on the relationship between speech and thinking when he made his claims about word meaning as a unit of analysis) as a matter of emphasis in his presentation, I still would assert that word meaning was intended by Vygotsky to serve a broader purpose. Hence in the final pages of *Thinking and Speech* (1934—or *Thought and Language* as translated into English) he asserted that the "word is the microcosm of human consciousness". For him, consciousness was a broad philosophical category. Hence I think his notion of word meaning had broader applications than for thinking and speech alone. He himself makes this point in *Thinking and Speech* when he argued for viewing thinking and speech as only one case of the broader issue of interfunctional relations. When considering this broader issue, he did not introduce other units of analysis.

Furthermore, as I have recently argued elsewhere (Wertsch in press), Vygotsky's account of word meaning cannot fulfil even the limited role he assigned to it. While it is a unit of *semiotic mediation* of higher mental functions and human consciousness, it is not a unit of analysis of either of these two phenomena *per se*.

While I think that your emphasis on dyadic interaction involving very young children is an absolutely essential aspect of Vygotsky's ideas, I would also argue that the dyad itself is not an appropriate unit of analysis. Rather, some mode of action or activity *of the dyad* (as well as of individuals) would seem most appropriate to me. (But here I am simply repeating what I said in my chapter.) With regard to your second comment, I again think that an element of differing interpretation of Vygotsky's voluminous texts is involved. I agree with you that social forces must be seen as playing a role in ontogenesis from birth onward.

However, as I have argued elsewhere (Wertsch in press), I also think that it is important to take seriously Vygotsky's comments about the natural line of development, elementary mental functioning, etc., in ontogenesis. Your comment was valuable to me in that you pointed out that I had presented my argument as if I thought Vygotsky posited a neat, separate period of non-social life before the onset of interpsychological functioning. This was not my intention. Rather, my goal was to produce a scheme in which the role of an individual's handicapping condition (e.g. lack of auditory discrimination skills) can be recognized and incorporated into a Vygotskian approach. My revisions reflect your critical insight on this point.'

Doise added:

'I do not want to intervene in a debate on the interpretation of Vygotsky, although I personally think that the notion of "individual pre-interpsychological functioning" is both not very central in his work and not very heuristic. But I want to stress the heuristic value of Wertsch's research procedure. It shows how the ideas of Vygotsky on the interpersonal and intrapersonal can be studied empirically. But such studies will probably ask for other experimental designs and especially for designs involving a pre-test and post-test in order to study the effects of different kinds of interpersonal functioning.'

Wertsch replied:

'With regard to your comments about pre-test and post-test designs in a Vygotskian approach, I have several observations (none of which may provide a definite response by itself). First, I have been primarily interested in documenting the kinds of interpsychological functioning that occur in various cultural and clinical contexts. Therefore, I have said less in detail than is eventually desirable about the individual forms of functioning that precede and follow such interaction. Given the overall claims of my approach, however, such functioning does need to be addressed. Two points arise. First, as Vygotsky argued, virtually any assessment of "individual" functioning in pre-tests and post-tests actually involves social factors that can be expected to influence scores. This point has been made in more detail recently by investigators such as Ann Brown (1983) in her studies of the zone of proximal development. In accordance with this point it is often very difficult to know where treatment stops and measurement begins.

Second, if one is to deal with the difficult issues of pre-tests and post-tests (and I think one must), various alternatives present themselves. For example, careful measures before and after extended training are possible. More to my liking, however, is the kind of "microgenetic" approach used by investigators such as myself, Ann Brown and her colleagues, and others. In this method, changes in a dyad's and in an individual's performance are examined within a single training or experimental session. I prefer such a method because it allows one to trace in more detail the sources, content, and processes of the shift from interpsychological to intrapsychological functioning.'

Bryant, addressing Ignajatovic-Savic, noted:

'About the relations between Piaget and Vygotsky, an important difference between them was that Piaget stressed conflict and Vygotsky stressed co-operation and what you call role division. If Piaget's theory is applied to the social domain, as

Perret-Clermont and Doise have done, then it seems to me, these theories dwell on contradiction and disagreement between people and Vygotsky's does not.'

Doise added:

'For the description of the sensori-motor development one should add a reference to the thesis of Hammes (University of Tilburg 1982) who compares the theory of Piaget and of Werner and Kaplan (1963) on the origin of symbolic thinking. To some extent his conclusion, based on experiments with autistic and other children, is that Piaget would fit perfectly for the autistic child, but Werner and Kaplan for both the autistic and the normal.

A theoretically important concept of "constructivism" is essential in the work of Piaget, but is it in Vygotsky? And how can we integrate it with the notion of the zone of proximal development? Does such a notion not necessarily involve modelling processes, whereas constructiveness does not?'

Ignjatovic-Savic commented:

'I would say that, for Vygotsky as well as for Piaget, the child is an active subject of his own development. He is not shaped by the environment but is actively formed through interaction. But Vygotsky's view on the influence of sociocultural factors upon development does not allow any notion of the universalism of cognitive structures nor of the invariable sequence of developmental steps. It follows from Vygotsky's theory that there is no one and unique direction of development but more divergent lines instead.'

21

Social regulations in cognitive development
WILLEM DOISE*

Probably because of the growing institutionalisation of educational practices, of which the raising of the school-leaving age is only one aspect, a certain postulate has come to pervade various areas of the human sciences. This concerns the existence of a causal link between the climate of human relationships in a given context and the cognitive functioning of the individuals living in that context. This conception is so general as to allow opposing forces, those that advocate restrictive and authoritarian methods and those that defend non-directive ideas or even spontaneity, to base their case on it. To a certain extent, these forces always find their expression in educational theories founded on the assumption that there is a positive or negative link between a set of characteristics in the sphere of social relationships and a set of characteristics appearing in the cognitive or affective functioning of the individual.

It is not my purpose here to review these conceptions systematically. This job remains to be done, though we are attempting elsewhere to investigate the social representations of the intellect and of its development (see Poeschl *et al.* in press).

The aim of this contribution is to specify the scientific status of conjectures about the causal intervention of social dynamics in individual cognitive development. However, I shall not demonstrate the scientific value of such conjectures once and for all; they fall in the realm of grand theories which will remain, almost by definition, impossible to verify as such. My goal is more modest but also more specific—namely to describe experimental illustrations of some of the processes at work in certain important aspects of individual cognitive functioning when known social regulations are introduced under appropriate conditions. Given the significance of the Piagetian contribution to the study of cognitive development, it seems appropriate to start with a brief account of the positions held by Piaget and by some of his critics concerning the weight of social factors in cognitive

---

* Translated by Evelyn Aeschlimann.

development. Many of the texts referred to are little known outside French speaking areas.

Piaget on the social development of the intellect

In an article first published in 1928, Piaget (1976a) considers the following philosophical problem: ' . . . do the operations by which we arrive at what rational consciousness calls the "true" depend on society, and in what way?' The point here, in reality, is to explain how cognitive operations can escape arbitrariness and come to enjoy a status of objectivity and universality. Two issues could impede this rationality of knowledge. Firstly, individual arbitrariness. Egocentrism could be seen as its proto-type, ' . . . anarchic thought subjected to feeling, as it is found in day-dreaming, in dreams, in certain states of the child's thought . . . ' (ibid., 66). Should society be the place to remedy this arbitrariness in the genesis of communicable thought? Is it not society that teaches the rules of thought and provides the means to reach agreements which are unamenable to indi-vidual arbitrariness? But conferring on society the power of imposing its rationality implies the intervention of a second arbitrariness. It amounts to replacing a fluctuating, individual arbitrariness with a stronger, unswerv-ing, social one. Both sorts of arbitrariness could well be mutually reinforc-ing: 'In a society where the generations thus heavily weigh upon each other, none of the conditions required to eliminate childish mentality can be met. There is no discussion, no exchange of views.' (ibid., 76). Whether such societies really exist or whether they are merely hypothesized for the sake of the argument is here beside the point.

But after all, it is thanks to a social process, which Piaget calls cooper-ation, that the subject frees himself from both individual and social arbi-trariness. Cooperation allows individuals gradually to acquire a sense of objectivity and rationality by confronting and coordinating their points of view. This argument is developed in Piaget's most frequently quoted book *The Moral Judgment of the Child*, which is indeed a book of social psychology. It often describes observations on social interactions and its theoretical approach is exclusively that of social psychology. The shift from heteronomy to autonomy in the child's moral concepts is explained by the many situations in which children have to cooperate with each other. For example, when explaining a change in mastering the rules of a game of marbles, ' . . . if, at a given moment, cooperation takes the place of constraint, or autonomy that of conformity, it is because the child, as he grows older, becomes progressively free from adult supervision' (Piaget 1965a).

Clearly, moral sense is considered to emerge from social interaction only. For Piaget it is obvious that the psychological analysis should be conducted within the framework of sociology for, as he states in the same book, one should not ' . . . return to the pre-sociological phase of psychology, . . . ' (ibid., p. 349, note 1). In other words, psychological and sociological explanations should be articulated in accounting for moral or cognitive development. It is true that Piaget did not subsequently continue in this direction, although the theme reappears all through his work in the form of brief remarks such as, ' . . . human intelligence develops in the individual in terms of social interaction—too often disregarded . . . ' (Piaget 1971, p. 224). And when Piaget enumerates the factors intervening in cognitive development, he places interindividual coordinations between biological development and equilibration on the one hand, and educational and cultural transmission on the other (Piaget 1974).

Although Piaget's thesis on the subject is expressed mostly in his earlier work, it was never denied in his subsequent writings. It clearly refers to a causal link between the dynamics of interindividual relationships and the onset of cognitive functioning. This thesis obviously fits a certain ideology: the kind of relationships that promote individual development and free societies from ideological relics are of a democratic type, 'To the residuum peculiar to the conforming attitude of the younger ones correspond the derivations "divine or adult origin" and "permanence in history". To the residuum peculiar to the more democratic attitude of the older children correspond the derivations "natural (childish) origin" and "progress".' (Piaget 1965a, p. 68).

But Piaget did not accept the idea of a causal influence of social events on cognitive development without further questioning. On the contrary, he set the matter in a more complex light from the start, ' . . . one may observe that precisely at the age when the child's social life develops, he acquires the possibility of entering into the point of view of others, he practises reciprocity and discovers how to handle the logic of relations. Must we then ask ourselves whether it is the logic of relations that leads to reciprocity or the reverse. This is the problem of the river and its banks. There are two aspects to the process: cooperation is the empirical fact of which reciprocity is the logical ideal.' (Piaget 1976a, p. 79s).

This mutual dependence of the social and the cognitive is most clearly stated in the *Etudes Sociologiques*. Here, Piaget (1965b, p. 143) returned once more to 'the question, so often discussed, of the social or individual nature of logic . . . '. Having published his famous studies on various forms of operational thought, Piaget approached again the problem of the links between social life and cognitive development 'while adding a new fact to the file: the existence of operational "groupements" distinguished by genetic psychology as playing a role in the forming of reason' (ibid., p. 143). It appears that the same logico-mathematical models which account for the cognitive activities that children between 7 and 10 are able to perform in relation to their physical environment also apply to their social capacities, such as the intellectual exchange of views during a debate or the exchange of qualitative values. No wonder that 'the period of operations properly speaking . . . corresponds to . . . a very marked progress in socialization: the child becomes capable of cooperation, he no longer thinks in terms of himself alone but in terms of actual or possible coordination of several points of view' (ibid., p. 157). In other words, a correlation between cognitive and social development seems to exist without any causality being attached to it: ' . . . if logical progress thus proceeds in parallel with that of socialization, is it necessary to say that the child becomes capable of logical operations because his social development qualifies him for cooperation or should one assert, on the contrary, that it is these logical acquisitions that allow him to understand others and thus lead him to cooperate? Since the two sorts of progress are on even terms, the question seems without answer except to say that they constitute the two indissociable aspects of a single and identical reality, at the same time social and individual.' (ibid., p. 158).

Questions and answers remain unchanged in Piaget's recent appreciation of our own work on the sociogenesis of cognitive operations. He wrote, ' . . . is it a matter of causality or of training (. . .), it remains clear that coordinations of actions and of operations are identical whether these liaisons are intra- or inter-individual and this more especially as the individual is himself socialized and that, reciprocally, joint performance would never function if each member of the group did not possess a nervous system and the psychobiological regulations which it comprises. In other words, the operational "structure" at stake is of a general or "common" nature, that is biopsychosociological, and that is why it is fundamentally logical.' (Piaget 1976b, p. 226).

Elsewhere, we (Doise 1982) have developed the idea that, for Piaget, interindividual coordination or cooperation was always a theoretical construction or even an idealized notion. Cooperation as having effects at the cognitive level was never studied empirically or experimentally by the master of the School of Geneva.

Finally, we must note that Piaget uses a definition of intelligence which disregards its social characteristics. The thesis we champion is, in essence, that individual cognitive action is but one moment of a complex process which is also of a social nature. As we have explained elsewhere (Doise and Mugny 1984), this thesis implies that the individual's cognitive coordinations be actualized by social coordinations. This means that the individual must coordinate his actions with those of others as a first step towards mastering individualized systems of coordination.

In the following paragraphs, we describe in more detail two kinds of social dynamics with considerable bearing on cognitive development.

Socio-cognitive conflict and cognitive operations

Experimental social psychology has long been studying the role of conflict in generating cognitive change in directions as diverse as cognitive dissonance or the effects of differences of opinion on group polarization. But our own research is based also on Piagetian ideas about the social origin of decentration.* Socio-cognitive conflict is said to exist when, in one and the same situation, different cognitive approaches to the same problem are socially produced. Given appropriate conditions, the confrontation of these different approaches may result in their being coordinated into a new approach, more complex and better adapted to solving the problem than any one of the previous approaches alone.

A first example of this can be drawn from the classical studies on the conservation of equal lengths (Inhelder *et al.* 1974). Five to six-year-old children rightly think that two identical rods placed parallel to each other, so that their extremities coincide perceptually, are of equal lengths. But if the perceived equality is disturbed by sliding one of the rods a few centimetres along the other, the same children will declare that one of the rods is longer because, in evaluating the respective lengths, they centre their attention on only one projecting end of a rod without taking into account that it is compensated by the projecting end of the other rod on the opposite side. In several experiments we have placed children who were non-conservers during an experimental phase in a situation of socio-cognitive conflict by contradicting them: when a child said that one of the rods was longer because it stuck out at one end, the experimenter answered that, in his opinion, the other rod was longer because it stuck out at the other end. We present here only a summary of the main conclusions of these experiments (Doise and Mugny 1984).

First many children, confronted with an incorrect centration proposed by an adult or by another child, adopt what we call a 'relational' solution to the conflict by saying that they agree with their contradictor. In this case, compliant responses prevail over innovative cognitive work; the children seem as unconcerned about the contradictory responses of others as they are about their own contradictions when one rod and then the other are shifted successively. Therefore, from the first we made sure that in situ-

---

* *Centration*: the young child naturally focuses attention on only one aspect of a problem at a time and may even behave as if ignoring the multidimensionality of an object or a problem. The term 'centration' is used to designate this internal focus of the child's attention. The developing child will learn to coordinate the different centrations that she/he experiences. Piaget saw this coordination as the source of the later operational structures that characterize the child's cognitive abilities. The child will also become aware that other persons may have centrations other than her/his own. Doise *et al.* (1975) suggest that such a discovery by the child of seemingly contradictory points of view is a potential source of cognitive growth.

ations of conflicting centrations both responses were kept in mind by having an associate experimenter remind the child of his initial answer whenever he gave in to the experimenter. With such an intensification of the socio-cognitive conflict, progress was systematically more often achieved than in another situation where the child was led to give a succession of incorrect and contradictory answers on his own. Furthermore, upon considering the results of a generalization test (using a test of conservation of unequal lengths), socio-cognitive conflict proved to be as efficient in generating progress as a 'modelling' condition in which the experimenter gave the correct answer and justified it by pointing to the compensations at the two opposite ends. Theoretically, the results obtained in the conditions of socio-cognitive conflict cannot be explained by a modelling effect as the child is never provided with the correct answer; on the other hand, the modelling situation may be reduced to a situation of socio-cognitive conflict. Indeed, in giving the correct answer, the experimenter socially emphasizes the existence of the two opposite projections. The correct answer, therefore, does imply a socially activated conflict of centration.

Here are the main conclusions that can be drawn from this experiment: having to keep one's point of view in mind while being obliged to take into account another incompatible one can lead to progress. This is also true when the conflict occurs in a more spontaneous manner between two children: only when they both give opposite answers do they progress.

In another series of experiments (Doise and Mugny 1984, Chapter 6), a paradigm of spatial transformation was reused. A model-village of several houses arranged on a cardboard base by the experimenter had to be reproduced by the subjects on an identical base but with a different orientation. However, the children involved were not yet capable of the spatial transformations required to conserve the intrafigural relations (left/right and front/back) and therefore produced an 'egocentric' copy of the village.

Having found in a first experiment that a couple of children accomplished the correct transpositions more easily than a child on his own, a further series of about twenty experiments were carried out to elucidate the superiority of collective performances over individual ones. Again, the socially created obligation to consider another point of view, different from one's own, was revealed as an important source of progress.

Consider such an experiment in more detail. Figure 21.1 shows the layout of a situation used during the experimental phase. The subject in the difficult position must transform front/back and right/left spatial relations, as he has been selected especially as unable to perform this task at pre-test. By contrast, the subject in the easy position has simply to rotate his visual plane by 90 degrees, something that most of the children in his age category can do.

But if the subject in the easy position does not experience any difficulty,

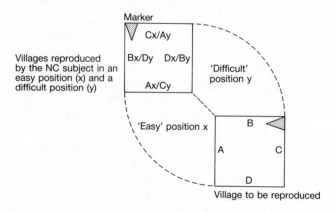

Fig. 21.1 Positions of houses as placed by two equally non-conserving subjects, one working from a 'difficult' position, and the other from an 'easy' position.

is he not likely to make no progress at all? No, not if he is confronted with another child who, being on the opposite side, is faced with a cognitive problem, provided however that the child in the difficult position proposes incorrect solutions. For any misplacement of a village house should create a problem for the child in the easy position. Such a conflict may oblige that subject to consider another point of view, to coordinate their differing points of view and eventually to grasp the error committed by the other subject, and even to convince the subject in the difficult position who originated this socio-cognitive conflict. Is it then not more than likely that the subject in the easy position will achieve cognitive progress from having dealt with this conflict?

This is the basic principle, but things do not always run smoothly. These experiments have given us further insight into the complex social dynamics which may take place between children and thwart the beneficial effects of socio-cognitive conflict. One major difficulty is due to the asymmetry generated by a dynamic of compliance, especially when one of the inter-acting members is overly sure of his answers.

We had already seen in an experiment with the same material (Doise and Mugny 1984, Chapter 6, exp. 2) that a very asymmetric interaction, between a subject who unhesitatingly adopted the correct solution and another who was not yet able to find it, very rarely ended in progress for the latter. The more advanced subject, sure of himself, simply imposed his solution without even considering the solution offered by the other, and there was no real confrontation of views.

Such an asymmetric situation also exists in the 'easy/difficult' version of the paradigm: there is an obvious solution for the 'easy' subject, but not for the 'difficult' subject. To overcome this problem, the 'difficult' subject must be given a chance to defend his point of view. This can be done by bringing a social support into the situation.

In the experimental situation presumed to be most propitious for socio-cognitive conflict, one 'easy' subject is confronted with two 'difficult' subjects, in the hope that they will assist each other in maintaining their initial centrations. In another experimental condition only one 'difficult' subject is opposed to one 'easy' subject.

The results were as expected: in the case where the 'easy' subjects were exposed more flagrantly to an incorrect solution, their progress at a post-test placing them too in a difficult position was definitely more marked.

We have tried to confirm these results with other experiments. Their purpose was to show that someone who already knows the correct answer can nevertheless learn from someone who, seeing the problem from a different angle, offers incorrect answers. So as to control these responses even further and to stress their impact, an adult collaborator was placed in the difficult position with the instruction systematically to offer incorrect solutions; or in another experiment, the child who had not yet mastered the transformations required in the difficult position had to execute the task before the child in the easy position was allowed to step in. In every instance, approximately half of the children progressed after having been confronted with such a clear-cut opposition to their own centration. More discriminating analyses again revealed the presence of two types of interaction: in one, dominated by compliance, the child simply applies the other persons' solution and so does not progress; in the other, over which socio-cognitive conflict prevails, the child progresses through having defended his own solution while having had to consider also the other erroneous one.

The same difference between regulation through compliance and regulation through cognitive work has been confirmed by other experimenters, including M. Levy (1981). In several of her experiments a child in the difficult position was confronted with different kinds of responses from an adult. The only answers that failed to lead to progress were obviously arbitrary ones: incorrect answers which were not haphazard but structured according to a logical plan did contribute to the child's progress. Similar effects were observed when the adult refrained from offering a solution but merely challenged each and every position that the child proposed for a house. However, as shown in yet another experiment, progress diminished if the credibility of the adult giving incorrect answers was questioned. This indicates once more that the child makes cognitive progress primarily when seriously confronted with a point of view challenging his own.

Other experiments (Ames and Murray 1982; Glachan and Light 1982; Silbereisen 1982) also have shown that socio-cognitive conflict can lead to progress provided that the cognitive work remains free of interfering compliance regulations. We would now like to show how cognitive progress can also come from exposure to the social norms and representations activated in certain situations.

Studies on social marking

The notion of social marking allows individual cognitive development to be studied beyond the limits of direct inter-individual encounters, placing it within the context of wider social relationships in which these encounters take on special meaning. A first definition of social marking was inspired by French studies on the isomorphic structures of communication networks and decision patterns which facilitate the execution of a specific task (Flament 1965; Faucheux and Moscovici 1960). The notion of social marking then postulated correspondences between the social relationships influencing the interaction of persons in a given situation and the cognitive relations bearing on certain properties of the things through which these social relationships materialize.

The experiment designed to illustrate this notion (Doise *et al.* 1978) was based on the idea that, given 'equal' socio-cognitive conflict, progress should appear only where the principles of cognitive regulation applicable to the task (in this case the conservation of unequal lengths throughout their various perceptual configurations) were homologous with principles related to the distribution of unequal objects among actors of unequal status. So we asked the child in the social marking situation of the experiment to share a short and a long bracelet between the experimenter and himself, after having expressed his judgment concerning the lengths (a judgment which was often incorrect as the subjects were non-conservers). This generated a socio-cognitive conflict capable of inducing the child to elaborate the correct answer for the conservation of unequal lengths. This did not happen in a control condition where the child was requested to share out the same objects between, this time, two cylinders, one with a circumference corresponding to a child's wrist and the other to the experimenter's wrist. Introducing a cognitive regulation into a well-established social regulation (that of unequal sharing between adult and child), conferred maximum efficiency to the socio-cognitive conflict, not merely because one functional relation happened to be more salient (since, when choosing the bracelets for distribution at the beginning of the experiment, there is no difference between conditions), but because the cognitive regulation obeyed a social rule which was made conspicuous.

Another way to study social marking was to invoke, or not, the child's vested interests (Doise *et al.* 1981). Children who were non-conservers on a conservation of liquids test participated in the interaction; in certain experimental conditions the experimenter told the children that because they had both done a good job they both deserved a drink, one as much as the other. Ensuring that the shares, to be poured in two glasses of different dimensions, were indeed equal was left to the children themselves. Being non-conservers (i.e. tending to evaluate the amount of liquid in the glasses only in terms of the level it reached without also taking into account the widths of the glasses) they usually made errors in estimating the equality of their respective shares. In the other condition, the experimenter did not stress the two partners' equal merits. Moreover, in one half of every condition sharing took place directly with an actual partner, in the other half with a 'symbolic' partner who allegedly was to join the child later. Results in fact showed that the subjects who believed they deserved equal shares progressed more (and also on generalization measures which vouch for the authenticity of their progress) than the subjects in the control condition where the requirement of sharing equally had not been stressed. Furthermore, even when the child was left alone to pour out and decide on the equality of the shares for himself and for a hypothetical future partner, the same effect was observed: an individual working on his own may still be faced with conflicting responses if he must abide by a social rule. Also, the fact that the children in the condition 'without merit' progressed considerably less indicates the necessity of emphasizing a social norm in order to generate the conflicts that lead children to cognitive progress. The need to activate such a norm was also recognized by Perret-Clermont (1980) in her experiments on sharing, where children during the experimental phase had to agree on 'fair' shares (of fruit drinks for the conservation of liquids or of 'smarties' for the conservation of number).

In continuing our experiments we extended the concept of social marking to other dynamics where norm structures and social representations could be articulated with processes of cognitive functioning. This recent extension also makes reference to socio-cognitive conflict in accounting for the role of social marking in cognitive development. The three following points summarize our present conception of social marking:

1) Social marking defines any socio-cognitive situation in which a correspondence exists between socially regulated responses (governing inter-individual relationships or relative positions in a social structure, in fact, the social representations of the child concerning various social relationships) and responses emanating from the organization of cognitive schemata at the child's disposal at the moment.

2) To achieve cognitive growth, this correspondence must be emphasized so that the subject is led actually to compare these responses of different natures.

3) The mechanism through which social marking ensures the elaboration of new cognitive responses is the confrontation of contradictory responses, in other words socio-cognitive conflict.

The notion of social marking, in this wider definition, does not necessarily imply that the norm be directly applicable to the interindividual relationship of the two interacting partners. For instance, an experiment carried out by D. Rilliet (Doise and Mugny 1984, Chapter 6), showed that, under certain circumstances, social interaction may be a place of choice for social marking to take on special emphasis without being of direct concern to the actual relationship going on between the actors. The paradigm of the village-copy was used at pre- and post-test. After pre-test, approximately half the subjects were given the opportunity to work on socially unmarked material (the village, as usual in this paradigm), and the others on socially marked materials (in this case, a 'classroom' with teacher and pupils). The idea was that the respective positions of the pupils' and teacher's desks were under much greater social constraint than the respective positions of the village houses. It was therefore expected that the incorrect answers furnished by the children at that cognitive level would be at odds with the socially demanded conservation of certain relations (either left/right, or front/back = near to/far from the teacher). As social interaction between partners of the same cognitive level was deemed to be particularly conducive to conflicts of this kind, merely because it increased the probability of cognitively produced answers being contradicted by socially produced ones, some of the children carried out the task in groups of two (placed on the same side of the board, a situation known to be unpropitious for socio-cognitive conflict when children of the same cognitive level work on socially unmarked material; cf. Mugny and Doise 1978); and the others worked on the task individually. Results showed two essential facts. Firstly, progress was significantly more frequent following the socially marked task, which confirms that the opposition between the answers that the child's cognitive level enable him to produce and those that appear to him through the emphasis placed on a social norm constitutes indeed a situation of social marking particularly favourable for cognitive growth; this greatly extends the range of situations to which our conceptualization applies. Secondly, as predicted, an order appears in the experimental conditions: the inter-individual condition with social marking being the most favourable for cognitive growth, followed by the individual conditions with social marking, the progress obtained in the two socially unmarked conditions being insignificant.

However the children could have been more familiar with the socially marked material used in this experiment. Furthermore, houses may be positioned in various ways in relation to a point of reference (swimming pool or lake), whereas the positioning of pupils' desks so that they face the teacher's desk seems far more obvious. In order to control such effects, which are inherent to all situations of social marking, P. De Paolis (see Doise and De Paolis 1984) designed two complementary researches. Their essential objectives were:

(i) to show how social marking can introduce socio-cognitive conflict, even into individual conditions, by opposing the regulations proceeding from social norms to the cognitive schemata that children appeal to with respect to given material; and

(ii) to determine whether the effects of social marking may be explained solely through familiarity with a given situation and material or whether, on the contrary, these effects depend on the salience of the social regulations that match the cognitive regulations needed to conserve certain relations between objects.

A first experiment was carried out, each of its two conditions involving material with the same degree of familiarity: the classroom (the usual condition of social marking) and a room where children watch television. Note that the direction being faced in looking at the teacher or looking at television is equally important from a functional point of view. The difference between the two conditions was that the social norm of hierarchy which determines the relative positions of pupils and teacher in the classroom was absent from the other setting. Results at post-test showed significantly more progress in the social marking condition. Post-testing for this and all other similar experiments involved the use of the village material which is free from social marking.

In another experiment, again on an individual task, we attempted to verify the salience of three different socially marked conditions compared to a control condition without social marking (the village). In the first condition, social marking was very explicit: the pupils in the classroom were to be positioned in accordance with the teacher's criterion that the best pupils should be nearest to her. The second condition represented an orchestra with several musicians whose point of reference was the conductor's podium; the rule for positioning the musicians was not so salient as in the classroom condition. The third condition represented a photographer taking a picture of the class; here there was an evident functional rule governing the respective positions but no obvious social rule. The results at post-test verified that the asymmetric social rules that command the spatial relationships in the classroom condition are more efficient in producing cognitive progress. The necessity to abide by these rules created a conflict with respect to the usual cognitive schemata that the child uses in effecting this task. In order to overcome this conflict, the child had to restructure his own cognitive schemata so as to find a solution that agreed with the social rules.

As with socio-cognitive conflict, the notion of social marking has been adopted by other researchers. Roux and Gilly (1984) applied the notion of social marking to a task designed for 12 to 13-year-olds which calls for rules to be applied according to two criteria of order. If several entities are identical according to one criterion and different according to the other, priorities will be set on the basis of the latter (for example, young people should take the initiative of greeting their elders). But if differences occur on both criteria at the same time, a rule must be introduced which specifies the relative importance of one and the other (for example, age precedes gender in determining who is to take the initiative of greeting the other). In the experimental phases of their experiment, Roux and Gilly either used tasks which did not call for social rules to be applied as organizational criteria, or tasks where social rules were applied either arbitrarily or according to well-established custom. The material used at post-test was devoid of social marking but the subjects who had progressed most were those who, during the experimental phase, had had the opportunity of practising alone or in pairs on socially marked material according to well-established customary rules. It is important to note that practising in a situation where the social rules are completely arbitrary is even more inefficient in producing progress than a situation where the criteria bear no social significance.

Conclusions

This paper does not attempt to give a complete overview of the various problems involved in studying the links between the social environment and individual functioning. Although sophisticated mathematical models have already been devised to account for these links, we believe that we have only just begun to shed light on some of the psycho-sociological processes that explain the ways in which social context and individual cognitive development interact. Our main hypothesis is that cognitive growth often arises from the contradiction between the responses stemming from the application of cognitive schemata at the child's disposal and the responses based on the application of a social rule (relative, as we have seen, to inter-individual relationships as in the case of equal sharing, and also to social relationships on a wider scale as found in the asymmetry of educational relationships). This dynamic model, while inverting the most frequently recognized direction of causality between cognitive development and social intelligence, the latter always being regarded as necessarily dependent on the former, restores the balance as it were, between the importance assigned respectively to social and situational factors and to factors at stake in individual dynamics. Basically, cognitive autonomy must be viewed as the result of social interdependence. But the progress that individuals achieve in situations of interdependence comes only if they can approach these situations with their own cognitive outlook. Moreover, social regulation can be built into a given situation with material evocative of certain social norms, rules, and representations.

Discussion

Conflict

Bryant noted:

'It seems to me that conflict is an inadequate explanation for cognitive change because it can be a signal to the child *only* that his cognitive system is not working. It does not tell him what to do about it. There is one exception to this and that involves space and perspective. Here the child needs to know that his and another person's point of view are different. This *is* the essential information for cognitive change. So your use of space may be a special case and does not rule out the possibility that you are just showing that giving the child the right information is sufficient. Hinde asked me to design an experiment which would sort that out. In a way I have done this already *vis-à-vis* measurement, in a recent paper (Bryant 1982). As far as space is concerned one needs a control group which is given the necessary information without conflict. There are a number of ways one might do that, but I think that I would show the child that the same scene viewed from different positions looks different and does so in systematic ways. No conflict would be involved.'

Doise replied:

'I agree that one of the effects of conflict is to draw attention to relevant features of the task. But that is just part of the problem of which social conditions further specific cognitive processes, which conditions lead to the child putting different centrations together. One such condition is just to make the opposite centrations salient through producing social conflict. This is very clear in the conservation of length experiments in which multiple intra-individual conflicts do not lead to the same progress as inter-individual conflict. And there the information is exactly the same in inter- and intra-individual conflict. See also the experiment on spatial relations in Doise and Mugny (1984, Chapter 6, experiment 3) in which subjects in intra-individual and inter-individual conditions have the same information. In relation with the Bryant (1982) experiment I think that it is rather an experiment on intra-individual conflict of responses.'

Vauclair added:

'If I understand correctly, a socio-cognitive conflict can lead either to the emergence of a relational response (that is, a response in conformity with pre-existing social norms) or to the elaboration of a new cognitive instrument. In this latter case I would like to know which is the nature of the factors(s) rendering necessary such a new structure?'

Doise replied:

'First of all normative responses and cognitively innovatory responses are not mutually exclusive as the social marking experiments show. Normative factors therefore can necessitate new cognitive answers as well as preventing their elaboration. It all depends on the relations between prevailing cognitive approaches, their conflict and the acceptable normative solution in a particular situation. Only the articulation of all these factors can lead to more precise hypotheses.'

Krappmann asked:

'Did you analyse the kind of communication between the children when they tried to find a better solution? Did you identify and distinguish strategies or patterns which the children use to solve the conflict?'

Doise replied:

'Silbereisen (1982) in Berlin, Bearison (1982) in New York, and myself studied communication patterns. But the problem is that communication is rather rare in the situation we have used. Children say yes, no, or fight with each other to place the houses. The only consistent pattern is that there should be a certain degree of symmetry between the intervention of the two partners in order to produce progress.'

22

On the interdigitation of social and cognitive processes
ANNE-NELLY PERRET-CLERMONT AND ALAIN BROSSARD

Mead, Vygotsky and Piaget have elaborated different views of the inter-dependence of cognitive processes and interpersonal relationships. In the first part of this paper we will be concerned with the *debate about the causality of cognitive development* that these different theories have postulated and we will refer to some empirical research carried out as a contribution to reconciling these different schools of psychological thought. However the experience gained from this empirical work in turn raises questions about the root of the debate: is it relevant to distinguish 'cognitive' vs. 'social' processes? And is it adequate to think in terms of a relatively mechanistic causal model relating univocally cognitive (or social) causes to cognitive (or social) consequences? In fact all these different aspects of reality are always incorporated within a larger set of meanings that is perceived differently by the subjects according to their previous and present socio-cognitive experience.

In the third part of this present contribution we will return to the 'founding fathers' and remember that their scope was much larger that the solving of a causal enigma and was concerned with the description of the integration of the different levels of verbal and nonverbal behaviour within a biopsychosocial model of development. This will call our attention to some aspects of behavioural interaction (e.g. gaze) that we have neglected up to now and that could open fruitful perspectives for a deeper understanding of the interdigitation of the cognitive and the relational factors in the elaboration of so-called 'cognitive' responses. Unfortunately the data available at present are not sufficient for an empirical validation of our hypothesis but we will put forward some case studies that illustrate our point in order to stimulate debate and ideas for further research on the integration of behaviour.

1 Cognitive processes and interpersonal relationships

Piaget's model of the development and functioning of thought was elaborated on the basis of his training as a biologist, his interest in epistemology and his view of the development of the child's cognitive processes. This orientation led him to focus on biological processes (self-regulation, *equilibration*) and processes of logical abstraction to the detriment of personal and social factors which, although acknowledged, were considered of secondary importance. The priority Piaget gave to the biopsychological dimension over social and symbolic interaction is found in many later studies, whose explanations imply a relatively mechanistic model of causality, whereby the individuals' social competences or sensitivities depend on their cognitive development. By contrast, the research to which we have contributed has explored an inverse causal model in which the structure of cognitive behaviours is seen as the result of adequate social interactions (Perret-Clermont 1980; Doise and Mugny 1981). This inversion of perspective finds its source in the work of authors such as Mead and Vygotsky. Let us say though, in order not to reduce the subject to a simplistic bipolarity, that Piaget himself in his first publications contributed to this orientation, and that Mead, on the oher hand, sometimes explicitly refers to the biologist Darwin (in particular to his study of emotion).

Piaget's, Mead's and Vygotsky's theories of cognitive development and their conceptions of the impact of social factors on intellectual growth

For Piaget, social factors are necessary for the completion of the structures of intelligence, *but they are not at the source of these structures*. In order to assimilate the contributions of his social experience, the child must already be endowed with mental structures which make this assimilation possible. Although Piaget allows for a possible incidence of social factors which might more or less facilitate the individual's development, he considers them only as supplementary variables which might affect individual behaviours. At one time the Piagetian school of cognitive psychology hoped to discover the basic mechanisms of intelligence, whose structure and functioning would be unaffected by social factors. But the ambition underlying this undertaking—*the search for universally valid explanations*—often leads to a failure to analyse the structuring effects of the *concrete* object's and partner's specific characteristics, to which the individual adapts his behaviour. There is a severe risk, then, to consider erroneously as general processes, phenomena which might be more accurately envisaged as *artefacts of particular social situations*. (This we will attempt to demonstrate later on.)

G. H. Mead (1934) proposed a conceptualisation of the links between social interaction and intellectual development via symbolic interactions stemming from *conversations with gestures*: even before the appearance of self-consciousness or of actual thought, the interaction between two individuals furnishes a base for the construction of symbolic thought. The

interaction between two individuals rests firstly on gestures. For Mead, gestures are not simply the expression of emotions in the Darwinian meaning of the word. They are the first stages of an individual's behaviour in response to another individual; they are thus the forerunners of the future stages of social behaviour, as they progressively become symbolic gestures. Animals also have conversations of gestures, but they are devoid of symbolic signification. Mead gives the example of a dog fight (Mead may have underestimated the complexity of a dog-fight, but it does not really affect his argument): each animal chooses his behaviour according to what the other does or will do. The behaviour of one of the two dogs forms a stimulus for the other, who will in turn be able to adjust his own behaviour in response to the other's. This adjusted behaviour becomes in turn a stimulus for the other dog and so on . . . Such acts, such behaviours, are a kind of communication. But if these gestures are to become significant symbols for the others, the individual must be able to anticipate by inference the response his gesture will elicit from the other, and thus to use this response of the other in order to determine his own future behaviour. Yet, according to Mead, the animal does not seem to be able to tell himself, 'If the animal comes from this direction he is going to spring at my throat and I will turn in such a way' (p. 43). Hence the following definition:
'Gestures become significant symbols when they implicitly arouse in an individual the same responses which they explicitly arouse, or are supposed to arouse, in other individuals, the individuals to whom they are addressed' (p. 47).

For Mead, construction of thought proceeds from the interiorization of conversations of gestures performed with other individuals. In this respect, Mead is in opposition to Wundt's theories, for whom the analyses of the communication between two individuals *presupposes* the existence of thoughts which can be communicated. The thus interiorized gestures (including verbal gestures, i.e. speech) are significant symbols because they have the same meanings for all the individuals of a given social group. For Mead, thought appears as the interiorization of the conversation of gestures. This interiorization occurs especially when verbal behaviours are mixed with the conversation of gestures.

Vygotsky (1934) developed a theory according to which the causal direction of the development of thought goes from the social to the individual. The instrument of thought is language which, before it is interiorized, is socialized, i.e. used to address the adult. But Vygotsky also refers to gesture in examples such as the one of the child who tries vainly to reach an object. This failure does not elicit a reaction from the object, but from another person. The repetition of this experience will soon lead the child to consider his act as an indicative act aimed at someone else. The function of this act will change: before, it was oriented towards an object, now it is meant for another person, and becomes therefore a means to initiate social interactions (the act of pointing which precedes deictic gestures, has been studied recently from an ontogenetic point of view by Murphy and Messer 1977; Clark 1978; Clark and Sengul 1978; Ochs and Schieffelin 1979; Masur 1982; and Wilcox and Howse 1982). The act of grasping becomes an act of indicating.

For Vygotsky, this transformation of an interpersonal process into an intrapersonal one can be generalized to all higher functions in Man, and to the cognitive ones in particular.

Some empirical investigations on the effects of social interaction on cognitive development in children

We thought it would be interesting to attempt to investigate further Mead, Piaget, and Vygotsky's conceptions by creating experimental paradigms which could ascertain more precisely the effect of social interactions between children or between adult and child.

Our hypothesis, contrary to those dominating the field of Piagetian research, can be expressed as follows: *cognitive co-ordinations between individuals* are at the basis of *individual cognitive co-ordinations*. Or in other words: the individual cognitive act is only a phase in a more complex process whose nature is also social. This thesis implies that cognitive co-ordinations in the individual are only made possible through social co-ordi-nations. It is by co-ordinating his actions with the actions of others that the individual gains individualized mastery of the co-ordination systems he will then later use on his own. This interactionist and constructivist conception takes from Piaget the idea that by acting on his environment the individual elaborates organizing systems. It also insists on the fact that the causality ascribed to social interaction is not unidirectional, but circular, or even more complex. Indeed, from a genetic point of view, it appears that inter-action enables the individual to master certain co-ordinations that enable him then to take part in more complex social interactions which, in turn, become the source of cognitive development. Social interactions are not only inductors, but also structuring causes of cognitive productions. And these can no longer be predicted when the psycho-social conditions of their elaboration are modified.

Our aim then was to illustrate experimentally, by using variations of clas-sical Piagetian tests, that operational capacities that were absent in the individual during the pre-tests could be elicited, or even fully mastered, during the post-tests following situations of social interaction where the partners had to take into account each other's view-point (Perret-Clermont 1980; Doise and Mugny 1981; Perret-Clermont and Schubauer-Leoni 1981).

We began by studying the conditions of social interaction during the elaboration of operational notions, such as the notion of liquid conservation. This research showed that when children aged 6 to 7 (which is the age during which this notion is constructed), could interact in the task with other children of the same age, they made more progress during the post-test* in the acquisition of this notion than children who were not given this opportunity.

These results suggest the importance of confrontations between peers. The non-conserving subject is led to *re-structure* his thinking as a result of the emergence of a conflict between his point of view and his partner's differing one, when the situation calls for common agreement or joint action. With Doise and Mugny we called these specific types of encounters 'socio-cognitive conflicts' and tried to determine their characteristics and impact:

a) First appeared the crucial role of *the difference between the cognitive view-points of the partners*. This might be due to the different stages of development of the children, but also to their different perspectives concerning the tasks (for example, in the research of Mugny, it is due to the respective spatial positions of the subjects during the test of building a village; see Doise, this volume). It appeared moreover that the 'right' answer given by the partner is not a necessary condition for cognitive progress: such progress is also observable when the partner's answer is different but wrong. It is therefore sometimes sufficient for cognitive progress that there be an opposition of view-points, a conflicting confrontation between partners.

b) There are developmental prerequisites necessary for the subjects to benefit from this situation of social interaction. Perret-Clermont and Schubauer-Leoni (1981) showed for instance that none of the subjects who did not show evidence of the notion of reversibility during the pre-test of liquid conservation progressed at the post-test. These results confirm former ones (Perret-Clermont 1980) concerning the conditions for the acquisition of number conservation. In other words, this means that the child must have reached a minimum operational level if the socio-cognitive conflict is to have any effect on his or her cognitive development. Some cognitive prerequisites are thus essential for the child to benefit from the situation of

---

* The experimental procedure is as follows: firstly a *pre-test*, in which the test for the conservation of liquids is given individually using a semi-directive clinical interview, in the Genevan tradition—this pre-test permits evaluation of the subject's operational level (other tests are sometimes added for a more precise assessment of the cognitive level). Then the phase of social *interaction*, which takes place after one week. Each non-conserving child interacts with another child (either non-conserving, intermediate, or conserving) by 'playing a game' whose purpose is the sharing of an equal quantity of juice. The game is over when the children both agree that they have the same quantity to drink. Finally the *post-test*, which takes place one week after the interaction session. Non-conserving subjects are re-tested on the conservation of liquids, as in the pre-test, and on other generalization problems.

social interaction. If the non-conserving child is unable to grasp the incompatibility of diverging viewpoints, he will not perceive the conflict and he will not be able to engage in the cognitive activity which would lead to its solution. Such an interaction will therefore have no effect on the operational structuring of the child's thought.

c) Let us also mention that there is no evidence of a causal link between the performance level observed during the interaction between peers and the cognitive progress of participants at the post-test. The *solution* of conflict during the phase of social interaction does not seem therefore to be a necessary condition for cognitive progress. The main element seems rather to be the *conflictual* aspect of the interactions as an urge for mental restructuring. But from a clinical point of view, it appears that this conflict should remain centred on the task and be perceived by the partners as a cognitive conflict, rather than as a sign of socio-affective tensions (unfairness, jealousy) or of incompetence ('Of course he gets it wrong, he is too young to understand'), in which case the subjects do not get involved in the search for a cognitive solution.

d) The cognitive restructuring that follows the interaction between peers influences the operational development of the child beyond the particular problem discussed: it can influence development on other conservation tests (matter and length, for instance), and results confirm the hierarchical order of genetic development according to Piaget's theory (Inhelder *et al.* 1974). However, the degree of generalization seems to depend on the type of relationship in the interaction situation: a complementary analysis of the data presented in Perret-Clermont and Schubauer-Leoni (1981) has shown that 'horizontal' (child-child) interactions provoke more generalization than 'vertical' (adult-child) interactions.

Some characteristics of the interpersonal relationships established between the subjects during the interaction are liable to render more or less salient the object of the socio-cognitive conflict, and hence the differing perceptions of the partners. This is one of the possible explanations for the results we have just mentioned (Perret-Clermont and Schubauer-Leoni 1981). Differences in cognitive performances have also been observed as a function of the repartition of the differently shaped glasses between the partners of the interaction who then seemed to be more or less favoured in the quantity of juice offered (Hendrix and Van de Voort 1982; Grossen and Perret-Clermont 1983; Rijsman 1984). Doise *et al.* (see his contribution to the present symposium) account for this type of phenomenon in terms of social marking effects.

However, in all these studies the causal system is complex: the social and cognitive processes are 'prerequisites' as well as 'consequences'. During the course of our experimentation, we have observed that subjects were sensitive to dimensions of the experiment other than those controlled by our paradigms. This is what we now propose to examine, postulating the hypothesis that the dissociation of processes into two distinct categories ('cognitive' and 'social') is an heuristic but arbitrary distinction which is liable to entrap the psychologist in causal dilemmas (see Bryant's contribution to the present symposium).

2 The interdigitation of social and cognitive processes

Data always came from socially marked situations

Traditionally, the 'psychological subject' has been viewed as if he had been raised free from any specific characteristic transmitted by his social and cultural origin. However every behaviour of the child is more or less revealing of his position in the socio-cultural field of which he is both inheritor and keeper. A certain number of observations have lead us to reconsider the testing situation as socially and culturally marked, and to take into consideration the fact that the socio-cultural conflicts we describe do not happen in a social vacuum. The impossibility of designing 'culture-free' tests is well known, but our point is precisely that the consequences of this impossibility have not yet been fully considered in cognitive psychology or in intercultural research. We are therefore concerned in this paper with the study of the interdigitation of these processes. This has lead us to suggest that the distinction between 'cognitive' and 'social' processes does have an heuristic value (as shown before), but that it could erroneously lead to an over-simplistic view of the phenomena and their underlying causation.

A sociological analysis of our data revealed differences in the performance of subjects which were linked to their social characteristics. These results are difficult to explain within the theory of genetic psychology. Considering these data together with others concerning the impact of various types of situations on social interaction, we were lead to study the relation of the children to the proposed tasks and to the style of interpersonal communication established by the experimenter.

In this perspective, we recognized that the pre-test is already a situation of social interaction that can provoke socio-cognitive conflicts between the child and his interlocutor and hence elicit learning right at the start of data collection.

a. *The operational level of pre-test subjects varies according to their social category*: analyses of Piagetian tests have often shown that the cognitive level of the child is influenced by his socio-cultural origin (for a description of these effects, see Perret-Clermont *et al.* 1984b). In the pre-test, children from privileged backgrounds present a significantly higher operational level than children of the same age who come from an underprivileged environment. Likewise, boys sometimes show a higher operational level on the pre-test than girls.

b. The cognitive effects of situations of social interaction seem to depend on which social category the children belong to: although the pre-tests often differentiate the subjects according to a hierarchy of cognitive levels which parallels the social hierarchy outside the test situation, post-test results often show that the participation of the subjects in the interaction session can reduce, or even eliminate, these differences of social origin and sex. The amount of cognitive progress shown in the post-test by children from underprivileged environments seems to be linked to the conditions of inter-action (Perret-Clermont and Schubauer-Leoni 1981) and to the child's developmental level at the time of the interaction (Mugny and Doise 1978).

These observations imply that we should acquire the means to go beyond mere sociological explanations of the differences between groups in order to examine the *relation* between these *sociological variables* and the *socio-psychological situation of testing* in which the cognitive level of the child is expressed in his behaviour. The fact that cognitive progress can be elicited experimentally through the experience of the subjects in interactions would suggest that individual psychological competences are not the only ones that are important, but that socio-psychological competences (i.e. the capacity to establish a certain type of discourse about a given object, with a given interlocutor and in a given type of situation) are also involved. Concepts like 'socio-cultural handicaps' are of little use for describing the subtlety of the processes which determine, in our research, these differences; differences which are sometimes (but not always . . .) liable to disappear after simple (but adequate!) experimental interaction situations lasting no more than ten minutes.

c. The pre-test situation in itself is already an occasion for operational and social learning: from the beginning of an operational test, the subject must take a stand socially and understand what is expected of him. Some tests, for example the conservation of liquids, may sometimes appear extremely ambiguous as to expected roles. Thus, in order to produce an operational answer, the child must understand in particular that:

  (i) the partners present must be considered as formally equal;

  (ii) the result of the pouring of the juice in spite of the illusion of non-conservation created by the differently shaped containers is legitimate within the framework of the experimental instructions;

  (iii) the object of the interview concerns only the abstract notion of *quantity*;

  (iv) and that after the pouring of the juice, the perceptual illusions purposefully created by the very choice of the material being used, must be perceived as such and not ignored, but rationalized in order to demonstrate to the experimenter the constancy of the quantities considered (even though the adult must be considered as aware of this, being the author of the experimental scenario!).

We have reviewed elsewhere (Perret-Clermont *et al.* 1984b) a series of studies (e.g. Rose and Blank 1974; McGarrigle and Donaldson 1974; Light 1979; Rommetveit 1979, among others) whose aim has been to understand how the subject, during the pre-test, is involved in intense cognitive activity in order to interpret the situation and to understand its meaning through the staging, the instructions, the questioning and the feedback inscribed in the choice of material, and the verbal productions and gestures of the child's interlocutor. From these cues, the child learns more or less quickly and with greater or lesser success, the kind of answer he must give. Results from the above mentioned studies suggest that children from different social backgrounds are unequally familiar with the type of interpersonal and cognitive exchanges required in an operational test. It seems that the children from more educated environments have better performances in Piaget's tests. Results from a study by Nicolet (1984) can be interpreted in a similar manner in that urban children perform better than rural children; among the latter group, it appears that children of farmers have the lowest level of operational performance. However, these sociological differences disappear in most of the experiments after the interaction session, indicating that they do not depend upon 'profound' differences. Data from the research using the conservation of liquid test (Perret-Clermont and Schubauer-Leoni 1981) also show that during the first exchanges of the standardized interview, and in particular after the fourth item of the interview procedure which consists in the traditional Piagetian 'counter-suggestion'* a certain number of subjects changed their answers by introducing operational answers among their initial non-conserving behaviours. It appears that for a certain number of subjects, this counter-suggestion could open the way for the elaboration of operational answers, perhaps because it makes explicit the nature of the reasoning expected by the adult. A close look at the data suggests that this counter-suggestion on its own is more effective with subjects who come from a socio-cultural background similar to that of the experimenter: their greater social ease when confronted with the staging of this test would given them a chance to *progress* more in terms of the operational level of their answers, than other children during the pre-test.

These observations call for an investigation of the effects of social variables and interactional processes involved in the realization of an operational notion. This is what we are going to attempt now, taking the three

---

* The strength of the subject's response is tested by suggesting him the answer opposite to his own.

situations (pre-test, interaction between peers, and post-test) as *three separate moments in which socio-cognitive conflicts are likely to take place*, three interactional situations which require both cognitive and social competences from the subjects.

How does the subject perceive these social situations?

In the course of the interactions which are the source of socio-cognitive conflicts, it appears that the participants must not only activate the cognitive skills necessary for solving the problems, but they must also draw on social knowledge which is essential to the understanding of the situation. Socio-cognitive conflicts will be fruitful only if the context and the object of the interaction are mutually understood. The crucial factor for mutual understanding is the sharing of a common framework of reference. The development of 'intersubjectivity' (Rommetveit 1979), i.e. the ascertaining of a shared and congruent social reality, is the *sine qua non* condition for signifying speech. This communality of significations depends on more or less explicit norms whose application is the condition for the success of the interaction; one of them being the acceptance of the limitations imposed by the adult experimenter in the structure of the task and of the interview. This means that when faced for the first time with a task of this kind, the main goal of the child will be to try to decode the adult's tacit assumptions concerning the definition of the situation, their respective roles, and the aim of the discussion. We will now consider the social knowledge required from the child to interpret the adult's discourse and intentions.

a. Perceptions of the experimental situation and task

For the child, the experimental situation is entirely new: the pre-test starts with an unknown adult who tells the child that 'we are going to play a game together'. However this game has nothing in common with the games the child is used to. This 'game' looks more like an exam situation in which the adult asks the child questions. Thus the situation can look quite ambiguous in the eyes of the child! One can therefore venture the hypothesis that the gap between the child's preconception of the concept of game and what the adult defines as a 'game' in this context can lessen the chances of the child answering correctly. For instance a situation requiring juice to be divided up between two dolls appeared to require a high degree of abstraction from the child who must understand that the game is about the equal distribution of juice and not about playing with dolls. This effect was found more marked for girls than for boys (Perret-Clermont and Schubauer-Leoni 1981).

One of the important elements in the perception of the situation is the understanding of the task. Faced with the material, the child must be able to distinguish which of its aspects are relevant to the resolution of the task. However the most evident clues of the situation—for instance the glass of juice the child would like to drink—are the very ones he must ignore! A testing situation implies an adult-child relation. In the experiment of the conservation of liquids, the adult asks the child to pour a quantity of liquid equal to the adult's. According to 'normal' social relations between two such participants, the child may find this demand absurd, because adults and children are not equals, neither in their social nor in their physical status! In order to succeed at the task, the child must therefore neglect relative status during the test situation. It seems moreover that operational reasoning is more efficient when the social relations structuring the tasks are isomorphic with the ones the child has to develop on a more abstract level (Doise *et al*. 1978; Doise and Mackie 1981).

The task concerning the conservation of liquids can be analyzed according to the expectancies that participants in the interaction have of their respective roles in the evolution of the situation. Each participant arrives at this situation with preconceived ideas, for instance about the manner in which the child—or the adult—must behave. Finn (1982) has studied the expectancies of children faced with a test situation by posing questions which transgress the usual rules of questioning: the questions the children were asked were nonsensical and were therefore impossible to answer. Yet almost all the subjects took the experimenter very seriously and answered his questions. Finn concludes that children create a context of intelligibility for the questions of an adult and their own answers depend upon their prior social knowledge. They strive to attribute meaning to the task situation.

b. The image of the partner

The experiments mentioned above show the importance of the child's perception of his partner in the interaction; this perception can determine the manner in which the socio-cognitive conflict will be solved. Other studies have measured the impact of various images of the experimenter on the cognitive performance of the child. Levy's results (1981) show that the induction of a negative image of the adult reduces the child's progress in a task of spatial transformation.

3 In search of nonverbal indicators of this interdigitation

In the first part of this paper we saw, refering to the importance of the socio-cognitive conflict for cognitive progress, that during the interaction between peers, some children do not always manifest behaviours corresponding to the operational level they had reached during the pre-test. It became clear to us that the expression of the operational level of the child depends, on the one hand, on the capacities of socio-cognitive adaption he must display in the experimental situation; and, on the other hand, on the social and cognitive background he has acquired previously in his interpersonal relationships. It is therefore misleading to consider (as we did at the beginning of our research endeavour) the expression of an operational level as an individual characteristic of the subject, given the number and complexity of cognitive and relational factors intervening in the realization of an operational behaviour. The operational level could thus be better understood as an index of the *testor-testee relationship* which is constructed during the interview.

Are there some *observable* signs in the child's behaviour which could inform us about this relationship, and hence about the interdigitation between cognitive and social elements? Two observable outputs can be analysed: the verbal behaviour, and the nonverbal behaviour of the child. We shall now but touch on the former in order to concentrate on the latter.

Verbal behaviour

The analysis of the verbal behaviour can be roughly divided along two main axes: a cognitive axis where verbal language becomes an index of the understanding (or of the lack of understanding) of the emitted or received utterances; and a social axis where verbal language appears as a means for regulating interindividual interactions.

a. Verbal behaviour as an index of understanding on the cognitive level

During the last thirty years, many studies in experimental psycholinguistics have tried to determine the cognitive processes involved in the understanding of verbal language, be it at lexical, syntactic, or semantic level. In the early fifties, a first generation of psycholinguists appeared in the United States (Osgood and Sebeok) who were much imbued with information and communication theory (Shannon and Weaver), behaviourism (Watson) and the theories of operant learning (Skinner). By the 1960s a second generation of psycholinguists was born, influenced by the work of Chomsky. Chomsky was mainly concerned with the syntactic aspects of language which he considered as an autonomous and genetically pre-formed tool; a conception opposed to Piaget, for whom language is linked to the general cognitive evolution of the individual, and is a result of a progressive construction.

We now find, at present, a third-generation in psycholinguistics, which tends rather to integrate language in a global activity unique to the subject (enunciation theories) and linked with the contextual communication strategies (pragmatics).

b. Verbal behaviour as a mean for regulating social interaction

The second function of verbal language was described during the 1960s. The ethnomethodological approach studies some verbal-vocal phenomena regulating conversational interactions, such as pauses, simultaneous speech, intonations, etc., in samples of conversations taken from daily situations: family conversations, in-the-street dialogues, etc. (Garfinkel 1967; Goffman 1973, 1974; Sacks and Schegloff 1974; Schegloff, Jefferson, and Sacks 1977; Sudnow 1972) or in therapeutic situations (Labov 1977) or even in school teaching situations (Sinclair and Coulthard 1975).

Nonverbal behaviour

There has been a prolific amount of research and publications in the last few years concerning nonverbal communication processes (cf. the impressive bibliography established by Key 1977). The subject of nonverbal communication covers a diversity of elements such as paralinguistics, facial expressions, gestures, proxemics, etc. Among these, *gaze* has already been studied in a number of ways (Cook 1977).

We shall concentrate on gaze, starting with the hypothesis that it might be an indicator of the changes in operational level if these are associated with a stronger need to interact with the experimenter. Our decision to study gaze was reinforced by the following considerations:

1. It is one of the nonverbal activities least controlled by the subjects, even in a formal experimental situation. Some studies have shown, also, that a modification of gaze in the direction of the partner (be it an increase or a decrease) rapidly provokes a reaction of the interlocutor, which could go as far as interrupting the interaction (Argyle and Williams 1969; Cook and Smith 1975).

2. The proxemic situation of our experiments with the adult and the child sitting side-by-side facilitates the analysis of gaze patterns: among others, it forces the subject explicitly to move his head in order to look at his partner. This is not the case in a face-to-face situation.

Systematic research on gaze since the 1960s, mainly in the United States (Nielsen 1962; Exline 1971) and in Great Britain (Kendon 1967) where a research group was formed with Argyle, Crossman, and Cook, has shown that gaze patterns can vary very much from one subject to another, and are closely linked to the structure of the individual's personality. Extrovert subjects look more at their interaction partner than introverts do (Cook 1977). An abnormally high or low level of gazes has been observed for asocial patients (Argyle *et al.* 1974). Depressive patients tend to avoid looking at others (Rutter and Stephenson 1972); the same goes for autistic children (Richer and Coss 1972, cited in Argyle and Cook 1976).

a. Gaze as an indicator of cognitive processes

Gaze has been studied primarily in connection with cognitive processes accompanying verbal speech. Kendon (1967) has shown for instance that spontaneous language activities are very often linked to gaze avoidance. Kendon hypothesizes that spontaneous verbal emissions require cognitive activity by the speaker. As the brain's capacity for simultaneous treatment of items of information is limited, temporarily irrelevant items which would interfere with the activity of verbal elaboration must be suppressed: hence the suppression of visual information.

b. Gaze as means for regulating social interaction

Several studies have demonstrated the role of gaze in regulating processes in communicative situations. The importance of gaze is manifest right from the first vocal and gestural interactions between the infant and its mother. Its regulating function is exerted from the very beginning through synchronous gazes connected with rhythmical vocalizations; these exchanges are the forerunners of future conversational interactions (Bateson 1975; Bruner 1975; Jaffe *et al.* 1973; Snow 1977; Stern *et al.* 1975).

With the appearance of language, the child extends the regulating function of gaze to simultaneous interaction with several partners. A study by Craig and Gallagher (1982) on groups of two or three interacting children shows that these American middle-class children are perfectly able to have a conversation as a threesome, and also that gaze patterns and proxemics are intimately linked to conversational turn-taking.

In conversations between adults, gaze is one of the behavioural elements that regulate interaction. A sophisticated system of conversational rules co-ordinates turn-taking between adults, offering regulating procedures which are necessary to start a conversation (Schegloff 1968), to maintain it (Schegloff and Sacks 1973), and to make progressive transitions from one interlocutor to the other. For instance, when the speaker wants to give the floor to his interlocutor, he punctuates the end of his speech with more frequent gazes aimed at him (Argyle *et al.* 1973; Duncan 1972, 1973; Duncan and Niederehe 1974). All these studies underline the fact that the regulation of speech between adults is the product of a complex interdigitation of verbal and nonverbal elements.

c. A few preliminary results based on case studies . . .

We have undertaken a preliminary study whose aim is to evaluate possible connections between development on the operational level, abilities to regulate interaction, and gaze patterns. Our hypothesis can be worded as follows: changes in the operational level of the subject might be associated with a stronger need to interact, which can be translated as 'a stronger need to look at the other person'. Will the gaze patterns of a subject announce and accompany the transition from non-conservation to conservation?

As of present, we have studied data (i.e. video-tapes) only from the pre-test (phase I), in which we analysed nine interactions between child and experimenter (the same experimenter interacted with all subjects). Among the nine subjects, three were already conservers, three became conservers during the post-test (phase III), and the last three remained non-conservers on the post-test. In order not to bias the analysis of gaze patterns, we thought it was methodologically important that the 'decoder' should not know which, among the six subjects who were non-conservers (NC) during the pre-test would remain NC, and which would become conservers (C) during the post-test. The gaze directions, both for the children and for the adult were recorded in three categories:

(i) towards the task;
(ii) towards the partner (from child to adult and from adult to child);
(iii) elsewhere.

These three gaze patterns in both children and adults were observed every five seconds resulting in approximtely 70–80 observations per subject. The data (expressed as relative percentages) are presented in Fig. 22.1. These results suggest the following observations:

1. Subjects who are already C seem to look most often at the task (respectively 83 per cent, 76 per cent and 67 per cent) in comparison to subjects NC who remain NC during phase III (respectively 83 per cent, 69 per cent and 57 per cent) and to those NC who become C during phase III (respectively 73 per cent, 69 per cent and 50 percent); conversely, it is precisely with these C subjects that the experimenter seems to look least often at the task (respectively 39 per cent, 52 per cent and 49 per cent).

2. Subjects NC who become C during phase III seem to look most often at the adult (respectively 16 per cent, 26 per cent and 46 per cent) in comparison with subjects NC who remain NC during phase III (respectively 6 per cent, 11 per cent and 39 per cent) and with subjects who are already C (respectively 15 per cent, 17 per cent and 21 per cent); conversely, the adult seems to look least often at these future C children (respectively 49 per cent, 34 per cent and 29 per cent).

3. Subjects NC who remain NC during phase III seem to look most often elsewhere (respectively 3 per cent, 10 per cent and 19 per cent) in

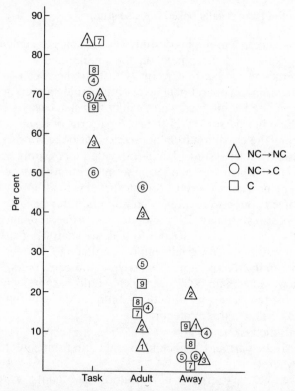

Fig. 22.1(a) Distributions of gaze for each child. [For example, during the pre-test, the gaze distribution of the child no ⑦ (a conserver, C) is respectively 83% towards the task (first column), 15% towards the adult (second column), and 2% elsewhere (third column).]

comparison with C subjects (respectively 1 per cent, 6 per cent and 11 per cent) and future C children (respectively 3 per cent, 3 per cent and 10 per cent); there is no noticeable difference between the three groups in the looking-elsewhere patterns of the adult.

The experimenter looks more often at the children than they do at her, and this lowers the proportion of gazes directed at the task. It seems that the adult's gazes are equally divided between the child and the task with the C group. This is the group where the adult seems to look most often at the children, but these C children look least often at the adult. On the contrary, the future C group looks most often at the adult, who looks least at them. These results confirm Cook's (1977) observation that mutual gazes are difficult to maintain and are hence avoided. We observe for our subjects the existence of rather marked individual differences in the distribution of gaze patterns. On the other hand, the experimenter's distribution of gazes is relatively stable from one group to the other and from one subject

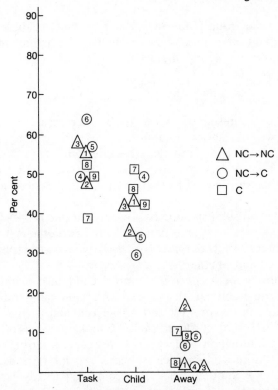

Fig. 22.1(b) Distributions of gaze for adult. [For example, during the pre-test, the gaze distribution of the adult with the child no ⑦ (a conserver, C) is respectively 39% towards the task (first column), 51% towards the child (second column), and 10% elsewhere (third column).]

to the other. This is concordant with Kendon and Cook's observation (1969) of an intra-individual stability of gaze.

The most interesting result, in our opinion, if confirmed by subsequent research, is the percentage of gazes directed at the adult: this seems to be the greatest for future C subjects. This would illustrate that it is during the elaboration of an operational concept that social interaction seems to have a special importance. It is as if these children were looking at the other person in a social quest for approval of their responses.

For a given task, the child's gaze towards the adult might serve different functions according to their developmental level. We suggest the following hypotheses, for each group of subjects:

(i) in the C group, these gazes function mainly as *regulators*, being shorter than 'gazes of understanding';

(ii) in the group of children who remain NC during phase III, these gazes would mainly indicate non-comprehension and they would hence tend to be less frequent but longer;

(iii) in the group of future C, the importance of these gazes would indicate both an unsecure understanding and a search for indications to regulate the course of social interaction.

d. . . . Which lead to others

Other results present some research perspectives:

1. If we consider the total duration of each experimental situation during the pre-test, we see that the three interactions with subjects who remain non-conservers during phase III (NC → NC) are systematically longer than the ones with other groups (with the exception of one future C subject).

2. For two subjects (one NC → NC and one NC → C), we had videotapes to analyse the gaze patterns in the two other phases of the experiment: interaction with a peer (phase II) and post-test (phase III). From this analysis, it appears that:

– as during the pre-test, the future C child looks mainly at the adult during the interaction with a C peer, with remarkably stable percentages (46 per cent in the pre-test and 47 per cent during the interaction with a peer). The subject who remains NC looks almost as often at the task during the interaction with a peer (62 per cent) as during the pre-test (69 per cent). Moreover, in each of these two interactions, peers look very rarely at each other. This result could be partially explained by the fact that, as in phase I and III, the two participants were sitting side-by-side.

During the post-test, the constancy of the distribution of gazes as compared with the pre-test is striking. For example, during the post-test the NC → NC child looks at:

– the task in 65 per cent of the time units (as compared to 69 per cent during the pre-test);
– the adult in 12 per cent of the time units (as compared to 11 per cent during the pre-test);
– elsewhere in 21 per cent of the time units (as compared to 19 per cent during the pre-test);

whereas the NC → C child looks at:

– the task in 50 per cent of the time units (as compared to 50 per cent during the pre-test);
– the adult in 44 per cent of the time units (as compared to 46 per cent during the pre-test);
– elsewhere in 5 per cent of the time units (as compared to 3 per cent during the pre-test).

These results are similar to observations made by Kendon and Cook (1969), who showed the stability of gaze patterns of a given individual across different situations. But further research is needed to test if this stability is not a function of the mastery of the notion discussed in the inter-action. Will these NC → C subjects not change their patterns when they will have become 'old faithful' C?

In conclusion, considering the scarcity of publications relating cognition wih nonverbal processes in interpersonal relationships, we think it is important to explore further the links between these two fundamental aspects of the psycho-sociological development of the individual. This implies for us that we must:

(i) augment the samples of subjects studied;

(ii) refine the time-cuts for gaze-patterns observations, taking into account the important variations in the duration of gazes (using for example intervals of one second instead of five);

(iii) search for other indicators.

Acknowledgements

We would like to acknowledge the support of the Fonds National Suisse de la Recherche Scientifique. Contrat No. 1.738.083. This paper was trans-lated from the French with the collaboration of Claude Béguin.

Discussion

Experimental procedure

Chandler drew attention 'to the relevance of your work to the topic of experimental procedures and intervention studies brought up by Bryant. With many children and especially children of low SES there is an improvement between pre- and post-tests, not because of the power of the independent variable or intervention strat-egy, but because of increased familiarity with the experimenters. This of course, cautions against controls which are simply no-treatment controls, which have little or no enduring contact with the experimenter.'

Perret-Clermont replied:
'We have controlled for the effect of familiarity with the task and the experimenter by comparing the cognitive development of groups of children submitted to the same experimental treatment (for instance: sharing juice or smarties) but with partners of different cognitive levels. We observed that those children who inter-acted only with peers displaying the same point of view did not progress in spite of their acquired familiarity with the setting. Instead the subjects who had been con-fronted with partners of different opinions due to their different cognitive level (for instance an intermediate subject having discussed with a non-conserver or a con-server, or a non-conserver having faced an intermediate or a conserver) were found more likely to progress.'

Epilogue

In trying to understand the nature either of relationships or of cognitive development we are forced to use concepts that are themselves products of the processes we are trying to understand. The result is a tangled skein, and our attempts to unravel it lead inevitably to the view that although the explanatory concepts may be well-suited to the conduct of our everyday affairs, they are not necessarily the best tools for guiding us towards an understanding which as yet we can only dimly perceive. But they are the only tools we have, so our best course must be to attempt to specify clearly their shortcomings, and perhaps then fashion new ones.

A basic issue concerns the nature of development. Attempts to define stages of cognitive, moral, emotional, and social development usually imply a unidimensional progression from a less advanced to a more advanced type. The criteria for progression used by psychologists include logical complexity, ability to delay self-gratification, autonomy, and so on, all envisaged as independent of the content, the culture, or the social demands of the situation. Four points arise. First, neither complexity nor any of the other criteria used are ends in themselves, but possible means to a variety of ends. In living beings in general evolution has led to complexity only if complexity brings greater reproductive success. Simple organisms may persist for hundreds of millions of years if they 'fit' their environment, and in many circumstances delay of gratification may be counter-productive. We must expect the same to be true of individual development, which will be geared to lead to a state making for better success.

And this is the second point, what makes for success depends on the context. The criteria chosen by psychologists are thus to be perceived not as absolutes, but relative to success in a particular context or range of contexts which are themselves created as a consequence of developmental progressions of the type under discussion. Indeed we may note in passing that there is a real risk of ethnocentricity in the circularity in the definitions of 'good performance' (that behaviour expectable and appropriate in a certain environment) and 'good environment' (that which produces good performances).

Therefore, thirdly, it follows that neither development nor performance can be considered independently of context. For example, even in the studies by Doise and Perret-Clermont demonstrating that cognitive conflict can lead to cognitive advances, the progress probably occurs primarily because the experimental situation and the task were set up in a particular way. If the situation were somewhat different other solutions (e.g. bargaining, giving up) might be found which were not associated with cognitive development (Perret-Clermont *et al.* 1984a).

Fourth, in considering the criteria of development, we must not forget that alternative means may lead to the same end, with the most effective depending on circumstances. Again a biological parallel may be helpful: the most effective way for an animal to find food depends on a variety of factors including its distribution and the costs of the various techniques that could be employed in searching for it, each of which may also vary with circumstances of weather, season, and so on. If 'good performance' is relative to a particular context, different behaviour may be necessary in another, and flexibility according to context may be the crucial desideratum. It may be here that complexity is important, for one concomitant of complexity may be the availability of alternative strategies and the ability to select between them according to circumstances. The importance of context is ubiquitous. We must face the fact that the individual's characteristics and actions cannot be described without reference to a context or range of contexts, and the relevant context cannot be specified without reference to the individual.

Yet development concerns change in the individual such that his or her behaviour, if not consistent across contexts, at least differs in a more or less consistent way between contexts. This implies that the cognitions developed in interaction with the real world become independent from the social interactions or relationships in which they were first developed. The development of social skills to regulate the behaviour of others leads to the development of symbolic mediators (among which language has a central role), which then seem to function autonomously. They then become tools, to be used in a variety of contexts, but at the same time they predispose the individual to operate preferentially in some contexts and to neglect or reject others. At every level, from the words of our language to the elaborated structures of scientific theories, concepts can be both tools that increase our scope and efficiency, and constraints on our vision. And, we reiterate, concepts useful for thinking about how others think in the course of managing our relationships may not be equally suitable for understanding the basic nature of thought or of relationships in general. There is an analogy (but no more) here to the way in which concepts of mass, energy, cause, and effect, essential to us for finding our way about the world of objects, cease to be useful for understanding the nature of sub-atomic particles.

As we stressed earlier (pp. 4–7, 312) there is also considerable conceptual fuzziness over the distinction between social and non-social. Within this volume, the distinction operates differently for the various authors, and we have deliberately avoided an attempt to impose uniformity. In some contexts the distinction seems heuristically valuable but whether it is to be seen in absolute terms is quite another issue. Even if it were desirable (which of course it is not) deliberately to replicate the Kaspar-hauser situation and study individuals brought up in total social but not physical isolation, we should still be dealing with beings whose evolu-tionary history *may* have predisposed them to perceive some entities in social terms, and we have no means of knowing the importance of this issue. In practical terms, physical reality is often socially mediated (for instance in early infancy, in school teaching, in all competitive or co-operative situations). Tests for abilities in the physical domain demand social skills for the subjects to display them. And are we dealing with social or non-social cognitive behaviour when someone helps someone else to remember something about the physical world by asking him questions?

A number of participants expressed the view that cognitive growth occurs especially in situations of conflict—that is, where participants in an interchange entertain different points of view, and the relevant infor-mation about the points of view and their resolution is available. It was further agreed that cognitive growth was more probable in the context of a continuing relationship than in a series of disconnected encounters. We would endorse this view. It seems reasonable to suppose that the conflict will be expressed and worked through only if the subject wishes to (or must) take into account the others' point of view, and this is most probable if he is interested in reaching agreement, or in validating his own position, and in general wishes either to maintain the relationship or re-establish a disrupted one.

However, there is a trap here, a trap which is perhaps ubiquitous and is certainly inherent in the title of this volume. When we discuss the effects of relationships on cognitive development, or of cognitive development on the nature of relationships, we speak as though cognition and rela-tionship were separate entities. In fact, of course, cognitive models, expec-tations, obligations, debts, hopes, and so on are part of the relationship.

Whilst at the behavioural level a relationship can be described as a series of interactions between participants known to each other, in another sense a relationship includes the models, expectations, and so on each has of the other and the relations between those cognitions (Hinde 1979). Indeed we see relationships with those we love (and hate) as continuing over long periods of separation. Thus what we are con-cerned with can be seen as cognitive change internalized by one or both participants that both is a change in the relationship itself and may affect behaviour outside the relationship. The former may involve a positive feed-back effect—cognitive growth inside the mother-child relationship changes the relationship in ways that make further cognitive growth possible. But it must not be forgotten that cognitive growth can also lead to the mutation or destruction of relationships (the success of the teacher-pupil relationship is the seed of its own destruction), and that some relationships may be inimical to cognitive growth.

But in considering the view that relationships include the cognitions that each participant has of the other, we must not forget that those cognitions often concern emotions—both those of the cognizer and those of the other participant. Furthermore we sometimes conceive of those cognitions as energized by emotions, and at the same time we may regard cognitions as the source of emotions. Lack of understanding or cognitive dissonance may be a source of stress (with its emotional concomitants) and stress may be a source of misunderstanding. Cognitions may control behaviour in such a way as to enhance or reduce emotional expression and/or the subjective experience of emotion. Furthermore the emotional content of a relationship may have a crucial influence on its effect on cognitive growth. For the young child security of attachment facilitates the development of cognitive abilities, yet the changes presented by potential discord may also be a source of cognitive growth. Of course the seeming contradiction here may be more apparent than real—secure attachment provides a forum for the harmonious resolution of conflicts that are inherent and inevitable in any mother-child relationship.

Finally, there is one other set of problems that must be mentioned. We are concerned with the development of individuals, and yet we have seen that the individual must always be seen in a social context, and indeed that in some senses it is not possible to separate the individual from the social context. In our view this is no excuse for an approach to the nature of the individual which sees the individual merely as part of a social whole, though it does question the goal of self-realization in so far as that implies independence of relationships. But in the current context it implies that we must seek to harden up our usage of terms like 'internalization' and 'autonomy' by seeking deeper understanding of the processes implied. This will involve also a clarification of what is involved in the change from an interpersonal process to an intrapersonal one. To what extent can we trace consistencies between the structures that we infer from an individual's behaviour in a social interaction, and those we infer from subsequent behaviour in other situations? And in so far as there are differences, what are the nature of the differences? We suspect that further progress in understanding will need further precision in the use of the concept of cognitive 'structure'—a term that is used with a non-negligible range of meanings in this volume: 'structures' are both inferred from behaviour and have a causal role imputed to them. (A similar problem with the concept of 'role identity' would seem to confront the symbolic interactionist approach to relationships. See, e.g. McCall 1970, 1974).

In conclusion, we acknowledge that pointing to shortcomings in, and overlap between, the descriptive and explanatory concepts that we use is not a difficult enterprise. But we seek understanding of development, and that requires us to recognize that the apparently neat conceptual tools we use are man-made, and are not necessarily applicable to every aspect of nature. We believe it does not pay, in studying psychological development, to ape physics and search for ubiquitously applicable explanatory systems that will account for everything, like $e = mc^2$. Rather than seeking for a mountain-top where we can climb above the jungle to see the whole of reality, we must acknowledge the constraints on our vision, seeking smaller hillocks each of which may give us a different perspective on part of the whole.

References

Abramovitch, R. and Strayer, F. F. (1973). Preschool social organization: agonistic spacing and attentional behaviors. In: P. Pliner, T. Kramer, and T. Alloway, (eds.) *Recent advances in the study of communication and affect*, Vol. 6. Plenum Press, New York.

—— Carter, C., and Pepler, D. J. (1980). Observations of mixed-sex sibling dyads *Child Development* **51**, 1268–71.

Adorno, T. (1967). Sociology and psychology. *New Left Review* **46**, 67–80.

Ainsworth, M. D. S. (1979). Attachment as related to mother-infant interaction. *Advances in the study of behavior* **9**, 2–52.

—— (1982). Attachment theory: retrospect and prospect. In: J. Stevenson-Hinde and C. M. Parkes (eds.) *The place of attachment in human behavior*. Basic Books, New York.

—— and Wittig, B. A. (1969). Attachment and exploratory behavior of one-year-olds in a strange situation. In: B. Foss (ed.) *Determinants of infant behavior*, Vol. 4, pp. 111–36. Wiley, New York.

—— Blehar, M. C., Waters E., and Wall, S. (1978). *Patterns of attachment: a psychological study of the strange situation*. Erlbaum, Hillsdale, N.J.

Alexander, R. D. (1974). The evolution of social behavior. *Annual Review of Ecology and Systematics* **5**, 325–83.

Allen, R. E. and Oliver, J. M. (1982). The effects of child maltreatment on language development. *Child Abuse and Neglect* **6**, 299–305.

Altmann S. A. (1981). Dominance relations: the Cheshire cat's grin? *Behavioral and Brain Sciences* **4**, 430–1.

Ames, G. J. and Murray F. B. (1982). When two wrongs make a right: promoting cognitive change by social conflict. *Developmental Psychology* **18**, 894–97.

Anderson C. O. and Mason W. A. (1974). Early experience and complexity of social organization in groups of young rhesus monkeys (*Macaca mulatta*). *Journal of Comparative Physiology and Psychology* **87**, 681–90.

Andrews, S. R., Blumenthal, J. B., Johnson, D. L., Kahn, A. J., Fergusson, C. J., Lasater, T. M., Malone, P. E., and Wallace, D. B. (1982). The skills of mothering: a study of parent child development centers. *Monographs of the Society for Research in Child Development*, Serial No. 198, **47**(6).

Arend, R. (1984). Preschooler's competence in a barrier situation: patterns of adaptation and their precursors in infancy. Unpublished doctoral dissertation, University of Minnesota.

—— Gove, F. L., and Sroufe, A. L. (1979). Continuity of individual adaptation from infancy to kindergarten: a predictive study of ego-resiliency and curiosity in preschoolers. *Child Development* **50**, 950–59.

Argyle, M. and Cook, M. (1976). *Gaze and mutual gaze*. Cambridge University Press, Cambridge.

—— Ingham, R., Alkema, F., and McCallin, M. (1973). The different functions of gaze. *Semiotica* **7**, 19–31.

—— Trower, P., and Bryant, B. (1974). Exploration in the treatment of personality disorders and neuroses by social skills training. *British Journal Medical Psychology* **47**, 63–72.

—— Williams, M. (1969). Observer or observed? A reversible perspective in person perception. *Sociometry* **32**, 396–412.

Asher, S. R. and Gottman, J. M. (1981). *The development of children's friendships.* Cambridge University Press, Cambridge.

Attili, G. (1982). The development of preferred relationships in preschool children: child–child and child–adult relationships. In: R. Gilmour and S. Duck (eds.), *The emerging field of personal relationships.* Erlbaum, Hillsdale, N.J.

—— (1984). Preschoolers' secondary attachments to adults and to peers: adaptiveness and functional independence. *Rassegna di Psicologia* **1**, 17–38.

—— (in press). Concomitants and factors influencing children's aggression. *Aggressive Behaviour.*

—— and Cavallo-Boggi, P. (1983). Aggression and children's social skills, a semilongitudinal study. Paper presented at the Seventh Biennial Meeting of the ISSBD, München, West Germany.

—— Hold, B., and Schleidt, M. (1982). Relationships among peers in kindergarten: a cross-cultural study. In: M. Taub and F. A. King (eds.) *Current perspectives in primate social dynamics.* Van Nostrand Reinhold, New York.

Azmitia, M. (1984). Social interaction: missing mechanism in memory development? Unpublished manuscript, University of Minnesota.

Bachmann, C. and Kummer, H. (1980). Male assessment of female choice in hamadryas baboons. *Behavioral Ecology and Sociobiology* **6**, 315–21.

Baddeley, A. D., Ellis, N. C., Miles, T. R., and Lewis, V. J. (1982). Developmental and acquired dyslexia: a comparison. *Cognition* **11**, 185–99.

Bard, K. A. and Vauclair, J. (1984). The communicative context of object manipulation in ape and human adult-infant pairs. *Journal of Human Evolution* **13**, 181–90.

Barnard, K. E., Bee, H. L., and Hammond, M. A. (1984). Home environment and cognitive development in a healthy, low-risk sample: the Seattle study. In: A. W. Gottfried (ed.) *Home environment and early cognitive development.* Academic Press, New York.

Barrett, D. E. and Radke-Yarrow, M. (1977). Prosocial behavior, social inferential ability, and assertiveness in children. *Child Development* **48**, 475–81.

Bates, E., Benigni, L., Bretherton, I., Camaioni, L., and Volterra, V. (1977). From gesture to first words. On cognitive and social prerequisites. In: M. Lewis and I. Rosenblum (eds.) *Origins of behavior: communication and language.* Wiley, New York.

Bateson, G. (1955). A theory of play and fantasy. *Psychiatric Research Reports* **2**, 39–51.

Bateson, M. C. (1975). Mother-infant exchanges: the epigenesis of conversational interaction. In: D. Aronson and R. Rieber (eds.) *Developmental Psycholinguistics and Communication Disorders. Annals of the New York Academy of Sciences* **263**, 101–13.

Bateson, P. P. G. (1979). How do sensitive periods arise and what are they for? *Animal Behaviour* **27**, 470–86.

Baumrind, D. (1967). Child care practices anteceding 3 patterns of preschool behavior. *Genetic Psychology Monographs* **75**, 43–88.

—— (1971). Current patterns of parental authority. *Developmental Psychology Monograph* **4**(1, Part 1).

Bearison, D. J. (1982). New directions in studies of social interaction and cognitive growth. In: F. C. Serafica (ed.) *Social-cognitive development in context*, pp. 199–221. Guilford Press, New York.

Beck, B. B. (1973). Cooperative tool use by captive hamadryas baboons. *Science* **182**, 594–97.

—— (1974). Baboons, chimpanzees, and tools. *Journal of Human Evolution* **3**, 509–16.

Bell, N. and Perret-Clermont, A–N. (1985) The socio-psychological impact of school failure. *International Review of Applied Psychology*. **34**, 149–160.

Bell, S. M. (1970). The development of the concept of object as related to infant-mother attachment. *Child Development* **41**, 291–311.

—— and Ainsworth, M. D. (1972). Infant crying and maternal responsiveness. *Child Development* **43**, 1171–90.

Bell, R. Q. and Harper, L. V. (1977). *Child effects on adults*. Wiley, New York.

Bem, D. J. and Allen, A. (1974). On predicting some of the people some of the time. *Psychological Review* **81**, 506–20.

—— and Funder, D. C. (1978). Predicting more of the people more of the time. *Psychological Review* **85**, 485–501.

Benhar, E., Carlton, P. L., and Samuel, D. (1975). A search for mirror-image reinforcement and self-recognition in the baboon. *Contemporary primatology*, Proceedings 5th International Congress on Primatology, pp. 202–08. Karger, Basel.

Berman, C. M. (1982). The ontogeny of social relationships with group companions among free-ranging infant rhesus monkeys. I and II. *Animal Behaviour* **30**, 149–70.

—— (1983). Differentiation of relationships among rhesus monkey infants. In: R. A. Hinde (ed.) *Primate Social Relationships*, pp. 89–93. Blackwell Science Publications, Oxford.

Berrueta-Clement, J. R., Schweinhart, L. J., Barnett, W. S., Epstein, A. S., and Weikart, D. P. (1984). *Changed lives: the effects of the Perry Preschool Program on youths through age 19*. High Scope Press, Ypsilanti, Michigan.

Berscheid, E. (in press). Interpersonal modes of knowing. In: E. W. Eisner (ed.) *Learning the ways of knowing. The 85th Yearbook of the National Society for the Study of Education*. University of Chicago Press, Chicago.

Billings, A. G. and Moos, R. H. (1983). Comparisons of children of depressed and non-depressed parents: a social-environmental perspective. *Journal of Abnormal Child Psychology* **11**, 463–486.

Birch, H. G. and Gussow, J. D. (1979). *Disadvantaged children: health, nutrition and school failure*. Harcourt Brace & World, New York.

Blasi, A. (1980). Bridging moral cognition and moral action: a critical review of the literature. *Psychological Bulletin* **88**, 1–45.

Blau, P. M. (1964). *Exchange and power in social life*. Wiley, New York.

Block, J. H. and Block, J. (1971). *The California Child Q Set: a procedure for describing personological characteristics of children*. Department of Psychology, University of California, Berkeley. (Mimeo).

—— —— (1980). The role of ego-control and ego-resiliency in the organization of behavior. In: W. A. Collins (ed.) *Minnesota symposia on child psychology*, Vol. 13. Erlbaum, Hillsdale, N.J.

Blurton-Jones, N. G. (1972). Categories of child-child interaction. In: N. G. Blurton-Jones (ed.) *Ethological studies of child behavior*. Cambridge University Press, London.

Blyth, D. A., Simmons, R. G., and Carlton-Ford, S. (1983). The adjustment of early adolescents to school transitions. *Journal of Early Adolescence* **3**, 105–20.

Bohman, M., Sigvardsson, S., and Cloninger, R. (1981). Maternal inheritance of alcohol abuse: cross-fostering analysis of adopted women. *Archives of General Psychiatry* **38**, 965–969.

Bouchard, T. J. and McGue, M. (1981). Familial studies of intelligence: a review. *Science* **212**, 1055–1059.

Boucher, J. and Warrington, E. K. (1976). Memory deficits in early infantile autism: some similarities to the amnesic syndrome. *British Journal of Psychiatry* **67**, 73–87.

Bowlby, J. (1969). *Attachment and loss*, Vol. 1, *Attachment*. Basic Books, New York.

—— (1973). *Attachment and loss*, Vol. 2, *Separation: anxiety and anger*. Hogarth Press, London.

—— (1977). The making and breaking of affectional bonds. *British Journal of Psychiatry* **130**, 201–210.

—— (1982). *Attachment and loss*, Vol. 1, 2nd Edn., *Attachment*. Hogarth Press, London.

Bradley, L. and Bryant, P. E. (1978). Difficulties in auditory organisation as a possible cause of reading backwardness. *Nature* **271**, 746–7.

—— —— (1983). Categorising sounds and learning to read—a causal connection. *Nature* **301**, 419–21.

Bradley, R. H. and Caldwell, B. M. (1984). The first five years: results of the HOME study. In: A. W. Gottfried (ed.) *Home environment and early cognitive development*. Academic Press, New York.

—— —— (1985). The relation of infants' home environments to achievement test performance in first grade: a follow-up study. *Child Development* **55**, 803–809.

—— —— and Elardo, R. (1979). Home environment and cognitive development in the first two years. *Developmental Psychology* **15**, 246–250.

Brazelton, T. B. (1982). Joint regulation of neonate-parent behavior. In: E. Z. Tronick (ed.) *Social interchange in infancy*, pp. 7–22. University Park Press, Baltimore.

Bretherton, I. (1980). Young children in stressful situations. In: G. V. Coelho and P. I. Ahmed (eds.) *Uprooting and development*. Plenum Press, New York.

—— Bates, E., Benigni, L., Camaioni, L., and Volterra, V. (1979). Relationships between cognition, communication, and quality of attachment. In: E. Bates, L. Benigni, I. Bretherton, L. Camaioni, and V. Volterra (eds.) *The emergence of symbols: cognition and communication in infancy*. Academic Press, New York.

Brion-Meisels, S. and Selman, R. L. (1984). Early adolescent development of new interpersonal strategies: understanding and intervention. *School Psychology Review* **13**, 278–91.

Broughton, J. M. (1981). Piaget's structural developmental psychology, I. Piaget and structuralism. *Human Development* **24**, 78–109.

Brown, A. L., Bransford, J. D., Ferrara, R. A., and Campione, J. C. (1983). Learning, remembering, and understanding. In: J. H. Flavell and E. M. Markman (eds.) *Cognitive development* (Vol. 3 of P. H. Mussen (ed.) *Handbook of child psychology*), pp. 77–166. Wiley, New York.

Brown, R. W. (1973). *A first language: the early stages*. Harvard University Press, Cambridge, Mass.

Brown, W. L., McDowell, A. A., and Robinson E. M. (1965). Discrimination learning of mirrored cues by rhesus monkeys. *Journal of Genetic Psychology* **106**, 123–28.

Brownell, C. A. (1982). Peer interaction among toddler aged children: effects of age and social context on interaction, competence, and behavioral roles. Unpublished doctoral dissertation, University of Minnesota.

—— and Nay, R. (1983). Combinatorial skills: converging developments in the second year. Paper presented at the meetings of the Society for Research in Child Development, Detroit, Mich.

Bruner, J. S. (1972). Nature and uses of immaturity. *American Psychologist*, **27**, 1–28.

—— (1975). The ontogenesis of speech acts. *Journal of Child Language* **2**, 1–19.

—— (1977). Early social interaction and language acquisition. In: H. R. Schaffer (ed.) *Studies in mother-infant interaction*, pp. 271–89. Academic Press, London.

—— and Tagiuri, R. (1954). The perception of people. In: G. Lindzey (ed.) *Handbook of social psychology*, Vol. 1, pp. 601–33. Addison Wesley, Cambridge, Mass.

——, Olver, R. R., and Greenfield, P. M. (1966). *Studies in cognitive growth.* Wiley, New York.

Bryant, P. (1982). The role of conflict and of agreement between intellectual strategies in children's ideas about measurement. *British Journal of Psychology* **73**, 243–251.

Busse, C. D. (1978). Do chimpanzees hunt cooperatively? *American Naturalist* **112**, 767–770.

Calhoun, J. (1971). Space and the strategy of life. In: A. H. Esser (ed.) *Behavior and environment*, pp. 329–87. Plenum Press, New York.

Camaioni, L., de Castro Campos, M. F. P., and de Lemos, C. (1984). On the failure of the interactions and paradigm in language acquisition: a re-evaluation. In: W. Doise and A. Palmonari (eds.) *Social interaction in individual development*, pp. 93–106. Cambridge University Press, Cambridge.

Campbell, F. (1979). How may we best predict the child's IQ? Paper presented at the Biennial Meeting of the SRCD, San Francisco, March 1979.

Campione, J. C., Brown, A. L., Ferrara, R. A. and Bryant, C. (1983). The zone of proximal development: implications for individual differences and learning. Paper presented at the meetings of the Society for Research in Child Development, Detroit, Mich.

Campos, J. and Barrett, K. C. (1984). Toward a theory of emotional development. In: C. Izard, J. Kagan, and R. Zajonc (eds.) *Emotion, cognition, and behavior*, pp. 229–62, Cambridge University Press, New York.

—— —— Lamb, M. E., Goldsmith, H. H., and Stenberg, C. (1983). Socioemotional development. In: M. M. Haith and J. J. Campos (eds.) *Infancy and developmental psychobiology* (Vol. 2 of P. H. Mussen (ed.) *Handbook of child psychology*), pp. 783–916. Wiley, New York.

Candland, D. K., Fell, J. P., Keen, E., Leshner, A. I., Plutchick, R., and Tarpy, R. M. (1977). *Emotion.* Brooks/Cole Publishing Co. California.

Carraher, T. N., Carraher, D. W., and Schliemann, A. D. (1985). Mathematics in the streets and in schools. *British Journal of Developmental Psychology* **3**, 21–29.

Carugati, F., Emiliani, F., and Palmonari, A. (1984). Re-socialization processes in institutionalized adolescents. In: W. Doise and A. Palmonari (eds.) *Social interaction in individual development.* Cambridge University Press, Cambridge.

Case, R. (1984). The process of stage transition: A neo-Piagetian view. In:

R. J. Sternberg (ed.) *Mechanisms of cognitive development*, pp. 19–44. Freeman, New York.

Cazden, C. (1983). Peekaboo as an instructional model: discourse development at home and at school. In: B. Bain (ed.) *The sociogenesis of language and human conduct*. Plenum Press, New York.

Centlivres, P. and Hainard, J. (eds.) *Les rites de passage aujourd'hui* L'Age d'homme (in press).

Chance, M. R. A. (1961). The nature and special features of the instinctive social bond of primates. *Viking Fund Publications in Anthropology* **31**, 17–33.

—— and Jolly, C. (1970). *Social groups of monkeys, apes, and men*. Jonathan Cape, London.

—— and Mead, A. P. (1953). Social behaviour and primate evolution. *Symposia of the Society for Experimental Biology* **7**, 395–439

Chandler, M. J. (1976). Social cognition: a selective review of current research. In: W. F. Overton and J. M. Gallagher (eds.) *Knowledge and development*, Vol. I. Plenum, New York.

—— (1982). Social cognition and environmental structure: a critique of assimilation-side developmental psychology. *Canadian Journal of Behavioural Science*, **14**, 290–306.

—— Koch, D., and Paget, K. F. (1977). Developmental changes in the responses of children to conditions of crowding and congestion. In: H. McGurk (ed.) *Ecological factors in human development*. North-Holland Publishing Co., New York.

—— Siegal, M. and Boyes, M. (1980). The development of moral behavior: Continuities and discontinuities. *International Journal of Behavior Development* **3**, 323–32.

Chapais, B. and Schulman, S. (1980). An evolutionary model of female dominance relations in primates. *Journal of Theoretical Biology* **82**, 47–89.

Charlesworth, W. R. (1978). Ethology: its relevance for observational studies of human adaptation. In: G. P. Sackett (ed.) *Observing behavior*, Vol. 1, *Theory and applications in mental retardation*, pp. 7–33. University Park Press, Baltimore, Md.

—— and LaFrenière, P. (1983). Dominance, friendship, and resource utilization in preschool children's groups. *Ethology and Sociobiology* **4**, 175–86.

Cheney, D. L. (1977). The acquisition of rank and the development of reciprocal alliances among free-ranging immature baboons. *Behavioral Ecology and Sociobiology* **2**, 303–18.

—— (1978a). The play partners of immature baboons, *Animal Behaviour 26*, 1038–50.

—— (1978b). Interactions of immature male and female baboons with adult females. *Animal Behaviour* **26**, 389–408.

—— (1983). Extra-familial alliances among vervet monkeys. In: R. A. Hinde (ed.) *Primate social relationships*, pp. 278–86. Blackwell, Oxford.

—— and Seyfarth R. M. (1980). Vocal recognition in free-ranging vervet monkeys. *Animal Behaviour* **28**, 362–67.

—— —— (1981). Selective forces affecting the predator alarm calls of vervet monkeys. *Behaviour* **76**, 25–61.

—— —— (1982a). Recognition of individuals within and between groups of free-ranging vervet monkeys. *American Zoologist* **22**, 519–29.

—— —— (1982b). How vervet monkeys perceive their grunts. *Animal Behaviour* **30**, 739–51.

—— —— (1983). Non-random dispersal in free-ranging vervet monkeys. *American Naturalist* **122**, 392–412.

—— —— (1985). Vervet monkey alarm calls: manipulation through shared information? *Behaviour* (in press).

—— Lee, P. C. and Seyfarth, R. M. (1981). Behavioral correlates of non-random mortality among free-ranging adult female vervet monkeys. *Behavioral Ecology and Sociobiology* **9**, 153–61.

—— Seyfarth, R. M., Andelman, S. J., and Lee, P. C. (1985). Reproductive success in vervet monkeys. In: T. H. Clutton-Brock (ed.) *Reproductive success*. University of Chicago Press, Chicago.

Christian, J., Flyer, V., and Davis, P. (1960). Factors in the mass mortality of a herd of sika deer *Cervus nippon*. *Chesapeake Science* **1**, 70–95.

Cicchetti, D. and Hesse, P. (1983). Affect and intellect: Piaget's contributions to the study of infant emotional development. In: R. Plutchik and H. Kellerman (eds.) *Emotion theory, research, and experience*, Vol. 2. Academic Press, New York.

Clark, E. V. (1978). From gesture to word: on the natural history of Deixis in language acquisition. In: J. Bruner and A. Garton (eds.) *Human growth and development*, pp. 85–120. Clarendon Press, Oxford.

—— (1983). Meanings and concepts. In: J. H. Flavell and E. M. Markman (eds.) *Cognitive development* (Vol. 3 of P. H. Mussen (ed.) *Handbook of child psychology*), pp. 787—840. Wiley, New York.

—— and Sengul, C. J. (1978). Strategies in the acquisition of Deixis, *Journal of Child Language* **5**, 457–75.

Clarke, A. M. (1982). Developmental discontinuities: an approach to assessing their nature. In: L. A. Bond and J. M. Joffe (eds.) *Facilitating infant and early childhood development*, pp. 58–77. University Press of New England, Hanover, New Hampshire.

—— (1984). Early experience and cognitive development. *Reviews of Research in Education* (in press).

—— and Clarke, A. D. B. (1976). *Early experience: myth and evidence*. Open Books, London.

Clarke-Stewart, K. A. (1973). Interactions between mothers and their young children: characteristics and consequences. *Monographs of the Society for Research in Child Development* (SRCD), Serial No. 153, **38** (6–7).

—— and Fein, G. G. (1983). Early childhood programs. In: M. M. Haith and J. J. Campos (eds.), *Infancy and developmental psychobiology* (Vol. 2 of P. H. Mussen (ed.) *Handbook of child psychology*), 4th edn., pp. 917–99. Wiley, New York.

Cloninger, R., Bonman, M., and Sigvardsson, S. (1981). Inheritance of alcohol abuse: cross-fostering analysis of adopted men. *Archives of General Psychiatry* **38**, 861–8.

Clutton-Brock, T. H. and Harvey, P. (1980). Primates, brains, and ecology. *Journal of Zoology, London* **190**, 309–23.

Cohn, J. F. and Tronick, E. Z. (1982). Communicative rules and the sequential structure of infant behavior during normal and depressed interaction. In: E. Z. Tronick (ed.) *Social interchange in infancy*, pp. 59–72. University Park Press, Baltimore.

Coie, J. D. and Dodge, K. A. (1983). Continuities and changes in children's social status: a five-year longitudinal study. *Merrill-Palmer Quarterly*, **29**, 261–82.

Cole, M. and Scribner, S. (1974). *Culture and thought: a psychological introduction*. Wiley, New York.

Collis, G. M. (1977). Visual co-orientation and maternal speech. In: H. R. Schaffer (ed.), *Studies in mother-infant interaction*, pp. 355–78. Academic Press, London.

Cook, M. (1977). Gaze and mutual gaze in social encounters, *American Scientist* **65**, 328–33.

—— and Smith, J. M. C. (1975). The role of gaze in impression formation *British Journal of Social and Clinical Psychology* **14**, 19–25.

Cooper, C. R., Ayers-Lopez, S., and Marquis, A. (1982). Children's discourse during peer learning in experimental and naturalistic situations. *Discourse Processes* **5**, 177–191.

Cooper, L. A. (1982). Internal representation. In: D. R. Griffin (ed.), *Animal mind–human–mind*, pp. 145–58. Springer-Verlag, Berlin.

Costanzo, P., Coie, J., Grumet, J., and Farnill, D. (1973). A reexamination of the effects of intent and consequence on children's moral judgments. *Child Development* **44**, pp. 154–61.

Costanzo, P. R. and Dix, T. H. (1983). Beyond the information processed: socialization in the development of attributional processes. In: E. Tory Higgins, D. N. Ruble, and W. W. Hartup (eds.) *Social cognition and social development*, pp. 63–81. Cambridge University Press, Cambridge.

Cowey, A. and Weiskrantz, L. (1975). Demonstration of cross-modal matching in rhesus monkeys (*Macaca mulatta*). *Neuropsychologia* **13**, 117–20.

Cox, J. K. and D'Amato, M. R. (1982). Matching to compound samples by monkeys (*Cebus apella*): shared attention or generalized decrement? *Journal of Experimental Psychology* **8**, 209–25.

Craig, H. K. and Gallagher, T. M. (1982). Gaze and proximity as turn regulators within three-party and two-party child conversations. *Journal of Speech and Hearing Research* **25**, 1, 65–75.

CRESAS (1978) (ouvrage collectif) *Le handicap socio-culturel en question*. Editions Sociales Françaises, Paris.

—— (1981) (ouvrage collectif) *L'échec scolaire n'est pas une fatalité*. Editions Sociales Françaises, Paris.

Cross, T. G. (1977). Mother's speech adjustment: the contribution of selected child listener variables. In: C. E. Snow and Ch. A. Ferguson (eds.) *Talking to children*, pp. 151–88. Cambridge University Press, Cambridge.

Cummings, E. M., Iannotti, R. J., and Zahn-Waxler, C. (in press). The influence of conflict between adults on the emotions and aggression of young children. *Developmental Psychology*.

—— Hollenbeck, B., Iannotti, R. J., Radke-Yarrow, M., and Zahn-Waxler, C. (in press). Early organization of altruism and aggression: developmental patterns and individual differences. In: C. Zahn-Waxler, E. M. Cummings, and R. J. Iannotti (eds.) *Altruism and aggression: social and biological origins*. Cambridge University Press, Cambridge.

Curtiss, S. (1977). Genie. A psycholinguistic study of a modern-day 'wild child'. *Perspectives in neurolinguistics and psycholinguistics*. Academic Press, New York.

Damon, W. (1977). *The social world of the child*. Jossey-Bass, San Francisco.

—— (1981). Exploring children's social cognition on two fronts. In: J. H. Flavell and L. Ross (eds.) *Social cognitive development: frontiers and possible futures*. Cambridge University Press, New York.

—— (1983a) The nature of social-cognitive change in the developing child. In:

W. Overton (ed.) *The relationship between social and cognitive development.* Erlbaum, Hillsdale, N.J.

—— (1983b). Five questions for research in social-cognitive development. In: E. Tory Higgins, D. N. Ruble, and W. W. Hartup (eds.) *Social cognition and social development*, pp. 371–93. Cambridge University Press, Cambridge.

Datta, S. B. (1983). Relative power and the acquisition of rank. In: R. A. Hinde (ed.) *Primate social relationships*, pp. 93–103. Blackwell, Oxford.

—— (1983). Patterns of agonistic interference. In: R. A. Hinde (ed.), *Primate social relationships*, pp. 289–97. Blackwell, Oxford.

Davie, C. E., Hutt, S. J., Vincent, E., and Mason, M. (1984). *The young child at home.* NFER-Nelson, Windsor, Berks.

Davies, B. (1982). *Life in the classroom and playground.* Routledge & Kegan Paul, London.

Deconchy, J. P. (1971). *L'orthodoxie religieuse. Essai de logique psychosociale.* Paris, Éditions Ouvrières.

—— (1980). *Orthodoxie religieuse et sciences humaines.* Mouton, La Haye.

DeLoache, J. S. (1983). Joint picture-book reading as memory training for toddlers. Paper presented at the meetings of the Society for Research in Child Development, Detroit, Mich.

Dennett, D.C. (1978). *Brainstorms: philosophical essays on mind and psychology.* Montgomery, Bradford Books, Vermont.

—— (1983). Intentional systems in cognitive ethology: the 'Panglossian paradigm' defended. *Behavioral and Brain Sciences* **6**, 343–90.

Dennis, M. and Whitaker, H. A. (1976). Language acquisition following hemidecortication: linguistic superiority of the left over the right hemisphere. *Brain and Language* **3**, 404–33.

Dennis, W. (1973). *Children of the crèche.* Appleton-Century-Crofts, New York.

Dixon, P. (1980). Personal communication cited in Rutter (1981).

Dodge, K. A. (1983). Behavioral antecedents of peer social status. *Child Development* **54**, 1386–99.

—— (in press). A social information processing model of social competence in children. In: M. Perlmutter (ed.) *Minnesota symposia on child psychology*, Vol. 18. Erlbaum, Hillsdale, N.J.

Döhl, J. (1968). Uber die Fähigkeit einer Schimpansin, Umwege mit selbständigen Zwischenzielen zu überblicken. *Zeitschrift für Tierpsychologie.* **25**, 89–103.

Doise, W. (1978). Soziale Interaktion und kognitive Entwicklung. In: G. Steiner (ed.) *Piaget und die Folgen*, pp. 331–47. Kindler, Zürich. (= Psychologie des 20. Jahr-hunderts. Vol. VII.)

—— (1980). Levels of explanation. *European Journal of Social Psychology.* **10**, 213–31.

—— (1982) *L'explication en psychologie sociale.* Presses Universitaires de France, Paris.

—— and de Paolis, P. (1984). Fattori sociali nello sviluppo cognitivo. La connotazione sociale del compito. *Età evolutiva* **19**, 5–10.

—— and Mackie, D. (1981). On the social nature of cognition. In: J. P. Forgas (ed.), *Social cognition.* Academic Press, London.

—— and Mugny, G. (1981). *Le développement social de l'intelligence.* Inter Editions, Paris.

—— —— (1984). *The social development of the intellect.* Pergamon Press, Oxford.

—— Dionnet, S. and Mugny, G. (1978). Conflit socio-cognitif, marquage social, et développement cognitif. *Cahiers de Psychologie* **21**, 231–43.

—— Meyer, G., and Perret-Clermont, A-N. (1976). Représentations sociales d'élèves enfui de scolasité obligatoire. *Cahiers de la section des science de l'education, Université de Geneve*, No. 2.

—— Mugny, G. and Perret-Clermont, A. N. (1974). Ricerce preliminari sulla sociogenesi delle strutture cognitive. *Lavoro educativo*, Vol. 1.

—— —— —— (1975). Social interaction and the development of cognitive operations. *European Journal of Social Psychology* **5**, 367–83.

—— Rijsman, J. B., Van Meel, J., Bressers, I., and Pinxten, L. (1981). Sociale markering en cognitieve ontwikkeling. *Pedagogische Studien* **58**, 241–8.

Donaldson, M. (1978). *Children's minds*. Fontana, London.

Dornbusch, S. M., Hastorf, A. H., Richardson, S. A., Muzzy, R. E., and Vreeland, R. S. (1965). The perceiver and perceived: their relative influence on categories of interpersonal perception. *Journal of Personality and Social Psychology* **1**, 434–440.

Douglas, J. W. B. (1964). *The home and the school*. McGibbon & Kee, London.

Duck, S. W. (1973). *Personal relationships and personal constructs: a study of friendship formation*. Wiley, New York.

—— (1983). *Friends, for life*. Harvester Press, Brighton.

Dunbar, R. I. M. (1980). Determinants and evolutionary consequences of dominance among female gelada baboons. *Behavioral Ecology and Sociobiology*, **7**, 253–65.

—— (1983) Structure of gelada baboon reproductive units. II. Social relationships between reproductive females. *Animal Behaviour* **31**, 556–64.

Duncan, S. (1972). Some signals and rules for taking speaking turns in conversations. *Journal of Personality and Social Psychology* **23**, 283–92.

—— (1973). Toward a grammar for dyadic conversation. *Semiotica* **9**, 29–46.

—— Niederehe, G. (1974). On signalling that its your turn to speak. *Journal of Experimental Social Psychology* **10**, 234–47.

Dunn, J. (1982). Comment, problems, and promises in the study of affect and interaction. In: E. Z. Tronick (ed.) *Social interchange in infancy*, pp. 197–206. University Park Press, Baltimore.

—— (1983), Sibling relationships in early childhood. *Child Development* **54**, 787–811.

—— and Dale, N. (1984). I a daddy: two-year-olds' collaboration to joint pretend with sibling and with mother. In: I. Bretherton (ed.) *Symbolic play: the representation of social understanding*, pp. 131–58. Academic Press, New York.

—— and Kendrick, C. (1982). *Siblings: love, envy, and understanding*. Harvard University Press, Cambridge, Mass.

Dunn, L. M. and Dunn, L. (1981). Peabody picture vocabulary task—revised. American Guidance Service, Circle Pines, Minnesota.

Easterbrooks, M. A. and Goldberg, W. A. (1984). Toddler development in the family: impact of father involvement and parenting characteristics. *Child Development* **55**, 740–52.

Eckerman, C. O., Whatley, J. L., and Kutz, S. L. (1975). Growth of social play with peers during the second year of life *Developmental Psychology*, **11**, 42–9.

Eisenberg, J. F. (1973). Mammalian social systems: are primates' social systems unique? In: E. W. Menzel (ed.), *Proceedings from the Symposia of the Fourth Congress of the International Primatological Society*, Vol. *1*. Karger, Basel.

Ekman, P., Friesen, W., and Ellsworth, P. (1972). *Emotion in the human face: guidelines for research and an integration of findings*. Pergamon Press, New York.

Elardo, R., Bradley, R., and Caldwell, B. (1975). The relation of infant's home environments to mental test performance from six to thirty six months: a longitudinal analysis. *Child Development* **46**, 71–6.

—— —— —— (1977). A longitudinal study of the relation of infants' home environments to language development at age three. *Child Development* **48**, 595–603.

Elias, M. J. (1978). The development of a theory-based measure of how children understand and attempt to resolve problematic social situations. Unpublished masters thesis, University of Connecticut, Storrs.

—— (1984). *Project AWARE: social problem solving as prevention model.* Progress Report No. 36828, National Institute of Mental Health, Washington, D.C.

Elkind, D. (1964). Discrimination, seriation, and numeration of size and dimensional differences in young children: Piaget replication study VI. *Journal of Genetic Psychology* **104**, 275–96.

Elmer, E. (1977). A follow-up study of traumatized children. *Pediatrics* **59**, 273–279.

Emde, R. (1980). Toward a psychoanalytic theory of affect. I. The organizational model and its propositions. II. Emerging models of emotional development in infancy. In: S. Greenspan and G. Pollack (eds.) *The course of life: psychoanalytic contributions toward understanding personality development.* US Government Printing Office, Washington, D.C.

Emiliani, F. and Zani, B. (1984). Behaviour and goals in adult-child interaction in the day nursery. In: W. Doise and A. Palmonari (eds.) *Social interaction in individual development*, pp. 78–90. University Press, Cambridge.

Emmerich, W. (1971). *Structure and development of personal-social behaviors in preschool settings: Head Start longitudinal study.* Educational Testing Service, Princeton, N.J.

Epstein, R., Lanza, R. P., and Skinner, B. F. (1981). Self-awareness in the pigeon. *Science* **212**, 695–6.

Erikson, E. H. (1950). *Childhood and society.* Norton, New York.

Esser, A. (1972). A biosocial perspective on crowding. In: J. Wohwill and D. Carson (eds.) *Environment and the social sciences: perspective and applications.* The American Psychological Association, New York.

Essock, S. M. and Rumbaugh, D. M. (1978). Development and measurement of cognitive capabilities in captive non-human primates. In: H. Markowitz and V. Stevens (eds.) *The behavior of captive wild animals*, pp. 161–203. Academic Press, New York.

Essock-Vitale, S. M. (1978). Comparison of ape and monkey modes of problem solution. *Journal of Comparative Physiology and Psychology* **92**, 942–57.

Exline, R. V. (1971). Visual interaction: the glances of power and preference. In: J. Cole (ed.) *Nebraska Symposium on Motivation*, pp. 163–206. University of Nebraska Press, Lincoln.

Fabricius, E. (1951). Zur Ethologie junger Anatiden. *Acta Zoologica Fennica* **68**, 1–178.

Fairbairn, W. R. D. (1952). *Psychoanalytic studies of the personality.* Basic Books, New York.

Fairbanks, L. A. (1980). Relationships among adult females in captive vervet monkeys: testing a model of rank-related attractiveness. *Animal Behaviour* **28**, 853–9.

Faucheux, C. and Moscovici, S. (1960). Etude sur la créativité des groups II: tâche, structure de communication, et réussite *Bulletin du CERP* **9**, 11–22

Fenson, L. and Ramsay, D. S. (1980). Decentration and integration of the child's play in the second year. *Child Development* **51**, 171–8.

Field, T., Woodson, R., Greenberg, R., and Cohen, D. (1982). Discrimination and imitation of facial expressions by neonates. *Science* **218**, 179–81.

Finn, G. P. T. (1982). Children's experimental episodes, or 'Ask a silly question but get a serious answer'. Unpublished manuscript, Department of Psychology, Jordanhill College of Education.

Flagg, S. F. and Medin, D. L. (1973). Constant irrelevant cues and stimulus generalization in monkeys. *Journal of Comparative Psychology* **85**, 339–45.

Flament, C. (1965). *Théorie des graphes et structures sociales*. Gauthier-Villars, Paris.

Flament, F. (1977). Quelques remarques sur la genèse de la communication non-verbale dans les interactions sociales entre nourrissons. In: *La genèse de la parole*. Symposium de l'Association de Psychologie Scientifique de Langue Française. Presses Universitaires de France, Paris.

Flavell, J. H. and Markman, E. M. (eds.) (1983). *Cognitive development*, Vol. 3 of P. H. Mussen (ed.) *Handbook of child psychology*. Wiley, New York.

Fogel, A., Diamond, G. R., Langhorst, B. H., and Demos, V. (1982). Affective and cognitive aspects of the 2-month-old's participation in face-to-face interaction with the mother. In: E. Z. Tronick (ed.), *Social interchange in infancy*, pp. 37–57. University Park Press, Baltimore.

Foot, H. C., Chapman, A. J., and Smith, J. R. (1980). *Friendship and social relations in children*. Wiley, New York.

Forbes, D. *et al.* (in press). Dinosaurs don't like rain: a dramatic analysis of children's fantasy play. In: E. Mueller and C. Cooper (eds.) *Process and outcome in peer-relationships*. Academic Press, New York.

Ford, M. (1982). Social cognition and social competence in adolescence. *Developmental Psychology* **18**, 323–40.

Forman, E. A. and Kraker, M. J. (in press). The social origins of logic: the contributions of Piaget and Vygotsky. In: M. W. Berkowitz (ed.) *Peer conflicts and psychological growth*. Jossey-Bass, San Francisco.

French, D. C., Waas, G., and Dopkins, A. L. (1984). *Leadership asymmetries in mixed-age children's groups*. Unpublished manuscript. University of Wisconsin.

Freud, S. (1933). *New introductory lectures*. Norton, New York.

Freud, A. and Dann, S. (1951). An experimental group up-bringing, Vol. 6: *Psychoanalytic study of the child*. Imago, London.

Frith, U. and Snowling, M. (1983). Reading for meaning and reading for sound in autistic and dyslexic children. *British Journal of Developmental Psychology* **1**, 329–42.

Frye, D. (1980). Developmental changes in strategies of social interaction. In: M. Lamb and L. Sherrod (eds.) *Infant social cognition: empirical and theoretical considerations*. Erlbaum, Hillsdale, New Jersey.

Furlong, V. (1976). Interaction set in classroom. In: M. Hammersley and P. Woods (eds.) *The process of schooling*, pp. 160–77. Routledge & Kegan Paul, London.

Furstenberg, F. F., Jr. (1976). *Unplanned parenthood*. Free Press, Glencoe, Ill.

Gaensbauer, T. J., Harmon, R. J., Cytryn, L., and McKnew, D. H. (1984). Social and affective development in infants with a manic-depressive parent. *American Journal of Psychiatry* **141**, 223–9.

Gallup, G. G. (1970). Chimpanzees: self-recognition. *Science* **167**, 86–7.

—— (1975) Towards an operational definition of self-awareness. In: R. H. Tuttle (ed.) *Socioecology and psychology of primates*, pp. 309–41. Mouton, The Hague, Paris.

—— (1977). Absence of self recognition in a monkey (*Macaca fascicularis*) following prolonged exposure to a mirror. *Developmental Psychobiology* **10**, 281–4.

—— (1983). Toward a comparative psychology of mind. In: Mellgren R. L. (ed.) *Animal cognition and behavior*, pp. 473–510. North-Holland, Amsterdam.

Garber, H. and Heber, F. R. (1977). The Milwaukee project: indications of the effectiveness of early intervention in preventing mental retardation, In: P. Mittler (ed.) *Research to practice in mental retardation. I. Care and intervention.* University Park Press, Baltimore.

—— and Heber, R. (1982). Modification of predicted cognitive development in high-risk children through early intervention. In: M. K. Detterman and R. J. Sternberg (eds.) *How and how much can intelligence be increased?*, pp. 121–37. Ablex, Norwood, N.J.

Garcia, H. S. and Ettlinger, G. (1978). Sorting of objects by the chimpanzee and monkey. In: D. J. Chivers and J. Herbert (eds.) *Recent advances in primatology*, pp. 957–9. Cambridge University Press, Cambridge.

Gardner, R. A. and Gardner, B. T. (1969). Teaching sign-language to chimpanzees. *Science* **165**, 664.

Garfinkel, H. (1967). *Studies in ethnomethodology*, Prentice Hall, Englewood Cliffs, N.J.

Garvey, C. (1974). Some properties of social play. *Merrill-Palmer Quarterly* **20**, 163–80.

—— (1979). An approach to the study of children's role play. *Quarterly Newsletters of the Laboratory of Comparative Human Cognition* **I**, 64–73.

—— (1982). Communication and the development of social role play. In: D. Forbes and M. Greenberg (eds.) *New directions in child development: children's playfulness*. Jossey-Bass, San Francisco.

—— and Berndt, R. (1977). Organization of pretend play. *Catalogue of selected documents in psychology* **7**, 15–89.

Garvin, J. B. and Sacks, L. S. (1963). Growth potential of preschool-aged children in institutional care: a positive approach to a negative condition. *American Journal of Orthopsychiatry* **33**, 399–408.

Gelman, R. (1979). Why we will continue to read Piaget. *The Genetic Epistomologist* **8**, 12–13.

—— and Baillargeon, R. (1983). A review of some Piagetian concepts. In: J. H. Flavell and E. M. Markman (eds.) *Cognitive Development* (Vol. 3 of P. H. Mussen (ed.) *Handbook of child psychology*), pp. 167–230. Wiley, New York.

—— and Spelke, E. (1981). The development of thoughts about animate and inanimate objects. In: J. H. Flavell and L. Ross (eds.) *Social cognitive development*, pp. 43–66. Cambridge University Press, Cambridge.

Gesten, E. L., Rains, M., Rapkin, B., Weissberg, R. G., Flores de Apodaca, R., Cowen, E. L., and Bowen, G. (1982). Training children in social problem-solving competencies: a first and second look. *American Journal of Community Psychology* **10**, 95–115.

Gillan, D. J. (1981). Reasoning in the chimpanzee: II. Transitive inference. *Journal of Experimental Psychology: Animal Behavioural Processes* **7**, 150–64.

—— Premack, D., and Woodruff, G. (1981). Reasoning in the chimpanzee. I. Analogical reasoning. *Journal of Experimental Psychology: Animal Behavioural Processes* **7**, 150–64.

Gilligan, C. (1982). *In a different voice. Essays on the psychological development of women.* Howard University Press, Cambridge, Mass.

Gilly, M. (1980) *Maitre-élève, rôles institutionnels et représentations*—Presses Universitaires de France, Paris.

Glachan, M. and Light, P. (1982). Peer interaction and learning: can two wrongs make a right? In: G. Butterworth and P. Light (eds.) *Social cognition: studies of the development of understanding*, pp. 238–62. University of Chicago Press, Chicago.

Gleitman, L., Newport, E. L., and Gleitman, H. (1984). The current status of the motherese hypothesis. *Journal of Child Language* **11**, 43–79.

Glick, J. (1978). Cognition and social cognition: an introduction. In: J. Glick and K. A. Clarke-Stewart (eds.) *The development of social understanding.* Gardner Press, New York.

Goffman, E. (1973). *La mise en scène de la vie quotidienne*, Editions de Minuit, Paris.

—— (1974a). *Les rites d'interaction*, Editions de Minuit, Paris.

—— (1974b). *Frame analysis*, Harper & Row, New York.

Goodall, J., Bandoro, A., Bergmann, E., Busse, C., Matama, H., Mpongo, E., Pierce, A., and Riss, D. (1979). Intercommunity interactions in the chimpanzee population of the Gombe National Park. In: D. Hamburg and E. R. McCown (eds.) *The great apes*, pp. 13–54. Benjamin Cummings, Menlo Park.

Goodnow, J. (in press). Some lifelong everyday forms of intelligent behavior: organizing and reorganizing. In: R. Sternberg and R. Wagner (eds.) *Practical intelligence: origins of competence in the everyday world.*

—— Knight, R. and Cashmore, J. (in press). Adult social cognition: implications of parents' ideas for approaches to development. In: M. Perlmutter (ed.) *Minnesota Symposia on Child Psychology*, Vol. 18. Erlbaum, Hillsdale, N.J.

Gotlib, I. and Asarnow, R. F. (1979). Interpersonal and impersonal problem-solving skills in mildly and clinically depressed university students. *Journal of Consulting and Clinical Psychology* **47**, 86–95.

Gottfried, A. W. (ed.) (1984). *Home environment and early cognitive development: longitudinal research.* Academic Press, Orlando and London.

—— and Gottfried, A. E. (1984). Children of middle S.E.S. families. In: A. W. Gottfried (ed.) *Home environment and early cognitive development.* Academic Press, New York.

Gottman, J. M. (1977). Towards a definition of social isolation in children. *Child Development* **48**, 513–517.

—— (1983). How children become friends. *Monographs of the Society for Research in Child Development.* Serial No. 201, **48**(3).

Granville, A. C., McNeil, J. T., Meece, J., Wacker, S., Morris, M., Shelly, M., and Love, J. M. (1976). *A process evaluation of project developmental continuity, interim report IV.* Vol. 1: *Pilot year impact study—instrument characteristics and attrition trends.* No. 105–75–1114, Office of Child Development, Washington, D.C.

Gray, S. B., Ramsey, B. K., and Klaus, R. A. (1982). *From 3 to 20—the Early Training Project.* University Park Press, Baltimore, Md.

Greenough, W. T. and Schwark, H. D. (1984). Age-related aspects of experience: effects upon brain structure. In: R. N. Emde and R. J. Harmon (eds.) *Continuities and discontinuities in development*, pp. 69–91. Plenum Press, New York.

Grossen, M. and Perret–Clermont, A. N. (1983). *Interactions sociales, représentations de la tâche et rôle du contexte dans l'épreuve de la conservation du nombre.* Séminaire de Psychologie, Université de Neuchâtel.

Grossman, K. and Grossman, K. (1981). Parent-infant attachment relationships in Bielefeld. A research note in: K. Immelman *et al.* (eds.) *Behavioral development: the Bielefeld Interdisciplinary Project.* Cambridge University Press, New York.

Guntrip, H. (1961). *Personality structure and human interaction.* International Universities Press, New York.

Hainard, J. and Kaehr, R. (eds) (1983). *Le corps enjeu.* Musé d'Ethnographie, Neuchâtel.

Halliday, M. A. K. (1975). *Learning how to mean: explorations in the development of language.* Edward Arnold, London.

Hallinan, M. T. (1979). Structural effects on children's friendships and cliques. *Social Psychology* **42**, 43–54.

Hare, R. M. (1964). *Freedom and reason.* Oxford University Press, New York.

Harlow H. F. (1949). The formation of learning sets. *Psychological Reviews* **5**, 51–65.

—— and Harlow, M. K. (1965). The affectional system. In: A. M. Schrier, H. F. Harlow, and F. Stollnitz (eds.) *Behaviour of non-human primates 2.* Academic Press, New York.

Harré, R. (1984). *Personal being.* Harvard University Press, Cambridge, Mass.

Harris, P. L. (1984). Infant cognition. In: P. Mussen (ed.). *Handbook of child psychology,* Wiley, New York.

Harris, P. L. and Heelas, P. (1979). Cognitive processes and collective representations. *European Journal of Sociology* **20**, 211–41.

Harter, S. (1983). Developmental perspectives on the self-system. In: E. M. Hetherington (ed.) *Socialization, personality, and social development* (Vol. 4 of P. H. Mussen (ed.) *Handbook of child psychology*), pp. 275–386. Wiley, New York.

Hartmann, H. (1939). *Ego psychology and the problem of adaptation.* International Universities Press, New York.

—— (1964). *Essays on ego psychology.* International Universities Press, New York.

Hartup, W. W. (1978). Children and their friends. In: H. McGurk (ed.) *Issues in childhood social development,* pp. 130–70. Methuen, London.

—— (1979). Peer relations and the growth of social competence. In: M. W. Kent and J. E. Rolf (eds.) *The primary prevention of psychopathology.* Vol. 3, *Promoting social competence and coping in children.* University Press in New England, Hanover, N.H.

—— (1983). Peer relations. In: E. M. Hetherington (ed.) *Socialization, personality, and social development* (Vol. 4 of P. H. Mussen (ed.) *Handbook of child psychology*), pp. 103–96. Wiley, New York.

Hay, D. F. (1984). Social conflict in early childhood. In: G. Whitehurst (ed.) *Annals of child development,* Vol. 1. JAI Press, Greenwich, Conn.

Hayes, K. J. and Nissen, C. H. (1971). Higher mental functions of a home-raised chimpanzee. In: A. M. Schrier and F. Stollnitz (eds.) *Behavior of non-human primates,* pp. 60–115. Academic Press, New York.

Heber, R., Garber, H., Harrington, S., Hoffman, C., and Falender, C. (1972). *Rehabilitation of families at risk for mental retardation,* December 1972 Progress Report. University of Wisconsin, Madison.

Hecaen, H. (1983). Acquired aphasia in children: revisited. *Neuropsychologia* **21**, 581–7.

Hendrix, J. M. and Van de Voort, M. (1982). *Social interaction and cognitive devel-*

opment: problems with the dis- and advantage. Department of Psychology, Tilburg University.

Herdt, G. H. (1981). *Guardians of the flute: idioms of masculinity*. McGraw-Hill, New York.

Herrnstein, R. J. (1979). Acquisition, generalization, and discrimination reversal of a natural concept. *Journal of Experimental Psychology: Animal Behavioural Processes* **5**, 116–29.

—— and DeVilliers, P.A. (1980). Fish as a natural category for people and pigeons. In: G. H. Bower (ed.) *The psychology of learning and motivation*, pp. 59–95, Academic Press, New York.

—— and Loveland, D. H. (1964). Complex visual concepts in the pigeon. *Science* **146**, 549–51.

—— —— and Cable, C. (1976). Natural concepts in pigeons. *Journal of Experimental Psychology: Animal Behavioural Processes* **2**, 285–311.

Hetherington, E. M., Cox, M., and Cox, R. (1982). Effects of divorce on parents and children. In: M. E. Lamb (ed.) *Non-traditional families*, pp. 233–88. Erlbaum, Hillsdale, N.J.

Hewison, J. and Tizard, J. (1980). Parental involvement and reading attainment. *British Journal of Educational Psychology* **50**, 209–15.

Heyns, B. (1978). *Summer learning and the effects on schooling*. Academic Press, New York.

Higgins, E. Tory and Parsons, J. E. (1983). Social cognition and the social life of the child: stages as subcultures. In: E. Tory Higgins, D. N. Ruble, and W. W. Hartup (eds.) *Social cognition and social development*, pp. 15–62. Cambridge University Press, Cambridge.

Hinde, R. A. (1972). Concepts of emotion. In: *Physiology, emotion and psychosomatic illness*. Ciba Foundation Symposium No. 8. Elsevier, Amsterdam.

—— (1974). *Biological basis of human social behaviour*. McGraw-Hill, New York.

—— (1976). Interactions, relationships, and social structure. *Man* **11**, 1–17.

—— (1979). *Towards understanding relationships*. Academic Press, London.

—— (1982). Attachment: some conceptual and biological issues. In: C. M. Parkes and J. Stevenson-Hinde (eds.) *The place of attachment in human behaviour*. Tavistock Publications, London.

—— (1982). *Ethology*. Fontana, London.

—— (1983a). Feedback in the mother-infant relationship. In: R. A. Hinde (ed.) *Primate social relationships*, pp. 70–73. Blackwell Scientific Publications, Oxford.

—— (1983b). Development and dynamics. In: R. A. Hinde (ed.) *Primate social relationships*, pp. 65–70. Blackwell Scientific Publications, Oxford.

—— (1983c). La comunicazione non-verbale nel contesto delle interazioni e delle relazioni: In: G. Attili and P. E. Ricci-Bitti (eds.) *Comunicare senza parole*, pp. 13–26. Bulzoni, Roma.

—— (1983d). Ethology and child development. In: Haith, M. H. and Campos, J. J. (eds) *Infancy and developmental psychobiology* (Vol. 2 of P. H. Mussen (ed.) *Handbook of child psychology*), 4th edn., pp. 27–99. Wiley, New York.

—— (1984). Why do the sexes behave differently in close relationships? *Journal of Social and Personal Relationships* **1**, 471–501.

—— (1985). Expression and negotiation. In: G. Zivin (ed.) *The development of expressive behavior: biology-environment interactions*, pp. 103–16. Academic Press, New York.

—— (1985). Biological bases of the mother-child relationship. In: J. Call, E.

Galensen, and R. L. Tyson (eds.) *Frontiers of infant psychiatry*, Vol. 2. Basic Books, New York.

—— (in press). Relazioni interpersonali e sviluppo infantile. *Eta Evolutiva*.

—— and Dennis, A. (in press). Categorizing individuals: towards linking developmental and clinical psychology.

—— and Spencer-Booth, Y. (1967). The behaviour of socially-living rhesus monkeys in their first two and a half years. *Animal Behaviour* **15**, 169–96.

—— and Stevenson-Hinde, J. (1973). *Constraints on learning*. Academic Press, New York.

—— —— (1976). Towards understanding relationships: dynamic stability. In: P. P. G. Bateson and R. A Hinde (eds.) *Growing points in ethology*. Cambridge University Press, Cambridge.

—— and Tamplin, A. (1983). Relations between mother-child interaction and behaviour in preschool. *British Journal of Developmental Psychology* **1**, 231–257.

—— Easton, P. F., Meller, R. E. and Tamplin, A. (1983). Nature and determinants of preschoolers differential behaviour to adults and to peers. *British Journal of Developmental Psychology* **1**, 3–19.

—— Thorpe, W. H. and Vince, M. A. (1956). The following response of young coots and moorhens. *Behaviour* **9**(2–3), 214–42.

Hiskey, M. S. (1966). *Hiskey-Nebraska test of learning aptitude*. Union College Press, Lincoln, Nebraska.

Hoffman, M. L. (1971). Father-absence and conscience development. *Developmental Psychology* **4**, 400–406.

—— (1981). Perspectives on the difference between understanding people and understanding things: the role of affect. In: J. H. Flavell and L. Ross (eds.) *Social cognitive development*, pp. 67–81. Cambridge University Press, Cambridge.

—— (1983). Empathy, guilt, and social cognition. In: W. F. Overton (ed.) *The relationship between social and cognitive development*, pp. 1–51. Erlbaum, Hillsdale, N.J.

—— (1984). Interaction of affect and cognition in empathy. In: C. E. Izard, J. Kagan and R. Zajonc (eds.) *Emotion, cognition, and behavior*. Cambridge University Press, New York.

Holst, E. von and Mittelstaedt, H. (1950). Das Reafferenzprinzip. *Naturwissenschaften* **37**, 464–76.

Homans, G. C. (1961). *Social behaviour: its elementary forms*. Routledge & Kegan Paul, London.

Honig, W. K. (1978). On the conceptual nature of cognitive terms: an initial essay. In: S. H. Hulse, H. Fowler, and W. K. Honig (eds.) *Cognitive processes in animal behavior*, pp. 1–14. Erlbaum, Hillsdale, N.J.

Hood, L. and Bloom, L. (1979). What, when, and how about why. *Monographs of the Society for Research in Child Development* **44**, 1–41.

Horn, J. M. (1983). The Texas adoption project: adopted children and their intellectual resemblance to biological and adoptive parents. *Child Development* **54**, 268–75.

Hulse, S. H., Fowler, H. and Honig, W. K. (eds.) (1978). *Cognitive processes in animal behavior*. Erlbaum, Hillsdale, N.J.

Humphrey, N. K. (1976). The social function of intellect. In: P. P. G. Bateson and R. A. Hinde (eds.) *Growing points in ethology*, pp. 303–17. Cambridge University Press, Cambridge.

Inhelder, B., Sinclair, H., and Bovet, M. (1974). *Apprentissage et structures de la connaissance*. Presses Universitaires de France, Paris.

Izard, C. E. (1984). Approaches to developmental research on emotion-cognition relationships. In: C. E. Izard, J. Kagan, and R. B. Zajonc (eds.) *Emotions, cognition, and behavior*. Cambridge University Press, Cambridge.

Jackson, E., Campos, J. J., and Fischer, K. W. (1978). The question of décalage between object permanence and person permanence. *Developmental Psychology* **14**, 1–10.

Jacquard, A. (1982). *Au péril de la science?* Interrogations d'un généticien. Seuil.

Jaffe, J., Stern, D., and Peery, J. (1973). 'Conversational' coupling of gaze behavior in prelinguistic human development. *Journal of Psycholinguistic Research* **2**, 321–9.

Jarrard, L. E. (ed.) (1971). *Cognitive processes in non-human primates*. Academic Press, New York.

Jarvis, M. J. and Ettlinger, G. (1977). Cross-modal recognition in chimpanzees and monkeys. *Neuropsychologia* **15**, 499–506.

Jaspars, J. M. F. and Leeuw, J. A. de (1980). Genetic-environment covariation in human behaviour genetics. In: L. J. T. van der Kamp *et al.* (eds.) *Psychometrics for educational debate*. Wiley, Chichester.

Jencks, C., Smith, M., Acland, M., Bane, M. J., Cohen, D., Gintis, H., Heyns, B., and Michelson, S. (1973). *Inequality: a reassessment of the effect of family and schooling in America*. Allen Lane, London.

Jensen, A. R. (1973). *Educability and group differences*. Methuen, London.

Johnson, D. L., Breckenridge, J. N., and McGowan, R. J. (1984). Home environment and early cognitive development in Mexican-American children. In: A. W. Gottfried (ed.) *Home environment and early cognitive development*. Academic Press, New York.

Johnson, J. E., Yu, S., and Roopnarine, J. (1980). Social cognitive ability, interpersonal behaviors, and peer status within a mixed age group. Paper presented at the meeting of the Southwestern Society for Research in Human Development, Lawrence, Ka.

Johnston, T. D. (1981). Contrasting approaches to a theory of learning. *Behavioral and Brain Sciences* **4**, 125–73.

Jolly, A. (1966). Lemur social behavior and primate intelligence. *Science* **153**, 501–6.

Judge, P. G. (1982). Redirection of aggression based on kinship in a captive group of pigtail macaques. *International Journal of Primatology* **3**, 301.

Kagan, J. (1984). The idea of emotion in human development. In: C. E. Izard, J. Kagan and R. Zajonc (eds.) *Emotions, cognition, and behavior*. Cambridge University Press, New York.

Kamin, L. J. (1974). *The science and politics of IQ*. Erlbaum, Potomac, Maryland.

Kaplan, J. R. (1978). Fight interference and altruism in rhesus monkeys. *American Journal of Physical Anthropology* **49**, 241–50.

Karnes, M. B. (1973). Evaluation and implications of research with young handicapped and low-income children. In: J. C. Stanley (ed.) *Compensatory education for children, ages 2 to 8*. Johns Hopkins University Press.

Kaye, K. (1977). Toward the origin of dialogue. In: H. R. Shaffer (ed.) *Studies in mother–infant interaction*, pp. 89–117. Academic Press, London.

—— (1979). The social content of infant development. *Report to Spencer Foundation*, cited in Camaioni *et al.* (1984).

—— (1982). Organism, apprentice, and person. In: E. Z. Tronick (ed.) *Social interchange in infancy*. Baltimore University Park Press, Baltimore.

Keating, D. P. (1979). Adolescent thinking. In: J. P. Adelson (ed.) *Handbook of adolescence*. Wiley, New York.

Keller, M. (1984). Resolving conflicts in friendship: the development of moral understanding in everyday life. In: W. M. Kurtines and J. L. Gewirtz (eds.) *Morality, moral behavior, and moral development*, pp. 140–58. Wiley, New York.

Kelley, H. H. (1971). *Attribution in social interaction*. General Learning Press, New York.

—— Berscheid, E., Christensen, A., Harvey, J. H., Huston, T. L., Levinger, G., McClintock, E., Peplau, L. A., and Peterson, D. R. (1983). *Close relationships*. Freeman, New York.

Kendon, A. (1967). Some functions of gaze direction in social interaction, *Acta Psychologica* **26**, 1–47.

—— and Cook, M. (1969). The consistency of gaze patterns in social interaction, *British Journal of Psychology* **69**, 481–94.

Kenrick, D. T. and Stringfield, D. O. (1980). Personality traits and the eye of the beholder. *Psychological Review* **87**, 88–104.

Key, M. R. (1977). *Non-verbal communication. A research guide and bibliography*. The Scarecrow Press, Metuchen, New Jersey.

Klaus, R. and Gray, S. W. (1968). The early training project for disadvantaged children: a report after five years. *Monographs of the Society for Research in Child Development* Serial No. 120, **33**(4).

Kliegl, R. (1984). Ein Beitrag zur Binnenstruktur des Freundschaftsverständnisses. In: K. E. Grossmann and P. Lütkenhaus (eds.) *Bericht über die 6. Tagung Entwicklungspsychologie 1983*, pp. 283–87. Druckerei der Universität Regensburg, Regensburg.

Kohlberg, L. (1968). Stage and sequence: the cognitive developmental approach to socialization. In: D. A. Goslin (ed.) *Handbook of socialization theory and research*, pp. 374–480. Rand McNally, Chicago.

—— (1969). The cognitive-developmental approach to socialization. In: D. A. Gozlin (ed.) *Handbook of socialization theory and research*. Rand McNally, Chicago.

—— Levine, C., and Hewer, A. (1983). Moral stages: a current formulation and a response to critics. *Contributions to human development*, Vol. 10. Karger, Basel, Switzerland.

Krappmann, L. (1982). *Soziologische dimensionen der Identität*. Klett, Stuttgart.

—— and Oswald, H. (1983a) Beziehungsgeflechte und Gruppen von gleichaltrigen Kindern in der Schule. In: F. Neidhardt (ed.) *Gruppensoziologie. Kölner zeitschrift für soziologie und sozialpsychologie*, Sonderheft **25**, 420–50.

—— —— (1983b) Types of integration into peer society. Presented at the Biennial Meeting of the SRCD, Detroit.

Krashen, S. D. (1973). Lateralization, language learning, and the critical period: some new evidence. *Language Learning* **23**, 63–74.

Kuczynski, L., Radke-Yarrow, M., and Zahn-Waxler, C. (1984). Content and development of imitation in the second and third years of life: a socialization perspective. Unpublished manuscript. National Institute of Mental Health.

Kuhn, D. and Ho, V. (1977). The development of schemes for recognizing additive and alternative effects in a 'natural experiment' context. *Developmental Psychology* **13**, 515–16.

Kummer, H. (1957). Soziales Verhalten einer Mantelpavian-gruppe. *Beiheft Schweizerischen Zeitschrift für Psychologie* **33**, 1–91.

—— (1968). *Social organization of hamadryas baboons*, University of Chicago Press, Chicago.

—— (1971). *Primate societies*. Aldine, Chicago.

—— (1975). Rules of dyad and group formation among captive gelada baboons. In: S. Kondo, M. Kawai, A. Ehora, and S. Kawamura (eds.) *Proceedings from the Symposia of the Fifth Congress of the International Primatological Society*. Japan Science Press, Tokyo.

—— (1982). Social knowledge in free-ranging primates. In: D. R. Griffin (ed.), *Animal-mind—human-mind*, pp. 113–30. Springer-Verlag, Berlin.

—— Götz W., and Angst W. (1974). Triadic differentiation: an inhibitory process protecting pair bonds in baboons. *Behaviour* **49**, 62–87.

Labov, W., Fanshel, D. (1977). *Therapeutic discourse. Psychotherapy as conversation*. Academic Press, New York.

Ladd, G. W. and Emerson, E. S. (1984). Shared knowledge in children's friendships. *Developmental Psychology* **20**, 932–40.

Lamb, M. E. (1977). A re-examination of the infant social world. *Human Development* **20**, 65–85.

—— (1978). The development of sibling relationships in infancy: a short term longitudinal study. *Child Development* **49**, 1189–96.

Lave, J. (1977). Cognitive consequences of traditional apprenticeship training in West Africa. *Anthropology and Education Quarterly* **7**, 177–80.

Lawson, A. and Ingleby, J. D. (1974). Daily routines of preschool children: effects of age, birth order, sex, and social class and developmental correlates. *Psychosocial Medicine* **4**, 399–415.

Lazar, I. and Darlington, R. (1982). Lasting effects of early education: a report from the consortium for longitudinal studies. *Monographs of the Society for Research in Child Development*, Serial No. 195, **47**(2–3).

Lee, P. C. (1983). Effects of the loss of the mother on social development. In: R. A. Hinde (ed.) *Primate social relationships*, pp. 73–9. Blackwell Scientific Publications, Oxford.

Leiter, R. G. (1969). Leiter international performance scale. Stoelting Co., Chicago.

Lempers, J. D., Flavell, E. R., and Flavell, J. H. (1977). The development in very young children of tacit knowledge concerning visual perception. *Genetic Psychology Monographs* **95**, 3–53.

Lenneberg, E. (1967). *Biological foundations of language*. Wiley, New York.

Leont'ev, A. N. (1959). *Problemy razvitiya psikhiki*. Translated as: A. N. Leont'ev (1981), *Problems in the development of mind*. Progress, Moscow.

—— (1975). *Deyatel'nost'. Soznanie. Lichnost'*. Translated as: A. N. Leont'ev (1978). *Activity. Consciousness. Personality*. Prentice-Hall, Englewood Cliffs, N.J.

—— (1981). The problem of activity in psychology. In: J. V. Wertsch (ed.) *The concept of activity in Soviet psychology*. M. E. Sharpe, Armonk, New York.

Levitt, M. J., Antonucci, T. C., and Clark, M. C. (1984). Object-person permanence and attachment: another look. *Merrill-Palmer Quarterly* **30**, 1–10.

Lévy, M. (1981). La nécessité sociale de dépasser une situation conflictuelle générée par la présentation d'un modèle de solution de problème et par le questionnement d'un agent social. Thèse de doctorat en psychologie, Université de Genève.

Lewis, M. and Brooks-Gunn, J. (1979). *Social cognition and the acquisition of self.* Plenum Press, New York.

—— and Rosenblum, L. (eds.) (1975). *Friendship and peer relations.* Wiley, New York.

—— Sullivan, M. W., and Michalson, L. (1984). The cognitive-emotional fugue. In: C. E. Izard, J. Kagan, and R. Zajonc (eds.) *Emotions, cognition, and behavior.* Cambridge University Press, New York.

—— Young, G., Brooks, J., and Michalson, L. (1975). The beginning of friendship. In: M. Lewis and L. Rosenblum (eds.) *Friendship and peer relations.* Wiley, New York.

Light, P. H., Buckingham, N., and Robbins, A. H. (1979). The conservation task as an interactional setting. *British Journal of Educational Psychology* **49**, 304–10.

Livesley, W. J. and Bromley, D. B. (1973). *Person perception in childhood and adolescence.* Wiley, London.

Loevinger, J. (1976). *Ego development.* Jossey-Bass, San Francisco.

Longstreth, L. E. (1981). Revisiting Skeels' final study: a critique. *Developmental Psychology* **17**, 620–25.

—— Davis, B., Carter, L., Flint, D., Owen, J., Rickert, M., and Taylor, E. (1981). Separation of home intellectual environment and maternal IQ as determinants of child IQ. *Developmental Psychology* **17**, 532–41.

Lougee, M. D., Grueneich, R., and Hartup, W. W. (1977). Social interaction in same- and mixed-age dyads of preschool children. *Child Development* **48**, 1353–1361.

Luce, J. De and Wilder, H. T. (eds.) (1983). *Language in primates. Perspectives and implications.* Springer-Verlag, Heidelberg, New York.

Ludeke, R. J. and Hartup, W. W. (1983). Teaching behaviors of nine- and eleven-year-old girls in mixed-age and same-age dyads. *Journal of Educational Psychology* **75**, 908–914.

Luria, A. R. (1981). *Language and cognition.* Wiley, New York.

Lyman, D. R. and Selman, R. L. (1985). Peer conflict in pair therapy: clinical and developmental analyses. In: M. Berkowitz (ed.) *Peer conflict and psychological growth: new directions in child development.* Jossey-Bass, San Francisco.

Mackintosh, N. J. (1974). *The psychology of animal learning.* Academic Press, London.

MacNamara, J. (1982). *Names for things.* MIT Press, Cambridge, Mass.

McCall, G. J. (1970). The social organization of relationships. In: McCall, G. J., McCall, M., Denzin, N. K., Suttles G. D., and Kurth, S. B. (eds.) *Social relationships.* Aldine, Chicago.

—— (1974). A symbolic interactionist approach to attraction. In: T. L. Houston (ed.) *Foundations of interpersonal attraction.* Academic Press, New York.

McCall, R. B. (1981). Nature-nurture and the two realms of development: a proposed integration with respect to mental development. *Child Development* **52**, 1–12.

—— Parke, R. D., and Kavanaugh, R. D. (1977). Imitation of live and televised models by children one to three years of age. *Monographs of the Society for Research in Child Development*, Serial No. 173, **42**(5).

McCartney, K., Scarr, S., Phillips, D., Grajek, S., and Schwarz, J. C. (1982). Environmental differences among day care centers and their effects on children's development. In: E. F. Zigler and E. W. Gordon (eds.) *Day care: scientific and social policy issues*, pp. 126–51. Auburn House, Boston, Mass.

McGarrigle, J. and Donaldson, N. (1974). Conservation accidents. *Cognition* **3**, 341–50.

McGonigle, B. O. and Chalmers, M. (1977). Are monkeys logical? *Nature* **267**, 694–6.

McGowan, R. J. and Johnson, D. L. (1984). The mother-child relationship and other antecedents of childhood intelligence: a causal analysis. *Child Development* **55**, 810–20.

McGrew, W. C. (1972). *An ethological study of children's behaviour.* Academic Press, New York.

McKim, B. J., Weissberg, R. P., Cowen, E. L., Gesten, E. L., and Rapkin, B. D. (1982). A comparison of the problem-solving ability and adjustment of suburban and urban third grade children. *American Journal of Community Psychology* **10**, 155–69.

Maccoby, E. E. (1984). Socialization and developmental change. *Child Development* **55**, 317–28.

—— and Feldman, S. S. (1972). Mother-attachment and stranger reactions in the third year of life. *Monographs of the Society for Research in Child Development*, Serial No. 146, **37**(1).

—— and Martin, J. A. M. (1983). Socialization in the context of the family: parent-child interaction. In: E. M. Hetherington (ed.) *Socialization, personality, and social development.* (Vol. 4. of P. H. Mussen (ed.) Handbook of child psychology, 4th edn), pp. 1–101. Wiley, New York.

Magnusson, D. (1981). *Toward a psychology of situations.* Erlbaum, Hillsdale, N.J.

Main, M., Kaplan, K. and Cassidy, J. (1985). Security in infancy, childhood and adulthood: a move to the level of representation. In Growing points in attachment theory and research (eds. I. Bretherton and E. Waters). *Monographs of the Society for Research in Child Development* (in press).

—— and Weston, D. R. (1982). Avoidance of the attachment figure in infances: descriptions and interpretations. In: C. M. Parkes and J. Stevenson-Hinde (eds.) *The place of attachment in human behaviour.* Tavistock Publications, London.

Maisonnet, R. and Stambak, M. (1983). Jeux moteurs chez des enfants de 12 à 18 mois. In: M. Stambak *et al.*, *Les bébés entre eux* (eds.) pp. 135–151. Presses Universitaires de France, Paris.

Malone, Daniel R., Tolan, James C., and Rogers, C. M. (1980). Cross-modal matching of objects and photographs in the monkey. *Neuropsychologia* **18**, 693–7.

Mandler, J. M. (1983). Representation. In: J. H. Flavell and E. M. Markman (eds.) *Cognitive development* (Vol. 3 of P. H. Mussen (ed.) *Handbook of child psychology*), pp. 420–94. Wiley, New York.

Maratsos, M. (1983). Some current issues in the study of the acquisition of grammar. In: J. H. Flavell and E. M. Markman (eds.) *Cognitive development* (Vol. 3 of P. H. Mussen (ed.) *Handbook of child psychology*), pp. 707–86. Wiley, New York.

Marc, P. (1984) *Autour de la notion pédagogique d'attente.* Collection exploration. P. Lang, Berne.

Marsh, D. T. (1982). The development of interpersonal problem-solving among elementary school children. *Journal of Genetic Psychology* **140**, 107–18.

Marvin, R. S. (1972). *Attachment and co-operative behavior in two-, three-, and four-year-olds.* University of Chicago.

—— (1977). An ethological-cognitive model for the attenuation of mother-child attachment behavior. In: T. M. Alloway, L. Krames, and P. Pliner (eds.) *Advances in the study of communication and affect*, Vol. 3. Plenum Press, New York.

Mason, W. A. (1978). Social experience and primate cognitive development. In: G. M. Burghardt and M. Bekoff (eds.) *The development of behavior: comparative and evolutionary aspects*, pp. 233–51. Garland Press, New York.

—— and Berkson G. (1975). Effects of maternal mobility on the development of rocking and other behaviors in rhesus monkeys: a study with artificial mothers. *Developmental Psychobiology* **8**, 197–211.

—— and Kenney, M. D. (1974). Redirection of filial attachments in rhesus monkeys: dogs as mother surrogates. *Science* **183**, 1209–11.

Massel, K. H., Macias III, Salvador, Meador, D. M., and Rumbaugh, D. M. (1981). The learning skills of primates: the rhesus macaque in comparative perspective. *International Journal of Primatology* **2**, 9–17.

Massey, A. (1977). Agonistic aids and kinship in a group of pigtail macaques. *Behavioural Ecology and Sociobiology* **2**, 31–40.

Masur, E. F. (1982). Mother's responses to infant's object-related gestures: influences on lexical development. *Journal of Child Language* **9**, 23–30.

Matas, L., Arend, R. A., and Sroufe, L. A. (1978). Continuity of adaptation in the second year: the relationship between quality of attachment and later competence. *Child Development* **49**, 547–56.

Maughan, B., Gray, G., and Rutter, M. (1985). Reading retardation and antisocial behaviour: a follow-up into employment. *Journal of Child Psychology and Psychiatry* (in press).

Maynard-Smith, J. (1979). Games theory and evolution of behaviour. *Proceedings of the Royal Society*, Series B, **205**, 475–88.

Mead, G. H. (1934). *Mind, self, and society*. University of Chicago Press, Chicago.

Medrich, E. A., Rosen, J., Rubin, V., and Buckley, S. (1982). *The serious business of growing up*. University of California Press, Berkeley, Calif.

Mellgren, R. L. (ed.) (1983). *Animal cognition and behavior*. North-Holland, Amsterdam.

Meltzoff, A., and Moore, M. (1977). Imitation of facial and manual gestures of human neonates. *Science* **198**, 75–8.

Metzl, M. N. (1980). Teaching parents a strategy for enhancing infant development. *Child Development* **51**, 583–6.

Meyenn, R. J. (1980). School girls' peer groups. In P. Woods (ed.) *Pupils strategies—explorations in the sociology of the school*, pp. 108–142. Croom Helm, London.

Miles, L. (1982). Sign language studies with an orang-utan. Paper presented at the IXth Congress of the International Primatological Society, Atlanta, Georgia.

Miles, R. C. (1965). Discrimination-learning sets. In: A. M. Schrier, H. F. Harlow, and F. Stollnitz (eds.) *Behavior of nonhuman primates*, Vol. 1, pp. 51–95. Academic Press, New York.

Miller, W. B. (1958). Lower class culture as a generating milieu of gang delinquency. *Journal of Social Issues* **14**, 5–19.

Milton, K. (1981). Distribution patterns of tropical plant food as an evolutionary stimulus to primate mental development. *American Anthropologist* **83**, 534–48.

Minton, C., Kagan, J., and Levine, J. A. (1971). Maternal control and obedience in the two-year-old. *Child Development* **42**, 1873–94.

Minuchin, P. O. and Shapiro, E. K. (1983). The school as a context for social development. In: E. M. Hetherington (ed.) *Socialization, personality, and social development* (Vol 4. of P. H. Mussen (ed.) *Handbook of child psychology*), 4th edn, pp. 197–274. Wiley, New York.

Moessinger, P. (1978). Piaget on equilibration. *Human Development* **21**, 255–67.

Moscovici, A. and Andpaicheler, G. (1973). Individu, travail, groupe. In: S. Moscovici (ed.) *Introduction à la psychologie sociale*, Vol. 2. Larousse, Paris.

Mowrer, O. H. (1950). *Learning theory and personality dynamics*. Ronald Press, New York.

Mueller, E. and Brenner, J. (1977). The origins of social skills and interaction among playgroup toddlers. *Child Development* **48**, 854–61.

—— and Lucas, T. (1975). A developmental analysis of peer interaction among toddlers. In: M. Lewis and L. A. Rosenblum (eds.) *Friendship and peer relations*. Wiley, New York.

—— and Vandell, D. (1979). Infant/infant interaction. A review. In: J. D. Osofsy (ed.) *Handbook of infant development*, pp. 591–622. Wiley Interscience, New York.

Mugny, G. (ed.) (1985). *Psychologie sociale du developpement cognitif*. Collection exploration. P. Lang, Berne.

Mugny, G. and Doise, W. (1978a). Factores sociologicos y psicosociologicos del desarrollo cognitivo. *Anuario de psicologia* **18**, 22–40.

—— —— (1978b). Socio-cognitive conflict and structuration of individual and collective performances. *European Journal of Social Psychology* **8**, 181–192.

—— —— and Perret-Clermont, A.-N. (1976). Conflict de centration et progrès cognitif. *Bulletin de Psychologie* **29**, 199–204.

—— De Paolis, P., and Carugati, F. (1984). Social regulations in cognitive development. In: W. Doise and A. Palmonari (eds.) *Social interaction in individual development*, pp. 127–46. Cambridge University Press, Cambridge.

Munsinger, H. (1975). The adopted child's IQ: a critical review. *Psychological Bulletin* **82**, 623–59.

Murphy, C. and Messer, D. (1977). Mothers, infants and pointing: a study of a gesture. In: M. R. Shaffer (ed.) *Studies in mother-infant interaction*, pp. 325–54. Academic Press, New York.

Murray, F. (1972). Acquisition of conservation through social interaction. *Developmental Psychology* **6**, 1–6.

Mussen, P. and Rutherford, E. (1963). Parent-child relations and parental personality in relation to young children's sex-role preference. *Child Development* **34**, 589–607.

Nelson, J. and Aboud, F. E. (in press). The resolution of social conflict between friends. *Child Development*.

Newcomb, A. F. and Brady, J. E. (1982). Mutuality in boys' friendship relations. *Child Development* **53**, 392–5.

—— Brady, J. E. and Hartup, W. W. (1979). Friendship and incentive condition as determinants of children's task-oriented social behavior. *Child Development* **50**, 878–81.

Newson, J. (1977). An intersubjective approach to the systemic description of mother-infant interaction. In: M. R. Shaffer (ed.) *Studies in mother-infant interaction*, pp. 47–61. Academic Press, London.

—— and Newson, E. (1968). Four years old in an urban community. Aldine, Chicago.

Nicolet, M. (1984). *Marquage social, caractéristiques de la situation et origine des sujets*. Séminaire de Psychologie, Université de Neuchâtel.

Nicolich McCune, L. (1977). Beyond sensorimotor intelligence: assessment of symbolic maturity through analysis of pretend play. *Merrill Palmer Quarterly* **23**, 89–99.

Nielsen, G. (1962). *Studies in self confrontation*. Munksgaard, Copenhagen.

Ninio, A. and Bruner, J. S. (1978). The achievement and antecedents of labelling. *Journal of Child Language* **5**, 1–15.

Nisbett, R. and Ross, L. (1980) *Human inference: strategies and shortcomings of social judgement*. Prentice-Hall, Englewood Cliffs, N.J.

Noble, C. S. and Thomas, R. K. (1970). Oddity learning in the squirrel monkey. *Psychonomic Science* **19**, 305–7.

Ochs, E. and Schieffelin, B. B. (eds.) (1979). *Developmental pragmatics*. Academic Press, New York.

Olshan, K. (1970). The multidimensional structure of person perception in children. Unpublished doctoral dissertation. Rutgers University.

Olson, S. L., Johnson, J., Belleau, K., Parks, J., and Barrett, E. (1983). Social competence in preschool children: interrelations with sociometric status, social problem-solving, and impulsivity. Paper presented at the meetings of the Society for Research in Child Development, April. Detroit, Mich.

Olton, D. S. (1979). Mazes, maps, and memory. *American Psychologist* **34**, 583–96.

Oser, F. (1981). *Moralishes Urteil in Gruppen. Soziales Handeln Verteilungsgerechtigkeit*. Suhrkamp.

Osborn, A. F., Butler, N. R., and Morris, A. C. (1984). *The social life of Britain's five-year-olds*. Routledge & Kegan Paul, London.

Oswald, H. and Krappmann, L. (1983). Interaction among peers in middle childhood—affiliation with peer group formations and patterns of interaction. Presented at the Biennial Meeting of the ISSBD, Munich.

—— —— (1984). Konstanz und Veränderung in den sozialen Beziehungen von Schulkindern. *Zeitschrift für Sozialisationsforschung und Erziehungssoziologie* **4**, 271–86.

Overman, W. H. and Doty, R. W. (1980). Prolonged visual memory in macaques and man. *Neuropsychologia* **5**, 1825–31.

Packer, C. (1977). Reciprocal altruism in *Papio anubis*. *Nature* **265**, 441–3.

Page, E. B. and Grandon, G. M. (1981). Massive intervention and child intelligence. The Milwaukee project in critical perspective. *Journal of Special Education* **15**, 239–256.

Pain, S. (1981) *Les difficultés d'apprentissage, diagnostic et traitement*. Collection exploration. P. Lang, Berne.

Pancake, L. V. (1985). Continuity between mother-infant attachment and ongoing dyadic peer relationships in preschool. Paper presented at meetings of the Society for Research in Child Development, Toronto, Canada.

Parke, R. D. (1978). Children's home environments. In: I. Altman and J. F. Wohlwill (eds.) *Children and the environment*, Vol. 3. Plenum Press, New York.

Parker, S. T. and Gibson, K. R. (1979). A developmental model for the evolution of language and intelligence in early hominids. *Behavioural and Brain Sciences* **2**, 367–408.

Parsons, T. (1964). *The social system* (1951). Free Press, Glencoe, Ill.

Patterson, F. and Linden, E. (1981). *The education of Koko*. Holt Rinehart and Winston, New York.

Patterson, G. (1982). *Coercive family process*. Castalia Publishing Co., Eugene, Oregon.

Pedersen, E., Faucher, T. A., and Eaton, W. W. (1978). A new perspective on the effects of first grade teachers on children's subsequent adult status. *Harvard Educational Review* **48**, 1–31.

Peery, J. C. (1979). Popular, amiable, isolated, rejected: a reconceptualization of sociometric status in preschool children. *Child Development* **50**, 1231–4.

Pellegrini, D. (1985). Social cognition and competence in middle childhood. *Child Development* **56**, 253–64.

Pepler, D. J., Abramovitch, R., and Carter, C. (1982). Sibling interaction in the home: a longitudinal study. *Child Development* **52**, 1344–7.

Perret-Clermont, A.-N. (1980). *Social interaction and cognitive development in children*. Academic Press, New York.

—— and Schubauer-Leoni, M. L. (1981). Conflict and cooperation as opportunities for learning. In: P. Robinson (ed.) *Communication in development*. Academic Press, London.

—— Brun, J., Saada, E. H., and Schubauer-Leoni, M. L. (1984a). Learning: a social actualization and reconstruction of knowledge. In: H. Tajfel (ed.), *The social dimension*, Vol. 1. Cambridge University Press and Maison des Sciences de l'Homme.

—— —— —— —— (1984b). *Psychological processes, operatory level, and the acquisition of knowledge*. Interactions Didactiques 2 Bis, Universities of Geneva and Neuchâtel.

Philp, A. J. (1940). Stranger and friends as competitors and co-operators. *Journal of Genetic Psychology* **57**, 249–58.

Piaget, J. (1929). *The child's conception of the world*. Harcourt Brace, New York.

—— (1932). *The moral judgment of the child*. Free Press, Glencoe, Ill.

—— (1937). *La naissance de l'intelligence chez l'infant*, Delachaux and Niestlé (English translation: Routledge, London, 1953).

—— (1950). *The psychology of intelligence*. Harcourt Brace, New York.

—— (1951). *Play, dreams, and imitation in childhood*. Norton, New York.

—— (1952). *The child's conception of number*. Routledge & Kegan Paul, London.

—— (1963). *The psychology of intelligence*. International Universities Press, New York.

—— (1965a). *The moral judgment of the child*. Free Press, New York.

—— (1965b). *Etudes sociologiques*. Droz, Paris.

—— (1968). *Six psychological studies*. Vintage, New York.

—— (1970). Piaget's theory. In: P. M. Mussen (ed.) *Carmichael's manual of child psychology*, 3rd edn, Vol. I. Wiley, New York.

—— (1971). *Biology and knowledge*. University of Chicago Press, Chicago.

—— (1974). Need and significance of cross-cultural studies in genetic psychology. In: J. W. Berry and P. R. Dasen (eds.) *Culture and cognition*, pp. 299–309. Methuen, London.

—— (1976a). Logique génétique et sociologie. In: G. Busino (ed.) *Les sciences sociales avec et après*, pp. 44–80. Librairie Droz, Genève.

—— (1976b). Postface. *Archives de Psychologie*, **44**, 223–228.

—— (1981). *Intelligence and affectivity*. Annual Reviews, California.

—— and Inhelder, B. (1969). *The psychology of the child*. Basic Books, New York.

—— —— and Szeminska, A. (1960). *The child's conception of geometry.* Basic Books, New York.

Platt, J. J. and Spivack, G. (1972a). Problem-solving thinking of psychiatric patients. *Journal of Consulting and Clinical Psychology* **39**, 148–51.

—— ——. (1972b). Social competence and effective problem-solving thinking in psychiatric patients. *Journal of Clinical Psychology* **28**, 3–5.

—— —— (1973). Studies in problem-solving thinking of psychiatric patients: patient-control differences and factorial structure of problem-solving thinking. *Proceedings of the 81st Annual Convention of the American Psychological Association* **8**, 461–2. (Summary).

—— —— (1974). Means of solving real-life problems: I. Psychiatric patients vs. controls, and cross-cultural comparisons of normal females. *Journal of Community Psychology* **2**, 45–8.

—— —— Altman, N., Altman, D., and Peizer, S. B. (1974). Adolescent problem-solving thinking. *Journal of Consulting and Clinical Psychology* **42**, 787–93.

Plomin, R. and DeFries, J. C. (1983). The Colorado adoptive project. *Child Development* **54**, 276–89.

—— —— (in press). *Origins of individual differences in infancy: the Colorado Adoptive Project.* Academic Press, New York.

—— —— and Leohlin, J. C. (1977). Genotype-environment interaction and correlation in the analysis of human behavior. *Psychological Bulletin* **84**, 309–22.

Poeschl, G., Doise, W., and Mugny, G. (in press). *Les représentations sociales de l'intelligence et de son développement chez des jeunes de 15 à 22 ans.*

Poole, J. and Lander, D. G. (1971). The pigeon's concept of pigeon. *Psychonomic Science* **25**, 157–8.

Porter, R. and Collins, G. M. (eds.) (1982). *Temperamental differences in infants and young children.* Ciba Foundation Symposium 89. Pitman Books, London.

Premack, D. (1976). *Intelligence in ape and man.* Erlbaum, Hillsdale, N.J.

—— (1978). On the abstractness of human concepts: why it would be difficult to talk to a pigeon. In: S. H. Hulse, F. Fowler and W. K. Honig (eds.) *Cognitive processes in animal behavior* pp. 421–51. Erlbaum, Hillsdale, N.J.

—— (1983). The codes of man and beasts. *Behavioural and Brain Sciences* **6**, 125–67.

—— and Premack, A. (1982). *The mind of an ape.* Norton, New York.

—— and Woodruff G. (1978). Does the chimpanzee have a theory of mind? *Behavioural and Brain Sciences* **4**, 515–26.

Quinton, D. and Rutter, M. (1984). Parenting behaviour of mothers raised 'in care'. In: A. R. Nicol (ed.) *Longitudinal studies in child psychology and psychiatry: practical lessons from research experience.* Wiley, Chichester.

—— —— and Liddle, C. (1984). Institutional rearing, parental support, and marital difficulties. *Psychological Medicine* **14**, 107–24.

Radke-Yarrow, M. and Zahn-Waxler, C. (1984). Roots, motives, and patterning in children's prosocial behavior. In: E. Staub, D. Bar-Tal, J. Karylowski, and J. Reykowski (eds.) *The development and maintenance of prosocial behavior: international perspectives on positive morality.* Plenum Press, New York.

—— Cummings, E. M., Kuczynski, L., and Chapman, M. (in press). Patterns of attachment in two- and three-year-olds in normal families and families with parental depression. *Child Development.*

Rahe, D. F. (1984). Interaction patterns between children and mothers on teaching tasks at age 42 months: antecedents in attachment history, intellectual correlates and consequences on children's socio-emotional functioning. Unpublished doctoral dissertation, University of Minnesota.

Ramey, C. T. and Campbell, F. A. (1981). Educational intervention for children at risk for mild retardation: a longitudinal analysis. In: P. Mittler (ed.) *Frontiers of knowledge in mental retardation*, Vol. I. *Social, educational, and behavioral aspects*, pp. 47–57. University Park Press, Baltimore, Md.

—— —— (1984). Preventive education for high risk children: cognitive consequences of the Carolina Abecedarian project. *Amererican Journal of Mental Deficiency* **88**, 515–24.

—— Bryant, D. M., and Suarez, T. M. (1984). Preschool compensatory education and the modifiability of intelligence: a critical review. In: D. Detterman (ed.) *Current topics in human intelligence*. Ablex Publishing Co., Norwood, N.J.

—— MacPhee, D. and Yeates, K. (1982). Preventing developmental retardation: a general systems model. In: L. Bond and J. Joffe (eds.) *Facilitating infant and early childhood development*. University Press of New England, Hanover, New Hampshire.

Record, R. G., McKeown, T., and Edwards, J. H. (1969). The relation of measured intelligence to birth order and maternal age. *Annals of Human Genetics Lond.* **33**, 61–9.

Reese, H. W. and Overton, W. (1970). Models of development and theories of development. In: L. R. Goulet and P. B. Baltes (eds.) *Life-span developmental psychology: research and theory*. Academic Press, New York.

Reiser, J., Doxsey, J., McCarrell, N., and Brooks, P. (1982). Wayfinding and toddlers' use of information from an aerial view of a maze. *Developmental Psychology* **18**, 714–20.

Reitman, W. (1970). What does it take to remember? In: D. Norman (ed.) *Models of human memory*. Academic Press, New York.

Renshaw, P. D. and Asher, S. R. (1982). Social competence and peer status: the distinction between goals and strategies. In: K. H. Rubin and H. S. Ross (eds.) *Peer relationships and social skills in childhood*. Springer-Verlag, New York.

Research and Clinical Center for Child Development. Annual Report 1981–1982. Faculty of Education, Sapporo, Japan.

Reynolds P. C. (1981). *On the evolution of human behavior. The argument from animals to man*. University of California Press, Berkeley.

Richard, B. A. and Dodge, K. A. (1982). Social maladjustment and problem-solving in school-aged children. *Journal of Consulting and Clinical Psychology* **50**, 226–33.

Richer, J. M. and Coss, R. G. (1976). Social looking in autistic and normal children. Cited in: M. Argyle and M. Cook (eds.), *Gaze and mutual gaze*, pp. 130–1. Cambridge University Press, Cambridge.

Richman, N., Stevenson, J., and Graham, P. J. (1982). *Pre-school to school: a behavioural study*. Academic Press, London.

Rijsman, J. (1984). Variables cathectiques dans le développement social de l'intelligence, In: G. Mugny (ed.) *Psychologie sociale de l'apprentissage*, pp. 151–166. Collection exploration. P. Lang, Berne.

Rodman, P. S. (1977). Feeding behavior of orang-utans in the Kutai Nature Reserve, East Kalimantan. In: T. H. Clutton-Brock (ed.), *Primate ecology* pp. 384–414. Academic Press, New York.

Roff, M., Sells, B., and Golden, M. M. (1972). *Social adjustment and personality development in children*. University of Minnesota Press, Minneapolis.

Rogosa, D. (1979). Causal models in longitudinal research: formulas and interpretations. In: J. R. Nesselroade and P. B. Baltes (eds.) *Longitudinal research in the study of behaviour and development*. Academic Press, New York.

Rommetveit, R. (1979). On common codes and dynamic residuals in human communication. In: R. M. Blakar and R. Rommetveit (eds.) *Studies of language, thought, and verbal communication*. Academic Press, London.

Rose, S. A. and Blank, M. (1974). The potency of context in children's cognition: an illustration from conservation, *Child Development* **45**, 499–502.

Rosenthal, T. and Zimmerman, B. (1978). *Social learning and cognition*. Academic Press, New York.

Ross, H. S. and Goldman, B. M. (1976). Establishing new social relations in infancy. In: T. Alloway, L. Krames, and P. Pliner (eds.). *Advances in communication and affect*, Vol. 4. Plenum Press, New York.

Roux, J. P. and Gilly, M. (1984). Aide apportée par le marquage social dans une procédure de résolution chez des enfants de 12–13 ans: données et réflexions sur les mécanismes. *Bulletin de Psychologie* (in press).

Rowe, D. C. and Plomin, R. (1981). The importance of nonshared (E_1) environmental influences in behavioral development. *Developmental Psychology* **17**, 517–31.

Rowell, T. E., Hinde, R. A., and Spencer-Booth, Y. (1964). Aunt-infant interactions in captive rhesus monkeys, *Animal Behaviour* **12**, 219–26.

Roy, P. (1983). Is continuity enough? Substitute care and socialization. Paper presented at the Spring Scientific Meeting, Child and Adolescent Psychiatry Specialist Section, Royal College of Psychiatrists, London, March.

Rozin, P. (1976). The evolution of intelligence and access to the cognitive unconscious. In: J. N. Sprague and A. N. Epstein (eds.) *Progress in psychology*, Vol. 6, pp. 245–80. Academic Press, New York.

—— and Gleitman, L. R. (1976). The structure and acquisition of reading, II. In: A. S. Reber and D. Scarborough (eds.) *Reading: the CUNY conference*. Erlbaum, Potomac, Maryland.

Rubin, K. H. (1973). Egocentrism in childhood: a unitary construct? *Child Development* **44**, 102–10.

Rubin, K. H. (1978) Role taking in childhood: some methodological considerations. *Child Development* **49**, 428–33.

Rubin, K. H. and Daniels-Beirness, T. (1983). Concurrent and predictive correlates of sociometric status in kindergarten and Grade 1 children. *Merrill-Palmer Quarterly* **29**, 337–51.

—— and Krasnor, L. (1984). Social-cognitive and social behavioral perspectives on problem-solving. In: M. Perlmutter (ed.) *The Minnesota symposium on child psychology*, Vol. 18. Erlbaum, Hillsdale, N.J.

—— and Ross, H. S. (1982). *Peer relationships and social skills in childhood*. Springer-Verlag, New York.

Ruble, D. (1983). The development of social-comparison processes and their role in achievement-related self-socialization. In: E. Tory Higgins, D. N. Ruble, and W. W. Hartup (eds.), *Social cognition and social development*, pp. 134–57. Cambridge University Press, Cambridge.

Rumbaugh, D. M. (1977). *Language learning by a chimpanzee: the LANA Project*, Academic Press, New York.

Russell, J. (1979). The status of genetic epistemology. *Journal of the Theory of Social Behavior* **9**, 14–26.

—— (1979). Children deciding on the correct answer: social influence under the microscope. Paper presented at the Annual Conference of the Developmental Section of the British Psychological Society. Southampton, (cited in M. Glachan and P. Light, 1982, 249 f.).

Rutter, D. R. and Stephenson, G. M. (1972). Visual interaction in a group of schizophrenic and depressive patients. *British Journal of Social and Clinical Psychology* **11**, 57–65.

Rutter, M. (1981). *Maternal deprivation reassessed.* Penguin Books, Harmondsworth, Middx.

—— (1983a). Cognitive deficits in the pathogenesis of autism. *J. Child Psychology and Psychiatry* **24**, 513–31.

—— (1983b) (ed.). *Developmental neuropsychiatry.* Guilford Press, New York.

—— (1983c). School effects on pupil progress: research findings and policy implications. *Child Development* **54**, 1–29.

—— (1984a) Family and school influences: meaning, mechanisms, and implications. In: A. R. Nicol (ed.) *Longitudinal studies in child psychology and psychiatry: practical lessons from research experience.* Wiley, Chichester.

—— (1984b). Psychopathology and development: II. Childhood experiences and personality development. *Australian and New Zealand Journal of Psychiatry* **18**, 314–27.

—— (1984c). Continuities and discontinuities in socio-emotional development: empirical and conceptual perspectives. In: R. Emde and R. Harmon (eds.) *Continuities and discontinuities in development.* Plenum, New York.

—— and Madge, N. (1976). *Cycles of disadvantage: a review of research.* Heinemann Educational, London.

—— and Quinton, D. (1984). Parental psychiatric disorder: effects on children. *Psychological Medicine* **14**, 853–80.

—— Tizard, J. and Whitmore, K. (eds.) (1970). *Education, health and behaviour.* Longmans, London. (Reprinted, 1981, Krieger, Huntingdon, N.Y.).

—— Maughan, B., Mortimore, P. and Ouston, J., with Smith, A. (1979). *Fifteen thousand hours: secondary schools and their effects on children.* Open Books, London; Harvard University Press, Cambridge, Mass.

Sacks, H. and Schegloff, E. (1974). Two preferences in the organization of reference of persons in conversation and their interaction. In: N. H. Avison and R. J. Wilson (eds.) *Ethnomethodology: labelling theory and deviant behavior.* Routledge & Kegan Paul, London.

Sade, D. S. (1972a). Sociometrics of *Macaca mulatta* I: Linkages and cliques in grooming matrices. *Folia Primatologia* **18**, 196–223.

—— (1972b). A longitudinal study of social behavior of rhesus monkeys. In: R. Tuttle (ed.) *The functional and evolutionary biology of primates.* Aldine, Chicago.

Sammarco, J. G. (1984). Joint problem-solving activity in mother-child dyads: a comparative study of normally achieving and language disordered preschoolers. Unpublished doctoral dissertation, Northwestern University.

Samuel, J. and Bryant, P. E. (1984). Asking only one question in the conservation experiment. *Journal of Child Psychology and Psychiatry* **25**, 315–18.

Sands, S. F. and Wright, A. (1980). Serial probe recognition performance by a rhesus monkey and a human with 10- and 20-item lists. *Journal of Experimental Psychology: Animal Behavioural Processes* **6**, 386–96.

—— Lincoln C. E. and Wright A. A. (1982). Pictorial similarity judgments and the organization of visual memory in the rhesus monkey. *Journal of Experimental Psychology: General* **111**, 369–89.

Savage-Rumbaugh, E. S., Rumbaugh, D. M., and Boysen, S. (1978). Sarah's problems of comprehension. *Behavioural and Brain Sciences* **4**, 555–7.
—— —— Smith, S. T. and Lawson, J. (1980). Reference: the linguistic essential. *Science* **210**, 922–5.
Savin-Williams, R. C. (1976). An ethological study of dominance formation and maintenance in a group of human adolescents. *Child Development* **47**, 972–79.
Scarr, S. (1981). *Race, social class and individual differences in IQ*. Erlbaum, Hillsdale, N.J.
—— and Kidd, K. K. (1983). Developmental behavior genetics. In: M. M. Haith, and J. J. Campos (eds.) *Infancy and developmental psychobiology* (Vol. 4 of P. H. Mussen (ed.) *Handbook of child psychology*), pp. 345–433. Wiley, New York.
—— and McCartney, K. (1983). How people make their own environments: a theory of genotype-environment effects. *Child Development* **54**, 424–35.
—— and Weinberg, R. A. (1976). IQ test performance of black children adopted by white families. *American Psychology* **31**, 726–39.
—— —— (1978). The influence of 'family background' on intellectual attainment. *American Sociological Review* **43**, 674–92.
—— —— (1983). The Minnesota adoptive studies: genetic differences and malleability. *Child Development* **54**, 260–7.
Schachter, S. and Singer, J. (1962). Cognitive, social, and physiological determinants of emotional state. *Psychological Review* **69**, 379–399.
Schaffer, H. R. (ed.) (1977). *Studies in mother-infant interaction*. Academic Press, London.
—— and Emerson, P. E. (1964). The development of social attachments in infancy. *Monographs of the Society for Research in Child Development* Serial No. 94, **29**(3).
—— Collis, G. M., and Parsons, G. (1977). Vocal interchange and visual regard in verbal and preschool children. In: H. R. Schaffer (ed.) *Studies in mother-infant interaction*, pp. 291–324. Academic Press, London.
Schegloff, E. (1968). Sequencing in conversational openings. *American Anthropologist* **70**, 1075–95.
—— Sacks, H. (1973). Opening up closings. *Semiotica* **8**, 289–327.
—— Jefferson, G. and Sacks, H. (1977). The preference for self-correction in the organization of repair in conversation. *Language* **53**, 361–82.
Schiff, M., Duyme, M., Dumaret, A., and Tumkiewics, S. (1982). How much *could* we boost scholastic achievement and IQ scores? A direct answer from a French adoption project. *Cognition* **12**, 165–96.
Schiller, J. D. (1978). Child care arrangements and ego functioning: the effects of stability and entry age on young children. Unpublished doctoral dissertation, University of California, Berkeley, Calif.
Schrier, A. M., Angarella R., and Povar, M. L. (1984). Studies of concept formation by stumptailed monkeys: concepts, humans, monkeys and letter A. *Journal of Experimental Psychology: Animal Behavioural Processes* **10**, 564–84.
Schwartzman, H. B. (1976). Children's play: a sideways glance at make believe . In: D. F. Lancy and B. A. Tindall (eds.) *The anthropological study of play: problems and prospects*. Leisure Press, New York.
—— (1978). *Transformations: the anthropology of children's play*. Plenum Press, New York.
Schweinhart, L. J. and Weikart, D. P. (1980). *Young children grow up: the effects of the preschool program on youths through age 15*, Monographs of the High

Scope Educational Research Foundation No. 7. High Scope Press, Ypsilanti, Michigan.

Scribner, S. and Cole, M. (1981). *The psychology of literacy*. Harvard University Press, Cambridge, Mass.

Sears, R., Maccoby, E. E., and Levin, H. (1957). *Patterns of child rearing*. Harper & Row, New York.

Seligman, M. E. P. and Hager, J. L. (1972). *Biological boundaries of learning*. Appleton-Century-Crofts, New York.

Selman, R. L. (1980). *The growth of interpersonal understanding*. Academic Press, New York.

—— (1981a). The child as a friendship philosopher. In: S. R. Asher and J. M. Gottman (eds.) *The development of children's friendship*, pp. 242–72. Cambridge University Press, Cambridge.

—— (1981b). The development of interpersonal competence: the role of understanding in conduct. *Developmental Review* 1, 401–22.

—— and Demorest, R. P. (1984). Observing troubled children's interpersonal negotiation strategies: implications of and for a developmental model. *Child Development* 55, 288–304.

—— Schorin, M. Z., Stone, C., and Phelps, E. (1983). A naturalistic study of children's social understanding. *Developmental Psychology* 19, 82–102.

Serafica, F. C. (1982). Introduction. In: F. C. Serafica (ed.) *Social-cognitive development in context*, pp. 1–26. Guilford Press, New York.

Seyfarth, R. M. (1977). A model of social grooming among adult female monkeys. *Journal of Theoretical Biology* 65, 671–98.

—— (1980). The distribution of grooming and related behaviours among adult female vervet monkeys. *Animal Behaviour* 28, 798–813.

—— (1981). Do monkeys rank each other? *Behavioural and Brain Sciences* 4, 447–448.

—— (1983). Grooming and social competition in primates. In: R. A. Hinde (ed.) *Primate social relationships*, pp. 182–90. Blackwell Scientific Publications.

—— and Cheney, D. L. (1984). Social and non-social knowledge in vervet monkeys. *Philosophical Transactions of the Royal Society*, Series B (in press).

—— —— (1984). Grooming, alliances, and reciprocal altruism in vervet monkeys. *Nature* 308, 541–3.

—— —— and Marler, P. (1980). Vervet monkey alarm calls. *Animal Behaviour* 28, 1070–94.

Shantz, C. U. (1975). The development of social cognition. In: E. M. Hetherington (ed.) *Review of child development research*, Vol. 5. University of Chicago Press, Chicago.

—— (1983). Social cognition. In: J. H. Flavell and E. M. Markman (eds.) *Cognitive development* (Vol. 4 of P. H. Mussen (ed.) *Handbook of child psychology*) pp. 495–555. Wiley, New York.

Shatz, M. (1983). Communication. In: J. H. Flavell and E. M. Markman (eds.) *Cognitive development* (Vol. 4 of P. H. Mussen (ed.) *Handbook of child psychology*), pp. 841–89. Wiley, New York.

—— and Gelman, R. (1973). The development of communication skills: modifications in the speech of young children as a function of listener. *Monographs of the Society for Research in Child Development* Serial No. 152, 38(5).

—— —— (1977). Beyond syntax: the influence of conversational constraints on speech modifications. In: P. E. Snow and C. A. Ferguson (eds.) *Talking to children*. Cambridge University Press, Cambridge.

Shugar, C. W. (1979). Peer face-to-face interactions at age three to five. *International Journal of Psycholinguistics* **564**, 17–37.

Shure, M. B. (1980). Interpersonal problem-solving in ten-year-olds. Final report no. MH–27741. National Institute of Mental Health, Washington, D.C.

—— (1981). Social competence as a problem-solving skill. In: J. Wine and M. Smye (eds.) *Social competence.* Guilford Press, New York.

—— (1982). Interpersonal problem-solving: a cog in the wheel of social cognition. In: F. Serafica (ed.) *Social cognition and social development in context.* Guilford Press, New York.

—— (in progress). Problem-solving and mental health of 10- to 12-year-olds. Project No. 35989, National Institute of Mental Health, Washington, D.C.

—— and Selman, R. L. (1977). *Issues in social cognition.* Conversation hour at the meetings of the Society for Research in Child Development, New Orleans (March).

—— and Spivack, G. (1970). Cognitive problem-solving skills, adjustment, and social class. (Research and Evaluation Report No. 26). Department of Mental Health Sciences, Hahnemann Medical College and Hospital, Philadelphia.

—— —— (1971). *Interpersonal cognitive problem-solving (ICPS)—a mental health program for four-year-old nursery school children: Training script.* Department of Mental Health Sciences, Hahnemann University, Philadelphia.

—— —— (1972). Means-ends thinking, adjustment, and social class among elementary school-aged children. *Journal of Consulting and Clinical Psychology* **38**, 348–53.

—— —— (1974). *Interpersonal cognitive problem-solving (ICPS)—A mental health program for kindergarten and first grade children: Training script.* Department of Mental Health Sciences, Hahnemann University, Philadelphia.

—— —— (1978). *Problem-solving techniques in childrearing.* Jossey-Bass, San Francisco.

—— —— (1980). Interpersonal problem-solving as a mediator of behavioral adjustment in preschool and kindergarten children. *Journal of Applied Developmental Psychology* **1**, 29–43.

—— —— (1982a). *Interpersonal cognitive problem-solving (ICPS): a training program for the intermediate elementary grades.* Department of Mental Health Sciences, Hahnemann Medical College, Philadelphia.

—— —— (1982b). Interpersonal problem-solving in young children: a cognitive approach to prevention. *American Journal of Community Psychology* **10**, 341–56.

—— Newman, S., and Silver, S. (1973). Problem-solving thinking among adjusted, impulsive, and inhibited Head Start children. Paper presented at the May meeting of the Eastern Psychological Association, Washington, D.C.

—— Spivack, G., and Gordon, R. (1972). Problem-solving thinking: a preventive mental health program for preschool children. *Reading World* **11**, 259–73.

Sigg, H. (1980). Differentiation of female positions in hamadryas one-male units. *Zeitschrift für Tierpsychologie,* **53**, 265–302.

—— (1981) Entwicklung von Zweier-Beziehungen bei jung adulten Rhesus-Affenmannchen und ihre Beeinflussung durch Psychopharmaka. In M. Bloesch (ed.) *Die Beeinflussung angeborner Verhaltensweisen durch neurotrope Substanzen,* Erlanger Forschungen, Reihe B, Band II. Verlag Bund Universitäts-Bibliothek Erlangen Nurnberg EV., Erlangen.

—— and Stolba, A. (1981). Home range and daily march in a hamadryas baboon troop. *Folia Primatologia* **36**, 40–75.

Silbereisen, R. K. (1982). Untersuchungen zur Frage sozial-kognitiv anregender Interaktionsbedingungen. In: D. Geulen (ed.) *Perspektivenübernahme und soziales Handeln*, pp. 485–515. Suhrkamp, Frankfurt-am-Main.

Silk, J. B., Samuels, S. A., and Rodman P. S. (1981). The influence of kinship, rank, and sex on affiliation and aggression between adult female and immature bonnet macaques (*Macaca radiata*). *Behaviour* **78**, 111–37.

Silverman, I. W. and Geiringer, E. (1979). Dyadic interaction and conservation induction. *Child Development* **44**, 815–20.

Simon, H. A. (1979). Information processing modes of cognition. In: M. Rosenweig and L. Porter (eds.) *Annual Review of Psychology* **30**, 363–96.

Sinclair, H., Stambak, M., *et al.* (1982). *Les bébès et les choses*. Presses Universitaires de France, Paris.

Sinclair, J. and Coulthard, R. (1975). *Towards an analysis of discourse*. Oxford University Press, London.

Skeels, H. M. and Dye, H. (1939). A study of the effects of differential stimulation on mentally retarded children. *Proceedings of the American Association for Mental Deficiency* **44**, 114–36.

Skodak, M. and Skeels, H. M. (1949). A final follow-up study of one hundred adopted children. *Journal of Genetic Psychology* **50**, 427–39.

Skuse, D. (1984). Extreme deprivation in early childhood. II. Theoretical issues and a comparative review. *Journal of Child Psychology Psychiatry* **25**, 543–72.

Smedslund, J. (1966). Les origines sociales de la decentration. In: J. Grize and B. Inhelder (eds.), *Psychologie et epistemologie genetiques: thèmes Piagetiens*, pp. 159–67. Dunod, Paris.

Smuts, B. (1985). *Sex and friendship in baboons*. Aldine, Chicago.

Snow, C. E. (1977). The development of conversation between mothers and babies. *Journal of Child Language* **4**, 1–22.

Snow, M., Jacklin, C., and Maccoby, E. (1981). Birth order differences in peer sociability at 33 months. *Child Development* **52**, 589–95.

Snow, R. (1980). Intelligence for the year 2001. *Intelligence* **4**, 185–99.

Sommer, R. (1969). *Personal space: the behavioral basis of design*. Prentice-Hall, Englewood Cliffs, N.J.

Sophian, C. and Sage S. (1983). Developments in infants' search for displaced objects. *Journal of Experimental Child Psychology* **35**, 143–60.

Specht, W. (1982). *Die Schulklasse als Beziehungsfeld altershomogener Gruppen*. Arbeitsbericht 3, published by the Zentrum I Bildungsforschung, Sonderforschungsbereich 23, University of Konstanz.

Spivack, G. and Levine, M. (1963). Self-regulation in acting-out and normal adolescents. Report M–4531, National Institute of Health, Washington, D.C.

—— and Shure, M. B. (1974). *Social adjustment of young children*. Jossey-Bass, San Francisco.

—— Platt, J. J., and Shure, M. B. (1976). *The problem-solving approach to adjustment*. Jossey-Bass, San Francisco.

—— Standen, C., Bryson, J., and Garrett, L. (1978). Interpersonal problem-solving thinking among the elderly. Paper presented at the August meeting of the American Psychological Association, Toronto, Calif.

Sroufe, L. A. (1979). Socio-emotional development. In: J. Osofsky (ed.) *Handbook of infant development*, pp. 462–516. Wiley, New York.

—— (1979). The coherence of individual development. *American Psychologist* **34**, 834–41.

—— and Fleeson, J. (in press). Attachment and the construction of relationships.

In: W. W. Hartup and Z. Rubin (eds.) *Relationships and development*. Erlbaum, Hillsdale, N.J.

—— and Waters, E. (1977). Attachment as an organizational construct. *Child Development* **48**, 1184–99.

Staddon J. E. R. (1983). *Adaptive behavior and learning*. Cambridge University Press, Cambridge.

Stambak, M. *et al.* (1983). *Les bébés entre eux*. Presses Universitaires de France, Paris.

—— and Verba, M. (in press). Organization of some social plays among toddlers: an ecological approach. In: E. Mueller and C. Cooper (eds.) *Process and outcome in peer relationships*. Academic Press, New York.

—— Bonica, L., Maissonet, R., Reyna, S., and Verba, M. (1979). Modalités d'échanges entre enfants de moins de deux ans. In: *Les enfants dans les crêches*, SAESAS, **19**.

Stammbach, E. (1978). On social differentiation in groups of captive female hamadryas baboons. *Behaviour* **67**, 322–38.

—— and Kummer H. (1982). Individual contributions to a dyadic interaction: an analysis of baboon grooming. *Animal Behaviour* **30**, 964–71.

Steiger, J. H. (1980). Tests for comparing elements of correlation matrix. *Psychological Bulletin* **87**, 245–51.

Stern, D. N. (1974). Mother and infant at play: the dyadic interaction involving facial, vocal, and gaze behaviors. In: M. Lewis and L. A. Rosenblum (eds.) *The effect of the infant on its caregiver*. Wiley, New York and London.

—— Hofer, L., Haft, W., and Dore, J. (in press). Affect attunement. In: T. Field and N. Fox (eds.) *Social perception in infants*. Ablex, Norwood, N.J.

—— Beebe, B., Jaffe, J., and Bennett, S. L. (1977). The infant's stimulus world during social interaction: study of caregiver behaviours with particular reference to repetition and timing. In: H. R. Shaffer (ed.) *Studies in mother-infant interaction*. pp. 177–202. Academic Press, London.

—— Jaffe, J., Beebe, B., and Bennett, S. (1975). Vocalizing in unison and in alternation: two modes of communication within the mother-infant dyad. In: D. Aronson and R. Riber (eds.) *Developmental psycholinguistics and communication disorders*. *Annals of the New York Academy of Sciences* **263**, 89–100.

Sternberg, R. J. (1984). Mechanisms of cognitive development: a componential approach. In: R. J. Sternberg (ed.) *Mechanisms of cognitive development*, pp. 163–86. Freeman, New York.

Stockinger, S. and McCune-Nicolich, L. (1983). Shared pretend: sociodramatic play at three. In: I. Bretherton (ed.) *Symbolic play: the representation of social understanding*. Academic Press, New York.

Strayer, F. F. (1983). Social ecology of the preschool peer group. In: W. A. Collins (ed.) *Development of cognition, affect, and social relations*. Erlbaum, Hillsdale, N.J.

Strum, S. C. (1981). Processes and products of change: baboon predatory behavior at Gilgil, Kenya. In: R. Harding and G. Teleki (eds.) *Omnivorous primates*, pp. 255–302. New York.

Sudnow, C. E. (1972). *Studies in social interaction*. The Free Press, New York.

Sullivan, H. S. (1953a). *The interpersonal theory of psychiatry*, Norton, New York.

—— (1953b). *Conceptions of modern psychiatry*. Norton, New York.

Suomi, S. and Harlow, H. F. (1972). Social rehabilitation of isolate-reared monkeys. *Developmental Psychology* **6**, 487–96.

Sutton-Smith, B. (ed.) (1979). *Play and learning*. Gardner Press, New York.

Taylor, W. T. (1961). *Normative discourse*. Prentice Hall, Englewood Cliffs, N.J.

Teleki, G. (1974). Chimpanzee subsistence technology: materials and skills. *Journal of Human Evolution* **3**, 575–94.

—— (1981). The omnivorous diet and eclectic feeding habits of chimpanzees in Gombe National Park, Tanzania. In: R. Harding and G. Teleki (eds.) *Omnivorous primates*, pp. 303–43. Columbia Univ. Press, New York.

Terrace, H. S. (1979). *Nim*. Knopf, New York.

Thibaut, J. W. and Kelley, H. H. (1959). *The social psychology of groups*. Wiley, New York.

Thomas, R. K. and Kerr, S. R. (1976). Conceptual conditional discrimination in *Saimiri sciureus*. *Animal Learning and Behaviour* **4**, 333–6.

Thorpe, W. H. (1961). *Bird song*. Cambridge University Press, Cambridge.

Tinbergen, N. (1959). Einige Gedanken über Beschwichtigungs-Gebaerden. *Zeitschrift für Tierpsychologie* **16**, 651–5. English translation: N. Tinbergen, (1973) *The animal in its world*, Vol. 2. Harvard University Press, Cambridge, Mass.

Tizard, B. (1984). What Joyce learnt from her mother. *New Society* 13th September.

—— and Hodges, J. (1978). The effect of early institutional rearing on the development of eight-year-old children. *Journal of Child Psychology and Psychiatry* **19**, 99–118.

—— and Hughes, M. (1984). *Young children learning: talking and thinking at home and at school*. Fontana, London.

—— Schofield, W. N., and Hewison, J. (1982). Collaboration between teachers and parents in assisting children's reading. *British Journal of Educational Psychology* **52**, 1–15.

—— Carmichael, H., Hughes, M., and Pinkerton, G. (1980). Four-year-olds talking to mothers and teachers. In: L. A. Hersov, M. Berger, and A. R. Nicol (eds.) *Language and language disorders in childhood*. Book supplement to *British Journal of Educational Psychology* No. 2, pp. 49–76. Plenum Press, New York.

—— Hughes, M., Carmichael, H., and Pinkerton, G. (1982). Adults' cognitive demands at home and at nursery school. *Journal of Child Psychology and Psychiatry* **23**, 105–16.

—— —— —— —— (1983a). Children's questions and adults' answers. *Journal of Child Psychology and Psychiatry* **24**, 269–82.

—— —— —— —— (1983b). Language and social class: is verbal deprivation a myth? *Journal of Child Psychology and Psychiatry* **24**, 533–42.

—— —— —— —— (1983c). Children's questions and adults' answers. *Journal of Child Psychology and Psychiatry* **24**, 269–81.

Tizard, J. (1970). The role of social institutions in the causation, prevention, and alleviation of mental handicap. In: H. C. Haywood (ed.) *Socio-cultural aspects of mental retardation*. Appleton-Century-Crofts, New York.

Tomkins, S. S. (1962–1963). *Affect, imagery, consciousness*, 2 Vols. Springer-Verlag, New York.

Trevarthen, C. (1977). Descriptive analysis of infant communicative behaviour. In: H. R. Schaffer (ed.) *Studies in mother-infant interaction*, pp. 227–70. Academic Press, London.

—— (1979). Instincts for human understanding and for cultural co-operation: their development in infancy. In: M. von Cranach, K. Foppa, W. Lepenies, and D. Ploog, (eds.) *Human ethology*, pp. 530–71. Cambridge University Press, Cambridge.

Triseliotis, J. and Russell, J. (1984). *Hard to place: the outcome of late adoptions and residential care.* Heinemann Educational, London.

Trivers, R. L. (1971). Parental investment and sexual selection. In: B. Campbell (ed.) *Sexual selection and the descent of man.* Aldine, Chicago.

Tulkin, S. R. and Covitz, F. E. (1975). Mother-infant interaction and intellectual functioning at age 6. Paper presented at the meeting of the SRCD, Denver, April.

Turiel, E. (1982). *The development of social knowledge.* Cambridge University Press, New York.

—— (1983). Interaction and development in social cognition. In: E. T. Higgins, D. N. Ruble, and W. W. Hartup (eds.) *Social cognition and social development.* Cambridge University Press, Cambridge.

Uzgiris, I. and Hunt, J. McV. (1966). *Ordinary scales of infant development.* International Congress of Psychology, Moscow.

Vaitl, E. (1978). Nature and implications of complexly organised social systems in non-human primates. In: D. J. Chivers and J. Herbert (eds.) *Recent advances in primatology I.* Academic Press, London.

van Eerdewegh, M. M., Bieri, M. D., Parilla, R. H., and Clayton, P. (1982). The bereaved child. *British Journal of Psychiatry* **140**, 23–9.

Van Leishout, C. F. M. (1975). Young children's reactions to barriers placed by their mothers. *Child Development* **46**, 879–86.

Vandell, D. L. (1980). Sociability with peer and mother during the first year. *Developmental Psychology* **16**, 355–61.

Verba, M., Stambak, M., and Sinclair, H. (1982). Physical knowledge and social interaction in children from 18 to 24 months of age. In: G. E. Forman (ed.) *Action and thought.* Academic Press, New York.

Verdier, Y. (1979). *Façons de dire, façons de faire.* Gallimard.

Vernon, P. E. (1970). Intelligence. In: W. J. Dockrell (ed.) *On intelligence.* Methuen, London.

—— (1979). *Human intelligence: heredity and environment.* Freeman, San Francisco.

Verplanck, W. S. (1954). Burrhus F. Skinner. In: W. K. Estes, S. Koch, K. MacCorquodale, P. E. Meehl, C. G. Mueller, W. N. Schonfeld, and W. S. Verplanck (eds.) *Modern learning theory*, pp. 267–316. Appleton-Century-Crofts, New York.

Visalberghi, E. and Mason, W. A. (1983). Determinants of problem-solving success in *Saimiri and Callicebus. Primates* **24**, 385–96.

Vygotsky, L. S. (1934). *Myshlenie i rech': Psikhologicheskie issledovaniya* [Thinking and speech: psychological investigations]. Gosudarstvennoe Sotsial'no-Ekonomicheskoe Izdatel'stvo, Moscow-Leningrad.

—— (1934). *Thought and language.* M.I.T. Press, Cambridge, Massachusetts, 1962.

Vygotsky, L. S. (1978). In: M. Cole, V. John-Steiner, S. Scribner, and E. Souberman (eds.) *Mind in society: the development of higher psychological processes.* Harvard University Press, Cambridge, Mass.

—— (1981a). The instrumental method in psychology. In: J. V. Wertsch (ed.) *The concept of activity in Soviet psychology.* M. E. Sharpe, Armonk, New York.

—— (1981b). The genesis of higher mental functions. In: J. V. Wertsch (ed.), *The concept of activity in Soviet psychology.* M. E. Sharpe, Armonk, New York.

de Waal, F. (1977). The organization of agonistic relations within two captive groups of Java monkeys (*Macaca fascicularis*) *Zeitschrift für Tierpsychologie* **44**, 225–82.

—— (1977). The organization of agonistic relations within two captive groups of Java monkeys. *Zeitschrift für Tierpsychologie* **44**, 225–82.

—— (1982). *Chimpanzee politics.* Harper & Row, New York.

Wachs, T. D. and Gruen, G. E. (1982). *Early experience and human development.* Plenum Press, New York.

Walker, E. and Emory, E. (1985). Commentary: interpretive bias and behavioral genetic research. *Child Development* (in press).

Weber, M. (1964). *The theory of social and economic organization*, p. 118. Free Press, New York.

Weisbard, C. and Goy, R. (1976). Effect of parturition and group composition on competitive drinking order in stumptail macaques. *Folia primatologia* **25**, 95–121.

Wittenberger, J. (1981). *Animal social behavior.* Duxbury Press, Boston.

Weiskrantz, L. and Cowey, A. (1975). Cross-modal matching in the rhesus monkey using a single pair of stimuli. *Neuropsychologia* **13**, 257–61.

Werner, H. (1948). *Comparative psychology of mental development.* Follett, Chicago.

—— and Kaplan, B. (1963). *Symbol formation.* Wiley, New York.

Wertsch, J. V. (1978). Adult-child interaction and the roots of metacognition. *Quarterly Newsletter of the Institute for Comparative Human Development* **2**, 15–18.

—— (in press). The social formation of mind: a Vygotskian approach. Harvard University Press, Cambridge, Mass.

—— and Hickmann, M. (in press). A microgenetic analysis of problem-solving in social interaction. In: M. Hickmann (ed.) *Social and functional approaches to language and thought.* Academic Press, New York.

—— and Stone, C. A. (1985). The concept of internalization in Vygotsky's account of the genesis of higher mental functions. In: J. V. Wertsch (ed.) *Culture, communication, and cognition: Vygotskian perspectives.* Cambridge University Press, New York.

—— Minick, N., and Arns, F. J. (1984). The creation of context in joint problem-solving: a cross-cultural study. In: B. Rogoff and J. Lave (eds.) *Everyday cognition: its development in social context.* Harvard University Press, Cambridge, Mass.

—— McNamee, G. D., McLane, J. B., and Budwid, N. A. (1980). The adult-child dyad as a problem-solving system. *Child Development* **51**, 1215–21.

Whiting, B. B. and Pope-Edwards, C. (eds.) (1977). *The effects of age, sex and modernity on the behaviour of mother and children.* Report to the Ford Foundation.

—— and Whiting, J. W. M. (1975). *Children of six cultures. A psycho-cultural analysis.* Harvard University Press, Cambridge, Mass., and London.

Wilcox, M. J. and Howse, P. (1982). Children's use of gestural and verbal behavior in communicative misunderstandings. *Applied Psycholinguistics* **3**, 15–27.

Winters, K. C., Stone, A. A., Weintraub, S., and Neale, J. M. (1981). Cognitive and attentional deficits in children vulnerable to psychopathology. *Journal of Abnormal Child Psychology* **9**, 435–54.

Wolkind, S. N. and de Salis, W. (1982). Infant temperament, maternal mental state, and child behavioral problems. In: Ciba Foundation Symposium 89, *Temperamental differences in infants and young children.* Pitman, London.

Wood, D., McMahon, L., and Cranston, Y. (1980). *Working with under fives*. Grant McIntyre, London.

Woodhead, M. (ed.) (1976). *An experiment in nursery education*. NFER Publishing Co., Windsor.

Woodruff, G. and Premack, D. (1979). Intentional communication in the chimpanzee: the development of deception. *Cognition* **7**, 333–62.

Wozniak, R. (in press). Notes toward a co-constructive theory of the emotion/cognition relationship. In: D. Bearison and H. Zimiles (eds.), *Thought and emotion*. Erlbaum, Hillsdale, N.J.

Wrangham, R. W. (1975). Behavioral ecology of chimpanzees in Gombe National Park, Tanzania. Ph.D. Thesis, University of Cambridge.

—— (1977). Feeding behavior of chimpanzees in Gombe National Park, Tanzania. In: T. H. Clutton-Brock (ed.), *Primate ecology*, pp. 504–38. Academic Press, New York.

—— (1983). Ultimate factors determining social structure. In: R. A. Hinde (ed.) *Primate social relationships*. Blackwell, Oxford.

Yarrow, L. J., Rubenstein, J. L., and Pedersen, F. A. (1975). *Infant and environment*. Halstead Press, New York.

Yarrow, M., Scott, P. M., and Waxler, C. (1973). Learning concern for others. *Developmental Psychology* **8**, 240–60.

Yeates, K. O., MacPhee, D., Campbell, F. A., and Ramey, C. T. (1983). Maternal IQ and home environment as determinants of early childhood intellectual competence: a developmental analysis. *Developmental Psychology* **19**, 731–9.

Youniss, J. (1980). *Parents and peers in social development: a Sullivan-Piaget perspective*. University of Chicago Press, Chicago.

—— (1982). Die Entwicklung und Funktion von Freundschaftsbeziehungen. In: W. Edelstein and M. Keller (eds.), *Perspektivität und Interpretation*, pp. 78–109. Suhrkamp, Frankfurt-am-Main.

Zahn-Waxler, C. and Radke-Yarrow, M. (1982). The development of altruism: alternative research strategies. In: N. Eisenberg (ed.) *Development of prosocial behavior*. Academic Press, New York.

—— Chapman, M., and Cummings, E. M. (1986a). Cognitive and social development in infants and toddlers with a bipolar parent. *Child Psychiatry and Human Development* **15**, 75–85.

—— Radke-Yarrow, M., and King, R. A. (1979). Child rearing and children's prosocial initiations toward victims of distress. *Child Development* **50**, 319–30.

—— Cummings, E. M., McKnew, D. H., and Radke-Yarrow, M. (1984a). Altruism, aggression, and social interactions in young children with a manic-depressive parent. *Child Development* **55**, 112–22.

—— McKnew, D. H., Cummings, E. M., Davenport, Y. B., and Radke-Yarrow, M. (1984b). Problem behaviors and peer interactions of young children with a manic-depressive parent. *American Journal of Psychiatry* **141**, 236–40.

Zajonc, R. B. (1980). Feeling and thinking: preferences need no inferences. *American Psychologist* **35**, 151–75.

—— and Markus, H. (1984). Affect and cognition: the hard interface. In: C. E. Izard, J. Kagan, and R. B. Zajonc (eds.) *Emotions, cognition, and behavior*. Cambridge University Press, Cambridge.

Zigler, E. F. and Gordon, E. W. (eds.) (1982). *Day care: scientific and social policy issues*. Auburn House, Boston, Mass.

Zinchenko, P. I. (1939). Neproizvol'noe zapominanie. [Involuntary memory]. *Nauchnie zapiski Khar'kovskogo Pedagogicheskogo Instituta Inostrahnykh*

Yazykov, Tl. [Scientific notes of the Khar'kov Pedagogical Institute of Foreign Languages, Vol. 1]. Khar'kov.

Zinchenko, V. P. (1985). Vygotsky's ideas about units for the analysis of mind. In: J. V. Wertsch (ed.) *Culture, communication, and cognition: Vygotskian perspectives.* Cambridge University Press, New York.

—— and Smirnov, S. D. (1983). *Metodologicheskie voprosy psikhologi* (Methodological problems in psychology). Izdatel'stvo Moskovskogo Universiteta, Moscow.

Zivin, G. (1977). On becoming subtle: age and social rank changes in the case of facial gesture. *Child Development* **48**, 1314–21.

Author index

Subject index